Advances in Experimental Medicine and Biology

Volume 1521

Advances in Experimental Medicine and Biology provides a platform for scientific contributions in the main disciplines of the biomedicine and the life sciences. This series publishes thematic volumes on contemporary research in the areas of microbiology, immunology, neurosciences, biochemistry, biomedical engineering, genetics, physiology, and cancer research. Covering emerging topics and techniques in basic and clinical science, it brings together clinicians and researchers from various fields.

Advances in Experimental Medicine and Biology has been publishing exceptional works in the field for over 40 years, and is indexed in SCOPUS, Medline (PubMed), EMBASE, Reaxys, EMBiology, the Chemical Abstracts Service (CAS), and Pathway Studio.

2023 CiteScore: 5.9

Mustafa G. Aydogan
Editor

Biological Time Control

Autonomous Clocks in Cellular Organization, Dynamics, and Function

Springer

Editor
Mustafa G. Aydogan
Department of Genetics and Development
Vagelos College of Physicians and Surgeons, Columbia University
New York, NY, USA

ISSN 0065-2598 ISSN 2214-8019 (electronic)
Advances in Experimental Medicine and Biology
ISBN 978-3-032-36188-2 ISBN 978-3-032-36189-9 (eBook)
https://doi.org/10.1007/978-3-032-36189-9

This Springer imprint is published by the registered company Springer Nature Switzerland AG
The registered company address is: Gewerbestrasse 11, 6330 Cham, Switzerland

Preface

How do cells perceive time, and how do they use this information to shape their physiology? For decades, our understanding of biological time control has centered on canonical mechanisms that mark and relay the passage of time, including the CDK–Cyclin complexes that time cell division and the molecular oscillators that govern circadian rhythmicity. However, advances over the past decade have revealed that cells are not organized simply by master timing systems such as the canonical cell cycle or circadian clock, but rather by coupled networks of "autonomous clocks" that can operate independently to govern the timing of specific subcellular events. In this sense, an autonomous clock or cycle is one that has evolved with the intrinsic capacity to time and execute a specific subcellular process, while remaining capable of coupling to broader cellular timing systems. In proliferating cells, for example, such sub-cellular autonomous oscillators appear to be entrained to CDK–Cyclin activity so that their pace aligns with DNA replication and segregation. In post-mitotic cells, they instead appear to couple to the circadian clock, allowing their otherwise autonomous cyclic processes to adjust to daily changes in cellular energy availability—active during periods of metabolic abundance and suspended when resources are limited.

The concept of autonomous clocks is increasingly recognized across diverse areas of cell biology, spanning organelle biogenesis, cytoskeletal dynamics, redox metabolism, growth, proteostasis, transcription, nuclear envelope remodeling, and even the division of the cytoplasm independently of nuclei and the classical cell cycle oscillator, as my lab was fortunate to discover. This book, envisioned as the inaugural textbook devoted to this emerging area, arose less from the personal ambition of its editor than from a broader need to synthesize a decade of rapid developments in this nascent field. The idea for the volume began at the 2023 annual meeting of the American Society for Cell Biology during a particularly cold Boston winter, when Merry Stuber, the tireless executive editor at Springer, approached me with a suggestion from an anonymous colleague (I am yet to figure out!): that it was time for those challenging long-standing dogmas in the field—particularly the notion that CDKs and circadian clocks are the master clocks governing the timing of "everything" in the cell—to formalize this emerging area in a collective work. I remain humbled by the suggestion that I take the helm in bringing this book together, and I am deeply indebted to the many colleagues who generously contributed their time, insight, and scholarship to make this book possible.

In Chap. 1, Prof. Steven B. Haase—a dear friend and close colleague—introduces the concept of autonomous oscillators through the lens of his laboratory's pioneering discoveries on the autonomous transcriptional network that regulates the needs of a dividing cell. This network normally couples to the CDK machinery, yet retains the capacity to run independently. In Chap. 2, with the help of my student Cindy Ow, I turn to the biology of centriole duplication to illustrate how decades of work on the timing of centriole biogenesis during the cell cycle culminated in the discovery of an autonomous clock controlling the centriole biogenesis cycle. The "autonomous centriole clock" provides perhaps the most molecularly mature and functionally grounded example of this concept. By revealing that a dedicated molecular clock drives the cycles of centriole biogenesis, this work transformed what might otherwise have been viewed as "epiphenomenal observations" on autonomous cycles into a defined molecular framework with clear physiological relevance. In doing so, it also offered an experimental paradigm for how phenomenological descriptions of autonomous cellular cycles can be converted into mechanistic principles based on molecular mechanisms. In Chap. 3, Dr. Adel Al Jord and colleagues extended such centriole biology into the post-mitotic context, describing how the canonical cell cycle machinery can conversely be repurposed to pace aspects of cilia formation and maturation in multiciliated cells in human physiology.

Whereas the first three chapters center on foundational aspects of cell biology related to the cell division cycle, the next three shift toward homeostatic processes—redox metabolism, proteostasis, and immunity—that are essential for maintaining human physiology across the circadian cycle. In Chap. 4, Prof. Xinnian Dong and Dr. Sargis Karapetyan offer a comprehensive account of autonomous redox rhythms, tracing their history while describing the emerging mechanistic basis for their coupling to circadian rhythms. In Chap. 5, another dear colleague and close friend Prof. Bokai Zhu uses the lens of his trailblazing studies on the autonomous proteostasis clock to illustrate the logic—and perhaps the necessity—of autonomous oscillators regulating nearly every step of the central dogma, thereby safeguarding the lifecycle of a cell over the circadian period. In Chap. 6, Dr. Myong-Hee "Mia" Sung presents a compelling case for emerging studies on the autonomous oscillations of NF-κB nuclear localization and how they are paving the way toward uncovering its still-elusive molecular function.

Thus far, the chapters offer a wide-ranging survey of biological discoveries that continue to shape the concept of autonomous clocks. Together, they illustrate the breadth of molecular processes through which this concept is being recognized across distinct areas of cell biology. The next four chapters turn toward the future, asking how autonomous clocks can be studied, understood, and ultimately leveraged to engineer biological systems with defined temporal behaviors—what might be thought of as "synthetic autonomous cycles."

In Chap. 7, my long-time colleague Prof. Qiong Yang and her students open our eyes to the world of synthetic biology approaches for studying the principles of autonomous oscillators *in vitro*. In Chap. 8, another good friend Prof. Sahand Jamal Rahi and his team use the lens of their recent tour-de-force bioengineering work to show how the fundamental principles governing autonomous cycles can be leveraged to build artificial circuits that serve the needs of biological systems. In Chap. 9, Prof. Evelyn Tang joins forces with Dr. Fabio Echegaray Iturra, a postdoctoral fellow in my laboratory, to provide a comprehensive outlook on the physical frameworks and theories needed to understand how autonomous biological cycles can synchronize without relying on a master clock such as the CDK or circadian systems—and they do this by directly drawing lessons from these time-honored cellular pacemakers. Finally, in Chap. 10, emerging leaders and established authorities in mathematical biology—Prof. Bree Cummins, Marcio Gameiro, Konstantin Mischaikow, and Tomas Gedeon—come together to offer a quantitative synthesis of the architecture of autonomous clocks as described through mathematics and computation.

I sincerely hope that current and future students of this field will benefit from the rich insights offered across these chapters, written by emerging leaders and established authorities in their respective fields. I know that I have benefited enormously—perhaps even selfishly—as a student of the field myself while editing each chapter line by line. Although my gratitude to the contributors is already reflected in the book itself, which stands as a testament to the discoveries that have made this emerging area possible, I must also explicitly thank the colleagues and friends whose encouragement was essential in bringing this volume to completion.

First, I am forever grateful to a friend in eternity, Dr. Mohammad Mofatteh, whose encouragement during our shared years in Oxford led me toward the kinds of studies that continue to define new frontiers in the concept of autonomous clocks. I am also sincerely grateful to Prof. Susan Dutcher, who generously shared personal anecdotes from her time as a student of the great Lee Hartwell. From her recollections, it was clear that there had long been suspicions—and anecdotal evidence, both in Hartwell's laboratory and elsewhere—that autonomous cycles might exist beyond the classical CDK–Cyclin system that dominated the cell cycle field for so long. Similar encouragement came from my close colleague and former UCSF office neighbor, Prof. Pat O'Farrell—the inventor of the 2D gel electrophoresis—who would often recall seeing multiple oscillatory proteins in his gels beyond the cyclins themselves. His reflections encouraged me to keep pursuing the idea that additional autonomous clocks might be hidden within cellular physiology. Readers less familiar with the history of the cell cycle field may also wish to revisit the classic review by Marc Kirschner and Andrew Murray, "Dominoes and Clocks" (PMID: 2683077), whose final paragraphs highlight potential autonomous cycles as a major mystery—and perhaps even a profound challenge—to the emerging models of the cell cycle in the 80s and 90s that would later be recognized by the 2001 Nobel Prize.

This book is composed of ten chapters, but the emerging field of autonomous clocks extends far beyond the scope of that. I would therefore like to express my sincere gratitude to the colleagues who, although unable to contribute to this book because of time constraints, generously offered their encouragement: Profs. Fred Cross, Matthias Heinemann, Joe Takahashi, Filipa Rijo-Ferreira, Galit Lahav, Moe Mahjoub, Moumita Das, Marc Kirschner, and Dr. Xili Liu. I am also grateful to those whose work and critiques have helped place this growing concept on a stronger foundation. In particular, I thank Prof. John Tyson and the late Béla Novák for their accounts of classical models of the cell cycle and for their thoughtful engagement with the emerging concept of autonomous clocks, which Novak wittily referred to as the "clockshop" hypothesis. Readers interested in these discussions may wish to consult their more recent articles and reviews (PMID: 39241924; PMCID: PMC9010849).

Finally, I am deeply grateful to the many people whose help, support, and encouragement during my lab's move from UCSF to Columbia University sustained me as I brought this book to completion. Foremost among them are the members of my lab who moved with me from San Francisco to New York City. Their patience—not only with the move itself, but also with the countless months I devoted to putting this volume together—is something I can never fully repay. I also owe sincere thanks to my chair, Prof. Gerard Karsenty; our former dean and a continuing leader at Columbia, Prof. Katrina Armstrong; and other leaders of CUIMC, including Profs. Stavros Lomvardas and Hashim Al-Hashimi, whose engagement with and support for the science we do at Columbia have made this transition all the more meaningful.

I conclude these acknowledgments by gratefully recognizing the generous support of the NIH Director's New Innovator Award (DP2GM154328) and the Howard Hughes Medical Institute (Freeman Hrabowski Program).

Most of all, however, I could not have done this without the continuous support, brilliance, and partnership of my wife, Alissa Kleinnijenhuis, who is the light of my inspiration and my life.

New York, NY, USA Mustafa G. Aydogan

Contents

Evolution of the Cell-Cycle Clock Models: Biochemical or Transcriptional?

1

Claudia A. Petrucco and Steven B. Haase

Abstract

Models of the central oscillating mechanism of the cell-division cycle have evolved over more than 40 years of investigation across multiple model systems. The initial oscillator models were based on observations from early embryonic systems and suggested that oscillations were driven by a small biochemical feedback loop comprised of mitotic cyclin, cyclin-dependent kinase (CDK) and the ubiquitin ligase complex, anaphase promoting complex/cyclosome (APC/C). The model was expanded to include kinase and phosphatase regulators of CDK (Wee1 and Cdc25) with experiments in *Xenopus* extracts. Biochemical models were supported by genetic studies in both *S. cerevisiae* and *S. pombe* systems where cell-cycle arrests were associated with mutations in the kinases, phosphatases, and ubiquitin ligases of the cell-cycle clock. Mathematical models also supported these biochemical oscillator models and were able to accurately predict the outcome of genetic experiments. Thus, there was no apparent need for new models until work showed that autonomous cell-cycle oscillations continued when mitotic cyclins were deleted in *S. cerevisiae*. The advent of transcriptomics technologies in the late 1990s revealed the existence of large programs of oscillating gene expression associated with cell-cycle progression. It was further shown that these programs of oscillating gene expression could function autonomously in *S. cerevisiae* mutants where mitotic cyclins were disrupted and cell-cycle progression was halted. Taken together, these new findings suggested that a *g*ene *r*egulatory *n*etwork (GRN) may function as the underlying cell-cycle oscillator with CDKs serving primarily as the effectors of these oscillations. Here, we explore the evolution of these models from early biochemical to more contemporary GRN models.

Keywords

Cyclin Dependent Kinase (CDK) · Gene regulatory network (GRN) · Transcription factor (TF) · Cell cycle · Checkpoints · Autonomous transcriptional oscillator · Biochemical oscillator · Mathematical models · Ubiquitin ligase

The cell-division cycle is a repeating collection of duplication and segregation events that are executed in a strict temporal order. Although much attention has been paid to the regulation of individual events (e.g., DNA replication or spindle assembly), some of the most foundational questions revolve around the machinery that coordinates this sequence of events and drives

C. A. Petrucco · S. B. Haase (✉)
Department of Biology, Duke University, Durham, NC, USA
e-mail: steve.haase@duke.edu

M. G. Aydogan (ed.), *Biological Time Control*, Advances in Experimental Medicine and Biology 1521,
https://doi.org/10.1007/978-3-032-36189-9_1

the cells through cycles of division. Models of the structure and function of this machinery, termed the "cell-cycle clock," have evolved since the early days of cell-cycle research.

The cell-cycle clock must be capable of driving cells through multiple phases, activating a variety of distinct, temporally ordered events. The cell cycle must progress in a unidirectional order, suggesting that transitions between phases may represent bistable states regulated by irreversible molecular switches. The clock machinery must be capable of driving system-wide oscillations, necessitating some form of negative feedback. However, these oscillations must be regulatable so that cells can pause the cycle if specific cell-cycle events are delayed, if DNA is damaged, or if nutrients are limited. Together, these observations imply that the regulatory network of the cell-cycle clock is likely to have a complex set of topological features that can execute a diverse set of functions.

1.1 Models of a Master Biochemical Clock

Some of the first models of the cell-cycle clock grew from studies of early embryonic division cycles in marine invertebrates. These embryonic systems enabled in vitro biochemical studies using isolated extracts of fertilized eggs. The discovery of cyclin proteins, cyclin-dependent kinase (CDK) and a ubiquitin ligase called the *a*naphase *p*romoting *c*omplex/*c*yclosome (APC/C) gave rise to the earliest elementary cell-cycle clock models, and was the basis of the Nobel Prize awarded to Hunt, Hartwell and Nurse in 2001. In early embryonic systems, a simple three-component network, CDK functions in a negative feedback loop with the APC/C that drives oscillation of the system, and also functions as the effector by phosphorylating target proteins that trigger DNA replication and then mitosis (Fig. 1.1a) (reviewed in [1, 2]).

This seminal work greatly influenced the thinking about the structure and function of the cell-cycle clock in yeast and somatic animal cells. However, the events of the early invertebrate embryonic cell cycle are distinctly different from what is observed for yeast and somatic animal cell cycles. First, early embryonic divisions in many bilaterians are extremely rapid and are completed on the time scale of minutes (~10 min depending on the organism) in contrast to somatic tissues where a cell cycle can take 15–24 h on average. The gap phases, G1 and G2, are not observed in early embryonic cycles and they do not appear to have the ability to induce checkpoint-dependent arrests. Rapid DNA replication is likely facilitated by the lack of transcription observed in these early embryos, so the components of the mitotic oscillator, like cyclin B, are directly translated from maternally deposited mRNAs rather than through transcriptional regulation. Not surprisingly, the architecture of the cell-cycle clock in yeasts and somatic animal cells is also different to accommodate these new features of the cycle. For example, both yeasts and somatic animal cells have multiple cyclin/CDK complexes that operate in a cell cycle phase-specific manner to regulate specific cell-cycle events, while the bulk of the embryonic cycle is driven by mitotic cyclin/CDK complexes (M-CDK).

The unique features of embryonic systems where the key attributes include rapid genome replication and cell cleavage absent of growth suggests that models constructed from these systems might not be fully representative of the cell-cycle clock machinery in opisthokonts at large. However, as time passed, additional components of this biochemical model of an embryonic cell-cycle clock were identified, which could account for some somatic cell-cycle functions. A CDK regulatory kinase, Wee1 and the opposing phosphatase, Cdc25 were identified in *S. pombe* by Paul Nurse and colleagues [3]. Work in *Xenopus* embryonic systems demonstrated these additional biochemical components contributed to the dynamic function of the CDK-APC/C negative feedback motif via positive feedback loops (Fig. 1.1b) [4]. Cdc25 and Wee1 were also found to be the regulatory targets of checkpoint pathways that inhibit CDK activity and block cell-cycle progression. Supporting the notion of a unified cell cycle architecture across different cell types and species

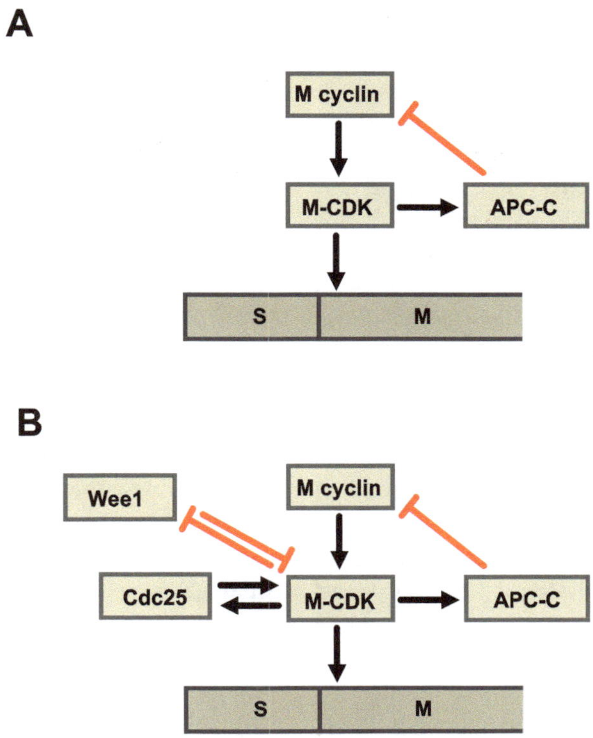

Fig. 1.1 Early model of the cell-cycle clock. (**a**) In early embryos, cell divisions consist of rapid cycles of DNA replication (S) followed by mitosis (M). Mitotic cyclin (M cyclin) binds to cyclin-dependent kinase (CDK) to form the active mitotic cyclin-CDK complex (M-CDK). M-CDK activity drives mitotic entry and triggers activation of the anaphase promoting complex (APC/Cdc20). Activation of the APC leads to degradation of M cyclin, which in turn inactivates M-CDK. As M-CDK levels drop, APC/Cdc20 activity also declines, allowing M cyclin to accumulate again and initiate the next cycle. (Adapted from Morgan [41]). (**b**) Both Wee1 and Cdc25 regulate, and are regulated by M-CDK. (Adapted from Pomerening et al. [64])

among opisthokonts, further work revealed that these core regulators were not exclusive to the early invertebrate embryo (reviewed in [5]).

Early genetic studies of the cell cycle in the *S. cerevisiae* and *S. pombe* model systems also suggested that cyclin/CDK complexes and APC/C were likely central components of the yeast cell-cycle clock [6–9]. As the counterpart regulatory components were also discovered in mammalian systems as homologs, observations remained consistent with the model that M-CDK and APC/C drove oscillations of the cell-cycle clock in somatic mammalian cells as well. Indeed, mathematical models of early embryonic clocks and relatively more complex biochemical models of the budding yeast cell-cycle clocks were capable of accurately simulating many of the cell-cycle behaviors that were observed experimentally [10, 11]. Together, both experimental and modeling studies in yeast systems were consistent with early biochemical models of the cell cycle, centered primarily on a negative feedback loop between M-CDK and APC/C. So, it is not surprising that these biochemical models pointing at a *master* cell-cycle clock has dominated the thinking in the field for many years.

1.2 Models of an Autonomous Transcriptional Network Oscillator

Yeast and somatic cells do not have maternal stores of mRNAs, thus a priori it was clear that cyclins, CDK, APC components, and other cell-cycle regulators must be synthesized de novo by transcription. Targeted expression studies demonstrated that different cyclins were transcribed periodically during the cell cycle just before they were needed to drive their cognate cell-cycle events [12]. For example, in *S. cerevisiae*, polarization, budding [13, 14]. and activation of the cell-cycle transcriptional program are initiated by G1 cyclins, while DNA replication and *s*pindle *p*ole *b*ody (SPB) duplication are controlled primarily by S-phase cyclins [15, 16] and mitosis is initiated by M-phase cyclins. In this regard, the extent and complexity of cell-cycle transcriptional regulation had not been fully understood until the development of microarray technologies.

In the late 1990s, early transcriptomic studies in synchronized populations of budding yeast cells demonstrated that ~15% of the yeast genome was transcribed in a cell-cycle phase-specific manner as cells progressed through the cell cycle [17]. This finding was strikingly similar to observations from circadian-clock systems that control large periodic transcriptional programs [18] and hinted at the potential for new models of the cell-cycle clock based on *g*ene *r*egulatory *n*etworks (GRNs).

Such extensive cell-cycle regulated transcription included genes encoding known cell-cycle regulators (e.g., cyclins), as well as the building blocks needed for SPB duplication, DNA synthesis, and mitotic entry [17]. Therefore, in addition to the core biochemical cell-cycle clock components, the cell-cycle transcriptional program appeared capable of regulating the periodic expression of genes encoding most of the proteins needed to execute cell-cycle functions, including the transcription factors themselves. These findings brought new attention to the importance of transcriptional regulation during the cell cycle and introduced an additional layer of control.

Globally, the mechanisms controlling the transcription of cell-cycle genes were poorly understood, but the emerging models hypothesized that cell-cycle transcription was largely under the control of cyclin/CDK complexes [19, 20]. The well-studied G1-CDK/Rb/E2F network motif that controlled genes required for S-phase entry became the archetype of cyclin/CDK regulatory mechanisms controlling periodic transcription during the cell cycle. To activate S-phase genes, G1-CDK complexes phosphorylate the transcriptional co-repressor, Rb, that represses genes activated by E2F. Phosphorylation of Rb at the promoters of S-phase genes, releases Rb from E2F complexes, consequently freeing E2F to activate transcription. Additional *t*ranscription *f*actor (TF) targets of cyclin/CDK have also been identified in multiple organisms [21–24], however, a novel 'omics approach in budding yeast suggested alternative regulatory models for the global regulation of the cell-cycle transcriptional program and for the cell-cycle clock.

New technologies enabled a genome-wide map of where particular TFs bind in the genome. Although there were caveats (e.g., binding in a promoter region does not always mean regulation) regulatory interactions can be inferred by locating TF binding to gene promoters. In the early 2000s, Rick Young's lab performed a ChIP-chip analysis on all annotated TFs (~200) in budding yeast and mapped their binding sites genome-wide [25, 26]. The findings from this approach indicated that a network of cell-cycle TFs transcribed in successive phases were likely regulating each other. As cells progressed through the cell cycle, TFs were activating each other in a cascade; when cells reached the end of mitosis, late mitotic TFs were activating TFs in early G1 of the next cell cycle. These observations led to the proposal that this network of TFs alone could produce an autonomously oscillating program of cell-cycle transcription (Fig. 1.2a). Because phase-specific transcription of cyclins and other cell-cycle regulators was also controlled by this network, it seemed plausible that a transcriptional network could function as the central oscillating mechanism driving the cell cycle in budding yeast. This bold new proposal, that an autonomous transcriptional oscillator could serve as the cell-cycle clock, was attractive; in that one of the best studied clocks, the circadian clock, is a gene regulatory network [27]. But, there were aspects of this model that did not fit the expectations for the cell-cycle-clock mechanism. One issue was the absence of any obvious negative feedback in

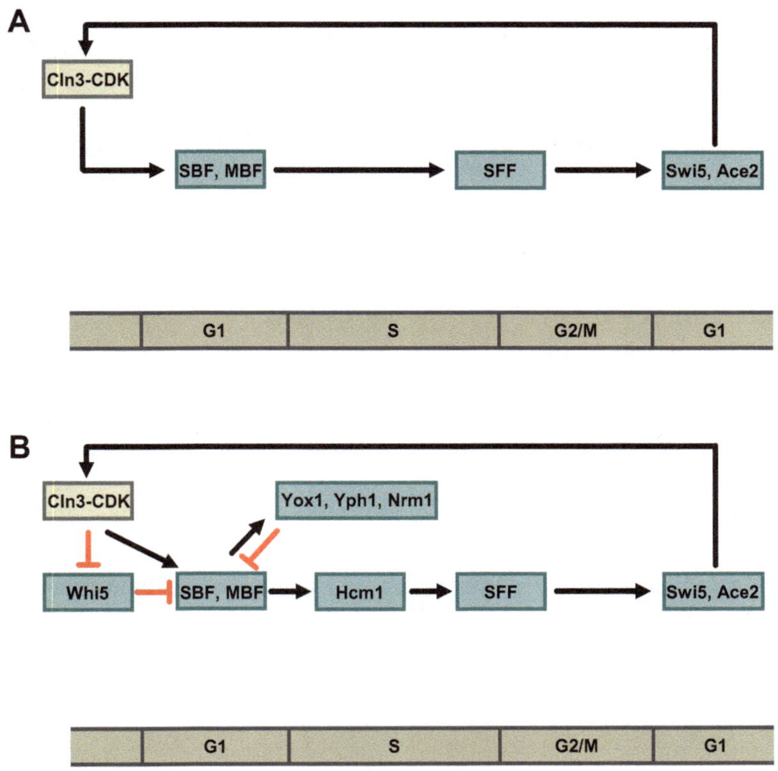

Fig. 1.2 Autonomously oscillating transcriptional clock in budding yeast. (**a**) Cell cycle transcription factors (blue boxes) active during one stage of the cell cycle activate transcription factors that function during the next stage. The G1 cyclin-CDK complex (Cln3-CDK) activates SBF (Swi4/Swi6) and MBF (Mbp1/Swi6) during late G1. SBF and MBF regulate SFF (Ndd1/Fkh2/Mcm1) which is responsible for activation of G2/M genes. SFF in turn regulates Swi5 and Ace2 which activate M/G1 genes including Cln3. (Adapted from Simon et al. [25]). (**b**) Identification of negative feedback in the cell cycle transcriptional regulatory network suggests the system could generate autonomous oscillations. The transcriptional repressors Whi5, Yox1, Yph1, and Nrm1 regulate SBF and MBF activity. Whi5 represses SBF and MBF in early G1. In late G1 Whi5 repression of MBF and SBF is relieved by Whi5 phosphorylation by G1-CDK. MBF and SBF induce Nrm1, Yox1, and Yhp1, which feedback and inhibit MBF- and SBF-activated transcription. (Adapted from Pramila et al. [29])

this transcriptional network that would be necessary to maintain stable oscillations. That issue was in part resolved by findings in 2006 that updated this network with three transcriptional repressors, including Yox1, Yhp1 [28], and Whi5 (the yeast ortholog of Rb) (Fig. 1.2b) [29]. With the addition of negative feedback there was now the possibility that a transcriptional network could oscillate autonomously under the right conditions, but there was still no experimental evidence, or a theoretical plausibility by mathematical modeling, that such a network could function in the absence of the CDK and APC/C—canonical components of the biochemical circuit postulated to govern the cell-cycle clock.

1.3 Coupling the Autonomous Transcriptional Clock to the CDK-APC/C Oscillator

Despite cell biological and genetic evidence supporting the kinds of physical interactions needed to constitute an autonomous transcriptional clock by a cascade of TFs (Fig. 1.2), postulates of autonomous transcriptional clock-based cell cycle models [14, 25, 29–33] did not garner immediate attention across the community and faced resistance from some cell-cycle researchers [34]. At the time there was no empirical evidence disputing the model that CDK-APC/C was the central oscillatory mechanism of the cell-cycle clock, so there was no impetus for developing an alternative model. Moreover, there was no experimental evidence that these transcriptional networks could drive oscillations in transcription independently of CDK-APC/C or cell-cycle progression.

Ironically, however, in the early 1970s Hartwell uncovered evidence that there might indeed be multiple autonomously oscillating cell-cycle circuits guiding cell division. Cells bearing a temperature-sensitive mutation in a gene called *CDC4* were found to arrest at the G1/S transition at the restrictive temperature. However, these arrested cells continued to undergo budding cycles independent of DNA replication and mitoses, which led Hartwell to postulate that budding was controlled by an oscillator that could function independently from the cell-cycle clock that drove progression through DNA replication and mitosis [35]. The function of Cdc4p was not known at the time, but today it is known to function as an F-box protein for an SCF ubiquitin ligase complex, and the central target of that SCF complex is a B-cyclin/CDK-specific inhibitor called, Sic1 [36, 37]. When Cdc4 is inactivated, Sic1 becomes hyperstabilized and blocks the function of all S-phase and Mitotic-CDK complexes in *S. cerevisiae*, but not G1-CDK complexes.

Nearly 30 years later in 1999, Haase and Reed analyzed cells that were disrupted for all S-phase and mitotic cyclin genes, and observed that, like *cdc4^{ts}* mutants, G1/S-arrested cells underwent budding cycles with a period similar to wild-type cells [13]. Oscillations of the mRNA encoding the G1-cyclin, Cln2 were also observed in arrested cells along with oscillations of Cln2/CDK kinase activity. These findings challenged the contemporary clock models and led to the proposal that an autonomous free-running oscillator, independent of M-CDK and APC/C, might function as the central clock of the budding yeast cell cycle. It was also the first report of transcriptional oscillations in arrested cells that lacked M-CDK activity. However, this work uncovering the transcript oscillations for this single transcript (Cln2) did not interrogate the full suite of regulators underlying the transcriptional clock models.

Evidence of genome-wide transcriptional oscillations in arrested cells lacking S- and M-CDK activities came from microarray studies in 2008. Orlando et al. demonstrated that the bulk of the cell-cycle-regulated transcriptional program continued to oscillate in concert with budding cycles in cells disrupted for all six S-phase and Mitotic cyclin genes [14]. Most of the cell-cycle-regulated genes were expressed in the proper order and oscillated with a period similar to normal wild-type cells. Data from cyclin mutants and wild-type cells were used to construct a network model, with network interactions drawn from data produced by previous ChIP-chip studies [25, 26, 29].

Dynamical model simulations suggested that the proposed transcriptional network could indeed give rise to autonomous transcriptional oscillations [38]. Simulations of the early transcriptional models [38] produced transcriptome dynamics consistent with those observed in experiments; however, the network models assumed that transcriptome dynamics were a suitable proxy for protein dynamics of the TFs. This underlying assumption was notable in that the dynamical models would fail to reproduce transcriptional dynamics if TF proteins were highly stable and did not oscillate in concert with their mRNAs. A subsequent study utilizing time-series proteomics measurements indicated that transcript levels (measured by microarray or RNA-seq) a reasonable proxy for the protein dynamics of the network components [39].

Although global transcriptional oscillations largely persisted in mutants lacking S-phase and mitotic cyclin activities, it was evident that loss of these cyclins did affect the expression of multiple genes. These mutant cells did have fully functional G1 cyclins, so it remained possible that G1 cyclins were driving a CDK-dependent oscillation. However, it was demonstrated in a subsequent study that transcriptional oscillations persisted even if oscillation of G1 cyclins was halted [30]. What clearly emerged from experiments with perturbed G1-CDK activity was that G1-CDK activities were essential for the high-amplitude transcriptional oscillations observed in wild-type cells. Thus, there was ample evidence that cyclin/CDKs were contributing to the robustness of the transcriptional program, but not required per se, to drive these transcriptional oscillations autonomously.

A complementary approach to analyzing cells mutated for cyclin/CDK complexes was to observe those depleted for Cdc20, a component of one of the APC/C complexes required for negative feedback in the biochemical clock models. Cells depleted for Cdc20 arrest in metaphase with stabilized levels of mitotic cyclin. A time-series microarray experiment employing synchronized cells depleted for Cdc20 demonstrated that large-scale transcriptional oscillations persist even in metaphase-arrested cells [31]. Oscillating gene sets in Cdc20-depleted cells and mitotic cyclin depleted cells were similar, suggesting that large programs of transcription could oscillate independently of either mitotic cyclin/CDK or APC^{Cdc20} functions. These findings provided further evidence that a transcriptional clock could function autonomously to drive oscillations independent of CDK or APC and independent of a cell-cycle arrest.

At this point of the debate regarding the cell-cycle clock models, there were two distinct networks that could oscillate autonomously in two distinct systems: one biochemical (early embryonic systems) and the other transcriptional (budding yeast). Taken together, it was possible either that there were two mutually exclusive cell-cycle clock mechanisms or that the two (what are otherwise) autonomous clocks were operating simultaneously in the budding yeast cell [40]. If the latter is true, the two systems must somehow be coupled under wild-type conditions to maintain synchrony between the transcriptional program and the progression through phases of the DNA replication and segregation cycles. That said, transcriptional oscillations continued even during a cell-cycle arrest at G1/S or mitosis when S-phase and mitotic cyclins (or Cdc20) were depleted, thus uncoupling transcriptional oscillations from cell-cycle progression. At that time, it was not yet clear how transcriptional oscillations could couple to normal cell-cycle progression in wild-type cells.

This question was resolved, at least in part, when time-series transcriptomics approaches were applied to cells arrested in response to checkpoint activation [31]. In two different checkpoint arrests, the dynamics of the transcriptional program halted along with the cell cycle, suggesting that checkpoint pathways are responsible for coupling progression through the transcriptional cycle with progression through the cell-division cycle. When the effector of the DNA replication checkpoint, *RAD53*, was mutated, cell-cycle progression remained arrested while global transcriptional oscillations persisted in checkpoint-arrested cells [31]. These findings indicated that Rad53 was a major checkpoint effector that regulated the transcriptional clock, likely via direct phosphorylation of TFs within the clock network. Together, these findings supported a mechanism by which an otherwise autonomous transcriptional-network clock could be coupled to a similarly autonomous CDK clock (Fig. 1.3).

1.4 A Gene Regulatory Network Clock as an Integrated Model for the Cell Cycle

The definitive network model for the budding yeast cell-cycle clock comes from the 2007 textbook *The Cell Cycle: Principles of Control* by David Morgan (Fig. 1.4a) [41]. The model was not intended to be comprehensive, capturing all the regulatory features of the cell cycle, but rather to illustrate how cells are guided through the successive stages of DNA replication and segregation, including how cells return from mitosis to G1. This model integrated transcription factors with CDKs and APC/C, but did not account for regulatory interactions between TFs. By updating the model with just a few additional TFs and their regulatory connections, the integrated network model captures the features of both the transcriptional (Fig. 1.4b, blue box), and biochemical clock models (Fig. 1.4a). At the top of the network is a transcriptional loop with negative feedback that can drive oscillations, including the

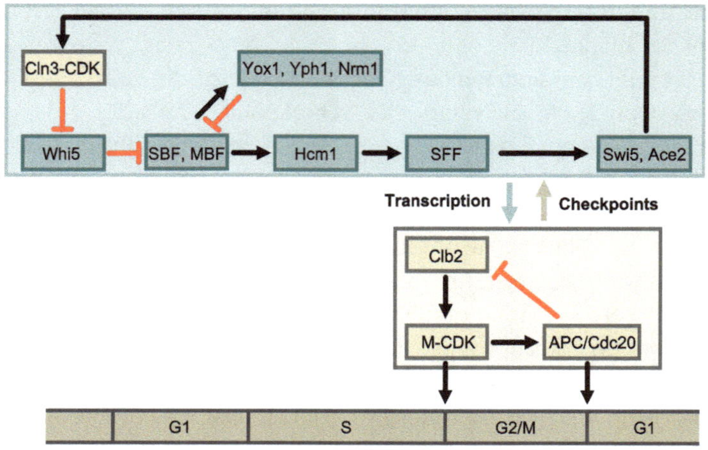

Fig. 1.3 Coupling of the transcriptional (blue box) and CDK/APC (tan box) oscillators in budding yeast. The transcriptional network (blue box) controls the ordered progression of cell-cycle regulated transcription including the synthesis of the M cyclin gene, CLB2. Clb2 being a component of the CDK/APC oscillator (tan box). Checkpoint pathways stall cell-cycle progression in response to perturbations to cell-cycle events. Checkpoint arrest also stalls transcriptional oscillator, coupling the CDK/APC clock with the transcriptional clock. (Adapted from Cho et al. [33])

periodic expression of cyclins even in the absence of M-CDK [32]. At the bottom of the network are the individual CDKs that drive cells through successive cell-cycle phases along with APC/C, which forms a biochemical negative feedback loop that drives oscillations in early embryonic cell-cycle clocks.

The integrated network clock model depicted in Fig. 1.4b is not simply a compromise between the conflicting models, but instead reflects the notion that, under normal conditions, the transcriptional loop and the CDK-APC/C loop exhibit multiple molecular connections and thus are not likely to be functioning as separate autonomous oscillators. The biochemical (kinases and ubiquitin ligases) and transcriptional components (transcription factors) in the model are not sequestered from one another in the cell; to the contrary, they are coupled to each other. Moreover, the integrated network model appears to capture wider functional spectrum of the cell cycle, particularly including the oscillatory transcriptional program observed in cyclin mutant cells arrested for cell-cycle progression [14].

In such an interconnected network it may be empirically difficult to assess whether system-wide oscillations are controlled by a small M-CDK/APC-C negative feedback loop, or whether the transcriptional loop drives oscillations, or both. However, phylogenetic reconstructions of cell-cycle genes suggest that cyclins and CDKs appeared later in the evolution of the cell-cycle clock [42], so it is likely that ancient eukaryotes relied on transcriptional negative feedback loops to drive cell-cycle oscillations. In support of this hypothesis, genetic experiments demonstrated that when M-CDKs are depleted from yeast cells, at least one of the transcriptional repressors of SBF/MBF (Yhp1, Yox1, or Nrm1) is required for the oscillating transcriptional program [32].

If the transcriptional loop is the central clock mechanism driving system-wide oscillations, then the main function of cyclin/CDK complexes may be to serve as the effectors of these oscillations to trigger DNA replication and segregation (i.e., mitosis). The cell cycle will not progress unless CDKs phosphorylate targets that initiate budding, DNA replication, and mitosis, so the cell-cycle arrest phenotypes observed when M-CDK activities are inhibited likely reflect the loss of effector function rather than the loss of an oscillator function per se. Note that in the textbook model (Fig. 1.4a) both G1 and S-phase cyclins are presumably effectors of the M-CDK/APC-C oscillator via controlling the activity of SBF/MBF transcription factors. So it is reasonable to suggest that M-CDK and APC/C are also effectors of a primarily transcriptional oscillator.

The effector functions of CDKs in the cell-cycle GRN are highlighted in studies assessing the phenotypes of yeast cells bearing mutations in different collections of cyclin or CDK mutations. In Cdk1 (*cdc28*) mutant yeast cells that lack all CDK activities, some low amplitude transcriptional oscillations are reproducibly observed, but no cycles of other cellular events persist [30, 32]. In yeast cells that have G1 cyclins but no S-phase or mitotic cyclins, oscillations in the global transcriptional program and budding cycles can be observed, consistent with G1-CDK controlling budding and amplitude of the transcriptional program [13, 14]. If yeast cells have G1 cyclins and one of the S-phase cyclins, but no mitotic cyclins, then cycles of global transcription, budding, and DNA replication are observed in the absence of mitosis [30]. In each case, oscillating expression of specific cyclins drives cycles of the corresponding cellular events triggered by those specific cyclins. Interestingly, these observations may suggest a mechanism for endoreduplication cycles that occur in many fungal, animal, and plant systems.

The regulatory logic of the cell-cycle clock model depicted in Fig. 1.4b can be inferred from the network structure, mathematical models, and from experimental results. In response to cell growth, G1/S transcription is activated by a double-negative regulatory motif in which

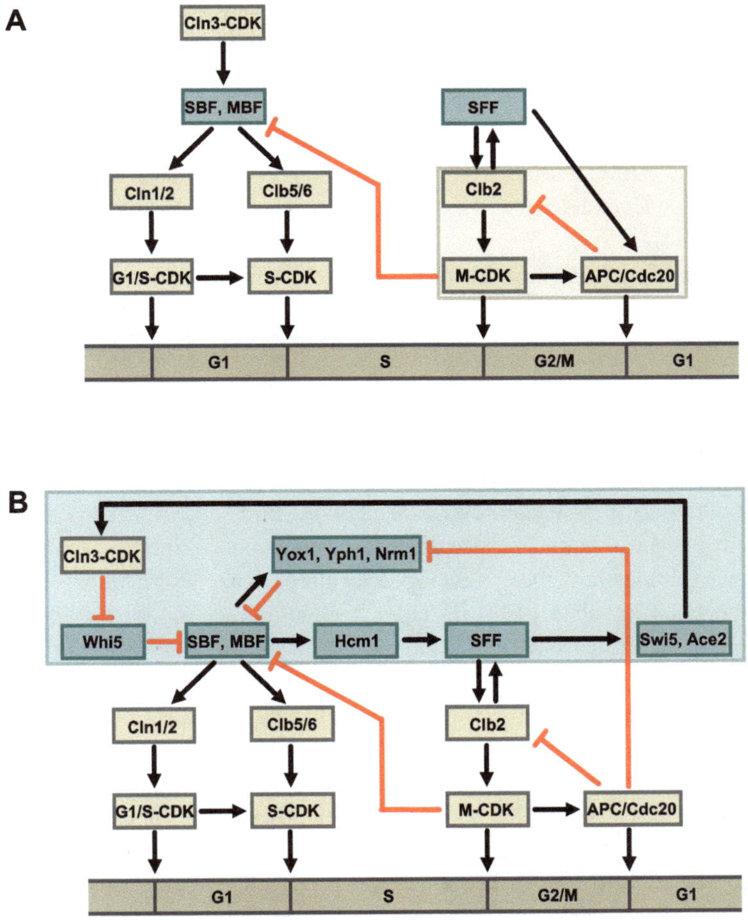

Fig. 1.4 Integrated network clock model for cell cycle control in budding yeast. (**a**) An overview of the budding yeast cell cycle control system. (Adapted from Morgan [41]). (**b**) An updated model that reflects greater functional diversity in the budding yeast cell cycle. Interactions between the core components of both the transcriptional (blue box) and biochemical CDK/APC (tan box) demonstrate how the two potentially autonomous clocks function as an integrated network

G1-CDK inhibits Whi5, thereby releasing the repression of SBF/MBF. The sequential propagation of transcription driven by cascades of TFs (such as SBF/MBF, Hcm1, SFF and Swi5/Ace2) is shaped into a transient transcriptional pulse by negative feedback from the transcriptional repressors Yox1, Yhp1, and Nrm1 [32]. This transcriptional pulse then propagates through the network to sustain oscillations and out from the network to cell-cycle-regulated genes to drive phase-specific expression. This transcriptional pulse also activates the phase-specific expression of distinct cyclins that target substrates required for activating budding, DNA replication, and mitosis. After G1/S, repression of SBF/MBF is maintained throughout the remainder of the cycle by both the transcriptional repressors and the mitotic cyclin, Clb2, to prevent a second transcriptional pulse from being generated. APC^{cdc20} and APC^{cdh1} complexes target Clb2 and the transcriptional repressors for destruction at the end of the cycle, which resets the system for activation when daughter cells arrive in the next G1.

1.5 Discussion and Future Directions

Much of the above integrated cell-cycle clock model is derived from the budding yeast model system, and it will be important to establish whether the integrated GRN clock model is applicable in other eukaryotic systems. Although networks are more difficult to identify and interrogate by genetics and genomics in larger eukaryotes, new technologies for network inference and validation are available. Computational technologies have been developed to infer GRN models from data [43–46]. These technologies often use data from transcriptomic experiments that tend to be accessible even in less tractable systems. Moreover, a new experimental technology using a highly specific estradiol-regulatable promoter system can input a transcription pulse into a GRN at a specified node [47, 48]. Following induction of the pulse, time-series transcriptomics can be used to follow this pulse as it propagates through the network. This approach has been used to map a gene regulatory network in the budding yeast and can distinguish between direct and indirect regulatory interactions (based on time) as well as providing information on whether the TF is a repressor or activator. This technology is likely extensible to other systems to aid in the inference and genetic validation of network models. Together, new technologies should accelerate the generation of robust, integrated GRN models in organisms beyond the budding yeast.

There are indications that the cell-cycle clocks of larger eukaryotes will have substantial similarity to the yeast GRN. Smaller subnetworks (motifs) within the larger clock network of *S. cerevisiae* (Fig. 1.4) are already known to be identical to their mammalian cell counterparts. The cyclin/CDK-Rb-E2F motif controlling S-phase entry in mammalian cells has the same arrangement as the cyclin/CDK-Whi5-E2F motif in budding yeast. This key motif controls the G1/S transition by initiating the transcriptional activation of S-phase genes. Although the arrangement of the cyclin/CDK-Rb-E2F motif is identical to what is observed in budding yeast, there is no sequence relationship between the cyclins and TFs in mammals and yeast [49, 50]. This surprising finding suggests that during clock evolution we may observe conservation of network structure because of functional selection, without conservation of gene sequences.

Although much remains to be fleshed out of the oscillatory systems of somatic cells of higher eukaryotes, there is some suggestive evidence of parallels to the yeast system. Several observations hint at autonomous control of transcriptional oscillations in more complex eukaryotes. When mitotic cyclins are inhibited in the budding yeast, the transcriptional loop of the integrated model can drive oscillations of G1 and S-phase cyclins that in turn drive repeated cycles of DNA replication without intervening mitoses. In *Drosophila*, mitotic cyclins are downregulated to inhibit mitosis in endocycling cells during development. This alternative cell cycle gives rise to large polyploid cells. During these endocycles, it has been proposed that oscillation of the S-phase cyclin

(cyclin E) is driven by an E2F-based oscillator [51]. Additional time-series experiments will be needed to interrogate the *Drosophila* transcriptome when mitotic cyclins are inhibited to determine whether global regulation of transcriptome oscillations is guided by a transcriptional component of the *Drosophila* cell-cycle clock. Despite the full conservation of the G1-CDK/Rb/E2F motif from budding yeast to humans, new evidence will be needed to identify the remaining components of the mammalian clock network, and to determine whether the transcriptional loop can function autonomously to facilitate system-wide oscillations.

During the development of multiciliated cells in humans, cells appear to utilize a different alternative cell cycle that has been called the multiciliation cycle [52]. During that cycle, DNA replication and cell division are inhibited, but cells continue to exhibit the periodic transcription of some cell-cycle regulators, while expression of others is damped. This observation is reminiscent of transcriptional and budding cycles in *S. cerevisiae* cells deleted for S-phase and mitotic cyclins [14]. The mechanisms that control this periodic program have not yet been elucidated, but it is possible that higher eukaryotes may have similar integrated GRNs that can drive periodic behavior in the absence of progression through some cell-cycle phases.

Furthermore, the integrated cell-cycle GRN clock model depicted in Fig. 1.4b shares some important structural and functional features with the circadian clock models for mammalian cells [53]. Both are moderately complex gene regulatory networks made up of TFs, kinases, and ubiquitin ligases, and have transcriptional negative feedback loops that can drive oscillation. Although cell-cycle and circadian clocks both serve different global functions in the cell, both the yeast cell-cycle clock and the mouse circadian clock drive large programs of temporally ordered periodic transcription. The components of the yeast cell-cycle clock and the mammalian circadian clock are unrelated at the sequence level, but they do share quantifiable features. A computational approach was developed to distinguish nodes of cell-cycle clock components from other cell-cycle regulated genes using time-series transcriptomics data. Remarkably, the metrics used to correctly identify the nodes from the budding yeast cell-cycle clock network also accurately identified components of the circadian clock from mouse data [54]. Taken together, similarities in the structure and molecular functions of these clock networks suggest they may share some evolutionary history. Moreover, there are reports of the circadian clock regulating the expression of important cell-cycle regulators such as Wee1 [55]. This regulatory connection suggests the possibility that the cell-cycle clock and the circadian clock may be coupled in some cells.

Given the growing number of discoveries characterizing novel autonomous cellular clocks beyond the canonical cell-cycle and circadian clocks, an important possibility that cells are regulated by multiple autonomous cellular clocks that interact with each other is emerging. For instance, autonomous clocks that control centriole biogenesis, cytoplasmic division, and organelle biogenesis have been reported [56–58]. Moreover, in budding yeast, a large oscillatory program of transcription along with oscillations in dissolved oxygen, called the yeast metabolic cycle (YMC), has been observed in cells grown in chemostat culture [59–61]. YMC oscillations and cell-cycle oscillations appear to have similar periods, indicating the two cycles are coupled in some way. Given the fundamental requirement for energy, cellular metabolism and its dynamic processes may be a foundation for the coordination of multiple oscillations.

A full understanding of the function of the cell-cycle clock will require a more complete understanding of the consequences of its coupling to other autonomous clocks. There is a deep literature from the quantitative sciences on behaviors of coupled oscillators (for example [62, 63]), so there are foundations for thinking about emergent properties of coupled oscillatory networks. To an outsider (and even many insiders) it may feel as though the mechanisms regulating the cell-division cycle have largely been solved. To the contrary, there are likely many more valuable fundamental lessons to be learned.

References

1. Morgan DO (1997) Cyclin-dependent kinases: engines, clocks, and microprocessors. Annu Rev Cell Dev Biol 13:261–291. https://doi.org/10.1146/annurev.cellbio.13.1.261. PubMed PMID: 9442875
2. Murray AW, Kirschner MW (1989) Dominoes and clocks: the union of two views of the cell cycle. Science 246:614–621
3. Nurse P, Thuriaux P (1980) Regulatory genes controlling mitosis in the fission yeast Schizosaccharomyces pombe. Genetics 96(3): 627–637. https://doi.org/10.1093/genetics/96.3.627. PubMed PMID: 7262540; PMCID: PMC1214365
4. Pomerening JR, Sontag ED, Ferrell JE Jr (2003) Building a cell cycle oscillator: hysteresis and bistability in the activation of Cdc2. Nat Cell Biol 5(4):346–351. https://doi.org/10.1038/ncb954. Epub 2003/03/12. PubMed PMID: 12629549
5. Murray AW (1989) Cell biology: the cell cycle as a cdc2 cycle. Nature 342(6245):14–15. https://doi.org/10.1038/342014a0. PubMed PMID: 2535628
6. Simanis V, Nurse P (1987) The cell cycle control gene *cdc2*$^+$ of fission yeast encodes a protein kinase potentially regulated by phosphorylation. Cell 49:569–576
7. Nurse P (1990) Universal control mechanism regulating onset of M-phase. Nature 344:503–508
8. Russell P, Moreno S, Reed SI (1989) Conservation of mitotic controls in fission and budding yeast. Cell 57: 295–303
9. Reed SI, Wittenberg C (1990) Mitotic role for the Cdc28 protein kinase of Saccharomyces cerevisiae. Proc Natl Acad Sci USA 87(15):5697–5701
10. Novak B, Csikasz-Nagy A, Gyorffy B, Chen K, Tyson JJ (1998) Mathematical model of the fission yeast cell cycle with checkpoint controls at the G1/S, G2/M and metaphase/anaphase transitions. Biophys Chem 72(1–2):185–200. https://doi.org/10.1016/s0301-4622(98)00133-1. Epub 1998/07/04. S0301-4622(98)00133-1 [pii]. PubMed PMID: 9652094
11. Chen KC, Calzone L, Csikasz-Nagy A, Cross FR, Novak B, Tyson JJ (2004) Integrative analysis of cell cycle control in budding yeast. Mol Biol Cell 15(8): 3841–3862. https://doi.org/10.1091/mbc.E03-11-0794. Epub 2004/06/01. E03-11-0794 [pii]. PubMed PMID: 15169868; PMCID: 491841
12. Haase SB, Wittenberg C (2014) Topology and control of the cell-cycle-regulated transcriptional circuitry. Genetics 196(1):65–90. https://doi.org/10.1534/genetics.113.152595. PubMed PMID: 24395825; PMCID: 3872199
13. Haase SB, Reed SI (1999) Evidence that a free-running oscillator drives G1 events in the budding yeast cell cycle. Nature 401(6751):394–397. https://doi.org/10.1038/43927. Epub 1999/10/12. PubMed PMID: 10517640
14. Orlando DA, Lin CY, Bernard A, Wang JY, Socolar JE, Iversen ES, Hartemink AJ, Haase SB (2008) Global control of cell-cycle transcription by coupled CDK and network oscillators. Nature 453(7197): 944–947. PubMed PMID: 18463633
15. Epstein CB, Cross FR (1992) *CLB5*: a novel B cyclin from budding yeast with a role in S phase. Genes Dev 6:1695–1706
16. Schwob E, Nasmyth K (1993) CLB5 and CLB6, a new pair of B cyclins involved in DNA replication in Saccharomyces cerevisiae. Genes Dev 7(7A):1160–1175
17. Spellman PT, Sherlock G, Zhang MQ, Iyer VR, Anders K, Eisen MB, Brown PO, Botstein D, Futcher B (1998) Comprehensive identification of cell cycle-regulated genes of the yeast Saccharomyces cerevisiae by microarray hybridization. Mol Biol Cell 9(12): 3273–3297
18. Zhang R, Lahens NF, Ballance HI, Hughes ME, Hogenesch JB (2014) A circadian gene expression atlas in mammals: implications for biology and medicine. Proc Natl Acad Sci USA 111(45):16219–16224. https://doi.org/10.1073/pnas.1408886111. PubMed PMID: 25349387; PMCID: 4234565
19. Koch C, Schleiffer A, Ammerer G, Nasmyth K (1996) Switching transcription on and off during the yeast cell cycle: Cln/Cdc28 kinases activate bound transcription factor SBF (Swi4/Swi6) at start, whereas Clb/Cdc28 kinases displace it from the promoter in G2. Genes Dev 10(2):129–141
20. Amon A, Tyers M, Futcher B, Nasmyth K (1993) Mechanisms that help the yeast cell cycle clock tick: G2 cyclins transcriptionally activate G2 cyclins and repress G1 cyclins. Cell 74(6):993–1007
21. Marais A, Ji Z, Child ES, Krause E, Mann DJ, Sharrocks AD (2010) Cell cycle-dependent regulation of the forkhead transcription factor FOXK2 by CDK·cyclin complexes. J Biol Chem 285(46): 35728–35739. https://doi.org/10.1074/jbc.M110.154005. Epub 20100901. PubMed PMID: 20810654; PMCID: PMC2975197
22. Darieva Z, Pic-Taylor A, Boros J, Spanos A, Geymonat M, Reece RJ, Sedgwick SG, Sharrocks AD, Morgan BA (2003) Cell cycle-regulated transcription through the FHA domain of Fkh2p and the coactivator Ndd1p. Curr Biol 13(19):1740–1745. https://doi.org/10.1016/j.cub.2003.08.053
23. Bulmer R, Pic-Taylor A, Whitehall SK, Martin KA, Millar JB, Quinn J, Morgan BA (2004) The forkhead transcription factor Fkh2 regulates the cell division cycle of Schizosaccharomyces pombe. Eukaryot Cell 3(4):944–954. https://doi.org/10.1128/ec.3.4.944-954.2004. PubMed PMID: 15302827; PMCID: PMC500873
24. Du W, Vidal M, Xie JE, Dyson N (1996) RBF, a novel RB-related gene that regulates E2F activity and interacts with cyclin E in Drosophila. Genes Dev 10(10):1206–1218. https://doi.org/10.1101/gad.10.10.1206. PubMed PMID: 8675008
25. Simon I, Barnett J, Hannett N, Harbison CT, Rinaldi NJ, Volkert TL, Wyrick JJ, Zeitlinger J, Gifford DK, Jaakkola TS, Young RA (2001) Serial regulation of transcriptional regulators in the yeast cell cycle. Cell 106(6):697–708. PubMed PMID: 11572776

26. Lee TI, Rinaldi NJ, Robert F, Odom DT, Bar-Joseph Z, Gerber GK, Hannett NM, Harbison CT, Thompson CM, Simon I, Zeitlinger J, Jennings EG, Murray HL, Gordon DB, Ren B, Wyrick JJ, Tagne JB, Volkert TL, Fraenkel E, Gifford DK, Young RA (2002) Transcriptional regulatory networks in Saccharomyces cerevisiae. Science 298(5594): 799–804. PubMed PMID: 12399584
27. Cox KH, Takahashi JS (2019) Circadian clock genes and the transcriptional architecture of the clock mechanism. J Mol Endocrinol 63(4):R93–R102. https://doi.org/10.1530/JME-19-0153. PubMed PMID: 31557726; PMCID: PMC6872945
28. Pramila T, Miles S, GuhaThakurta D, Jemiolo D, Breeden LL (2002) Conserved homeodomain proteins interact with MADS box protein Mcm1 to restrict ECB-dependent transcription to the M/G1 phase of the cell cycle. Genes Dev 16(23):3034–3045. PubMed PMID: 12464633
29. Pramila T, Wu W, Miles S, Noble WS, Breeden LL (2006) The Forkhead transcription factor Hcm1 regulates chromosome segregation genes and fills the S-phase gap in the transcriptional circuitry of the cell cycle. Genes Dev 20(16):2266–2278. PubMed PMID: 16912276
30. Simmons Kovacs LA, Mayhew MB, Orlando DA, Jin Y, Li Q, Huang C, Reed SI, Mukherjee S, Haase SB (2012) Cyclin-dependent kinases are regulators and effectors of oscillations driven by a transcription factor network. Mol Cell 45(5):669–679. https://doi.org/10.1016/j.molcel.2011.12.033. Epub 2012/02/07. S1097-2765(12)00043-3 [pii]. PubMed PMID: 22306294
31. Bristow SL, Leman AR, Simmons Kovacs LA, Deckard A, Harer J, Haase SB (2014) Checkpoints couple transcription network oscillator dynamics to cell-cycle progression. Genome Biol 15(9):446. https://doi.org/10.1186/s13059-014-0446-7. PubMed PMID: 25200947; PMCID: 4180952
32. Cho CY, Kelliher CM, Haase SB (2019) The cell-cycle transcriptional network generates and transmits a pulse of transcription once each cell cycle. Cell Cycle 18(4):363–378. https://doi.org/10.1080/15384101.2019.1570655. Epub 2019/01/23. PubMed PMID: 30668223; PMCID: PMC6422481
33. Cho CY, Motta FC, Kelliher CM, Deckard A, Haase SB (2017) Reconciling conflicting models for global control of cell-cycle transcription. Cell Cycle 16(20): 1965–1978. https://doi.org/10.1080/15384101.2017.1367073. PubMed PMID: 28934013; PMCID: PMC5638368
34. Rahi SJ, Pecani K, Ondracka A, Oikonomou C, Cross FR (2016) The CDK-APC/C oscillator predominantly entrains periodic cell-cycle transcription. Cell 165(2): 475–487. https://doi.org/10.1016/j.cell.2016.02.060. PubMed PMID: 27058667; PMCID: PMC4826480
35. Hartwell LH (1971) Genetic control of the cell division cycle in yeast. II. Genes controlling DNA replication and its initiation. J Mol Biol 59(1):183–194. PubMed PMID: 5283752
36. Schwob E, Bohm T, Mendenhall MD, Nasmyth K (1994) The B-type cyclin kinase inhibitor p40SIC1 controls the G1 to S transition in S. cerevisiae. Cell 79(2):233–244
37. Verma R, Feldman RM, Deshaies RJ (1997) SIC1 is ubiquitinated in vitro by a pathway that requires CDC4, CDC34, and cyclin/CDK activities. Mol Biol Cell 8(8):1427–1437
38. Sevim V, Gong X, Socolar JE (2010) Reliability of transcriptional cycles and the yeast cell-cycle oscillator. PLoS Comput Biol 6(7):e1000842. https://doi.org/10.1371/journal.pcbi.1000842. Epub 2010/07/16. PubMed PMID: 20628620; PMCID: 2900291
39. Kelliher CM, Foster MW, Motta FC, Deckard A, Soderblom EJ, Moseley MA, Haase SB (2018) Layers of regulation of cell-cycle gene expression in the budding yeast Saccharomyces cerevisiae. Mol Biol Cell 29(22):2644–2655. https://doi.org/10.1091/mbc.E18-04-0255. PubMed PMID: 30207828
40. Pomerening JR (2018) Better together: unifying discordant cell-cycle oscillator models. Cell Cycle 17(1): 9–10. https://doi.org/10.1080/15384101.2017.1389197. Epub 20180102. PubMed PMID: 29108455; PMCID: PMC5815437
41. Morgan DO (2007) The cell cycle: principles of control. New Science Press Ltd, London. 297 p
42. Krylov DM, Nasmyth K, Koonin EV (2003) Evolution of eukaryotic cell cycle regulation: stepwise addition of regulatory kinases and late advent of the CDKs. Curr Biol 13(2):173–177. PubMed PMID: 12546794
43. Bonneau R, Reiss DJ, Shannon P, Facciotti M, Hood L, Baliga NS, Thorsson V (2006) The Inferelator: an algorithm for learning parsimonious regulatory networks from systems-biology data sets de novo. Genome Biol 7(5):R36. https://doi.org/10.1186/gb-2006-7-5-r36
44. Cummins B, Motta FC, Moseley RC, Deckard A, Campione S, Gameiro M, Gedeon T, Mischaikow K, Haase SB (2022) Experimental guidance for discovering genetic networks through hypothesis reduction on time series. PLoS Comput Biol 18(10):e1010145. https://doi.org/10.1371/journal.pcbi.1010145. Epub 2022/10/11. PubMed PMID: 36215333; PMCID: PMC9584434
45. Hecker M, Lambeck S, Toepfer S, van Someren E, Guthke R (2009) Gene regulatory network inference: data integration in dynamic models—a review. Biosystems 96(1):86–103. https://doi.org/10.1016/j.biosystems.2008.12.004
46. Oates CJ, Mukherjee S (2012) Network inference and biological dynamics. Ann Appl Stat 6(3):1209–1235. https://doi.org/10.1214/11-aoas532. PubMed PMID: 23284600; PMCID: PMC3533376
47. Hackett SR, Baltz EA, Coram M, Wranik BJ, Kim G, Baker A, Fan M, Hendrickson DG, Berndl M, McIsaac RS (2020) Learning causal networks using inducible transcription factors and transcriptome-wide time series. Mol Syst Biol 16(3):e9174. https://doi.org/10.15252/msb.20199174. Epub 2020/03/18. PubMed PMID: 32181581; PMCID: PMC7076914

48. Hickman MJ, Petti AA, Ho-Shing O, Silverman SJ, McIsaac RS, Lee TA, Botstein D (2011) Coordinated regulation of sulfur and phospholipid metabolism reflects the importance of methylation in the growth of yeast. Mol Biol Cell 22(21):4192–4204. https://doi.org/10.1091/mbc.E11-05-0467. Epub 2011/09/09. mbc.E11-05-0467 [pii]. PubMed PMID: 21900497; PMCID: 3204079
49. Cross FR, Buchler NE, Skotheim JM (2011) Evolution of networks and sequences in eukaryotic cell cycle control. Philos Trans R Soc Lond Ser B Biol Sci 366(1584):3532–3544. https://doi.org/10.1098/rstb.2011.0078. Epub 2011/11/16. 366/1584/3532 [pii]. PubMed PMID: 22084380; PMCID: 3203458
50. Medina EM, Turner JJ, Gordan R, Skotheim JM, Buchler NE (2016) Punctuated evolution and transitional hybrid network in an ancestral cell cycle of fungi. elife 5. https://doi.org/10.7554/eLife.09492. PubMed PMID: 27162172; PMCID: PMC4862756
51. Edgar BA, Zielke N, Gutierrez C (2014) Endocycles: a recurrent evolutionary innovation for post-mitotic cell growth. Nat Rev Mol Cell Biol 15(3):197–210. https://doi.org/10.1038/nrm3756
52. Choksi SP, Byrnes LE, Konjikusic MJ, Tsai BWH, Deleon R, Lu Q, Westlake CJ, Reiter JF (2024) An alternative cell cycle coordinates multiciliated cell differentiation. Nature 630(8015):214–221. https://doi.org/10.1038/s41586-024-07476-z
53. Takahashi JS (2017) Transcriptional architecture of the mammalian circadian clock. Nat Rev Genet 18(3):164–179. https://doi.org/10.1038/nrg.2016.150. Epub 20161219. PubMed PMID: 27990019; PMCID: PMC5501165
54. Motta FC, Moseley RC, Cummins B, Deckard A, Haase SB (2022) Conservation of dynamic characteristics of transcriptional regulatory elements in periodic biological processes. BMC Bioinformatics 23(1):94. https://doi.org/10.1186/s12859-022-04627-9. Epub 2022/03/19. PubMed PMID: 35300586; PMCID: PMC8932128
55. Matsuo T, Yamaguchi S, Mitsui S, Emi A, Shimoda F, Okamura H (2003) Control mechanism of the circadian clock for timing of cell division in vivo. Science 302(5643):255–259. https://doi.org/10.1126/science.1086271. Epub 20030821. PubMed PMID: 12934012
56. Mofatteh M, Echegaray-Iturra F, Alamban A, Dalla Ricca F, Bakshi A, Aydogan MG (2021) Autonomous clocks that regulate organelle biogenesis, cytoskeletal organization, and intracellular dynamics. elife 10: e72104. https://doi.org/10.7554/eLife.72104
57. Bakshi A, Iturra FE, Alamban A, Rosas-Salvans M, Dumont S, Aydogan MG (2023) Cytoplasmic division cycles without the nucleus and mitotic CDK/cyclin complexes. Cell 186(21):4694–709.e16. https://doi.org/10.1016/j.cell.2023.09.010
58. Aydogan MG, Steinacker TL, Mofatteh M, Wilmott ZM, Zhou FY, Gartenmann L, Wainman A, Saurya S, Novak ZA, Wong SS, Goriely A, Boemo MA, Raff JW (2020) An autonomous oscillation times and executes centriole biogenesis. Cell 181(7):1566–81.e27. https://doi.org/10.1016/j.cell.2020.05.018. Epub 20200611. PubMed PMID: 32531200; PMCID: PMC7327525
59. Slavov N, Botstein D (2011) Coupling among growth rate response, metabolic cycle, and cell division cycle in yeast. Mol Biol Cell 22(12):1997–2009. https://doi.org/10.1091/mbc.E11-02-0132. Epub 2011/04/29. mbc.E11-02-0132 [pii]. PubMed PMID: 21525243; PMCID: 3113766
60. Tu BP, McKnight SL (2007) The yeast metabolic cycle: insights into the life of a eukaryotic cell. Cold Spring Harb Symp Quant Biol 72:339–343. https://doi.org/10.1101/sqb.2007.72.019. Epub 2008/04/19. PubMed PMID: 18419291
61. Tu BP, Kudlicki A, Rowicka M, McKnight SL (2005) Logic of the yeast metabolic cycle: temporal compartmentalization of cellular processes. Science 310(5751):1152–1158. https://doi.org/10.1126/science.1120499. Epub 2005/10/29. 1120499 [pii]. PubMed PMID: 16254148
62. Strogatz SH, Stewart I (1993) Coupled oscillators and biological synchronization. Sci Am 269(6):102–109. https://doi.org/10.1038/scientificamerican1293-102. PubMed PMID: 8266056
63. Gedeon T, Cummins B (2023) Oscillator death in coupled biochemical oscillators. Math Control Signals Syst 35(4):781–801. https://doi.org/10.1007/s00498-023-00348-3. Epub 20230421. PubMed PMID: 41178938; PMCID: PMC12574442
64. Pomerening JR, Kim SY, Ferrell JE Jr (2005) Systems-level dissection of the cell-cycle oscillator: bypassing positive feedback produces damped oscillations. Cell 122(4):565–578. https://doi.org/10.1016/j.cell.2005.06.016. PubMed PMID: 16122424

The Centriole Biogenesis Cycle: Temporal Mechanisms, Autonomy and Coupling to the Cell Cycle and Other Cellular Clocks

2

Cindy Ow and Mustafa G. Aydogan

Abstract

Centrioles are cytoskeletal organelles whose duplication must occur with high temporal precision to ensure faithful cell division, ciliogenesis and tissue organization. Traditionally, the centriole biogenesis cycle has been viewed as a surrogate to the nuclear division cycle, particularly to oscillations in cyclin-dependent kinase (CDK) activity. In this chapter, we revisit classic experimental observations demonstrating that centriole duplication can proceed independently of DNA replication, transcription and CDK oscillations, thereby revealing an intrinsic autonomy in this cycle. We then discuss emerging molecular evidence identifying oscillatory dynamics of Polo-like kinase 4 (Plk4) as the core autonomous clock mechanism that initiates and times centriole biogenesis. Building on this framework, we examine mechanisms that entrain or couple the centriole cycle to the mitotic cycle, ciliogenesis and circadian rhythms, highlighting shared regulatory modules such as ubiquitin ligases and phosphatases. We explore the physiological importance of maintaining proper coupling between these cycles through examples of cancer pathologies and developmental defects that arise when centriole timing becomes uncoupled. Finally, we outline outstanding questions regarding the temporal machinery that governs the centriole biogenesis cycle, its evolution, and emerging therapeutic opportunities, positioning this cycle as a model system for understanding how autonomous cellular clocks integrate into broader temporal control programmes in biology.

Keywords

Centriole biogenesis · Centriole duplication · Polo-like kinase 4 (Plk4) · Plk4 oscillator · Autonomous cellular clocks · Centrosome cycle · Cell cycle coupling · CDK-cyclin oscillations · Ciliogenesis · Circadian

C. Ow
Department of Genetics and Development, Vagelos College of Physicians and Surgeons, Columbia University, New York, NY, USA

Developmental and Stem Cell Biology Graduate Program, University of California, San Francisco, CA, USA

M. G. Aydogan (✉)
Department of Genetics and Development, Vagelos College of Physicians and Surgeons, Columbia University, New York, NY, USA

Howard Hughes Medical Institute, Columbia University, New York, NY, USA

Columbia Stem Cell Initiative, Columbia University Medical Center, New York, NY, USA

Herbert Irving Comprehensive Cancer Center, Columbia University, New York, NY, USA
e-mail: mga2157@columbia.edu

M. G. Aydogan (ed.), *Biological Time Control*, Advances in Experimental Medicine and Biology 1521,
https://doi.org/10.1007/978-3-032-36189-9_2

rhythms · CP110 · Ubiquitin ligases · Phosphatases · Centrosome amplification · Chromosomal instability · Cancer pathophysiology

2.1 Introduction

2.1.1 Centrioles and Their Biogenesis Cycle

Centrioles are microtubule-based cytoskeletal organelles that, during quiescence, serve as the structural foundation for cilia formation, and during proliferation, act as the core scaffold for the assembly of pericentriolar material that matures into a microtubule-organizing centre (MTOC), known as the centrosome [1, 2]. Centrosomes are essential for proper mitotic spindle assembly and the fidelity of chromosome segregation [3]. While centrioles can form de novo through self-assembly [4], their canonical mode of biogenesis in proliferating cells involves a pre-existing centriole (the mother centriole), which serves as a platform upon which a new centriole (the daughter centriole) is assembled—thereby generating a duplicate of the mother—a process thus termed centriole duplication. Although de novo centriole formation and centriole duplication occur through distinct mechanisms, they are not necessarily mutually exclusive: a centriole that initially forms de novo can subsequently mature into a mother centriole capable of initiating repeated rounds of duplication [5, 6]. As such, this chapter focuses on centriole duplication, its temporal regulation and the molecular basis of its repeated rounds, collectively referred to here as the centriole biogenesis cycle.

Centrioles are composed of a barrel-shaped, nine-fold symmetric cartwheel structure at their core, whose initiation of growth is closely followed by the nearly concomitant assembly of centriolar microtubule (MT) triplets that extend from each of the nine spokes of the cartwheel [7, 8] (Fig. 2.1). Centriole duplication first requires the maturation of newly formed daughter centrioles into mother centrioles—a process covered in extensive detail elsewhere. Newly matured mother centrioles recruit Asl/Cep152 in a ring-shaped halo surrounding the centriole, which in turn enables the recruitment of Polo-like kinase 4 (Plk4), initially in a similarly ring-shaped pattern [9, 10] (Fig. 2.1). Through a molecular symmetry-breaking event, Plk4 then resolves into a highly focused condensate [11], most likely mediated by phase-separation mechanisms, which rapidly concentrates the key cartwheel components Sas6 and STIL/Ana2 [9, 12–14] (Fig. 2.1). This event not only ensures the formation of a single procentriole site on the mother, thereby preserving the canonical

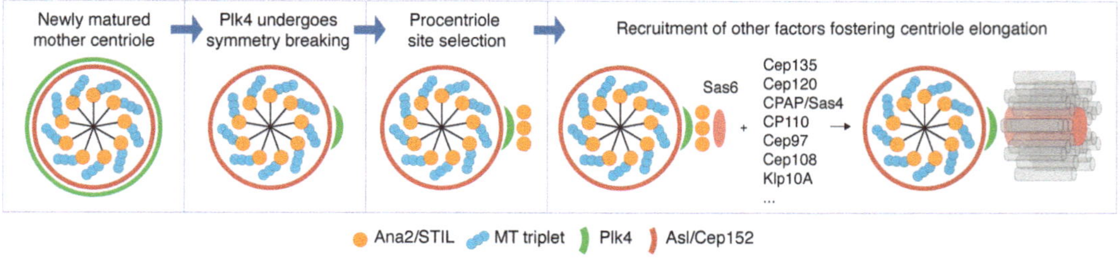

Fig. 2.1 Molecules involved in centriole duplication. Newly formed mother centrioles covered with Asl/Cep152 recruit Plk4 initially in a ring-shaped pattern. Plk4 then undergoes a molecular symmetry-breaking event and resolves into a highly focused condensate, which rapidly concentrates Sas6 and STIL/Ana2. This event not only ensures the formation of a single procentriole site on the mother and preserves the canonical one-daughter-per-mother duplication pattern but also enables the rapid and cooperative growth of the cartwheel structure. Soon after the onset of cartwheel assembly, proteins such as Cep135, Cep120 and CPAP/Sas-4 help tether centriolar MT triplets to the core cartwheel structure, while CP110, Cep97, Cep108 and Klp10A help mediate proper centriole elongation

one-daughter-per-mother duplication pattern, but also enables the rapid and cooperative growth of the cartwheel structure. As demonstrated by Gönczy and colleagues, while the cartwheel rings can self-assemble in a snap-like fashion due to locally high concentrations of Sas-6 in vitro [15], Plk4 regulates the kinetics of this process, accelerating or slowing cartwheel growth in response to the concentrations of additional regulatory proteins in vivo [16–18]. Consistent with this model, the cartwheel elongates from the proximal end of the growing daughter centriole—adjacent to the mother centriole, where Plk4 is localized—rather than from the distal end pointing away from the mother [16].

Soon after the onset of cartwheel assembly, proteins such as Cep135 help tether the cartwheel to nascent centriolar MT triplets [19], while elongation factors including Cep120 and CPAP/Sas-4 promote the regulated extension of centriolar MTs [20–22] (Fig. 2.1). As centriole elongation proceeds, the distal ends of the growing daughter centrioles become decorated by MT (+)-end–associated proteins such as CP110, Cep97, Cep104 and Klp10A, which collectively regulate centriole elongation and help fine-tune centriole length [18, 23–27]. Together, these processes couple the site selection and elongation of daughter centrioles on their mothers into a coordinated growth programme that defines both their number and ultimate structure.

The process of centriole duplication occurs once—and only once—during each cell cycle. For this reason, the centriole biogenesis cycle has long been considered a surrogate for the cell's DNA replication and segregation cycle [28]. In this chapter, we focus on the temporal mechanisms governing centriole biogenesis: first, we review classic evidence demonstrating that the centriole biogenesis cycle can proceed in the absence of DNA replication, the nucleus, and even oscillatory CDK–cyclin activity. Next, we examine emerging molecular insights into the centriole biogenesis cycle, centred on a newly discovered molecular clock capable of initiating and timing centriole biogenesis autonomously of the canonical cell cycle machinery. We then describe mechanisms that help entrain—and perhaps couple—such otherwise autonomous centriole biogenesis cycle to cycles of DNA replication and chromosome segregation. We also discuss molecular systems that synchronize alternating cycles of ciliogenesis and mitosis with the centriole biogenesis cycle. Moreover, we highlight emerging evidence suggesting that the centriole biogenesis cycle and ciliogenesis programme may be entrained in physiology beyond mitotic signals, including by circadian rhythms. Together, these perspectives allow us to explore the physiological and pathological implications of a centriole biogenesis cycle that is intrinsically autonomous yet normally coordinated with major temporal programmes such as the cell cycle and the circadian clock. We conclude with a forward-looking section outlining key outstanding questions in the field and the scientific opportunities that they present.

2.2 Autonomy of the Centriole Biogenesis Cycle from Major Temporal Programmes

2.2.1 Historical Evidence on the Autonomy of Centriole Biogenesis Cycle

Centrioles—when decorated with astral microtubules to form centrosomes—are easily observed across different phases of the cell division cycle, owing their birefringent properties under light microscopy. Accordingly, their duplication has long been considered a direct consequence of the progression of the cell division cycle, a view dating back to the late nineteenth century [29]. However, from the early days of modern cell biology, several classic observations have challenged this notion, suggesting that the centriole biogenesis cycle can proceed independently of key features of the cell division cycle.

Early studies from several laboratories reported that inhibition of DNA replication or enucleation in cultured mammalian cells prevents further rounds of centriole biogenesis [30, 31]. Reasoning that such perturbations might limit the production of precursors required for the centriole biogenesis cycle—perhaps due to stalled transcription—Sluder and Rieder turned to a system in which these components would be abundantly stockpiled. In their seminal 1986–1987 studies using sea urchin eggs, they showed that complete removal of the nucleus did not abolish centriole duplication (Fig. 2.2a); rather, centrioles underwent at least three to four successive rounds of duplication [32, 33]. Notably, this occurred with a periodicity slightly slower than that observed in control eggs undergoing normal cell cycles [33], suggesting that nucleus-associated processes may accelerate the centriole biogenesis cycle under physiological conditions. Importantly, as confirmed by electron microscopy, the newly formed centrioles were architecturally indistinguishable from those formed in control conditions [32], indicating that the centriole biogenesis process can proceed in periodic cycles autonomously.

These findings by Sluder and Rieder were pivotal to reinforcing other much earlier observations by Joan Lorch, who reported nearly three decades prior that enucleated sea urchin blastomeres can sustain repeated 'multiplication' of astral centres [34]. Although these earlier studies lacked the resolution and conceptual framework to definitively identify centrioles as the templates for centrosomes and astral microtubules—at least certainly by Lorch who regarded centrioles as 'fixation artefacts'—they nonetheless pointed to a nucleus-independent capacity for centrosome-associated structures to proliferate. Together, these findings provided early and compelling evidence that the centriole biogenesis cycle (provided that its building blocks are sufficiently abundant) can proceed independently of the nucleus, mRNA transcription

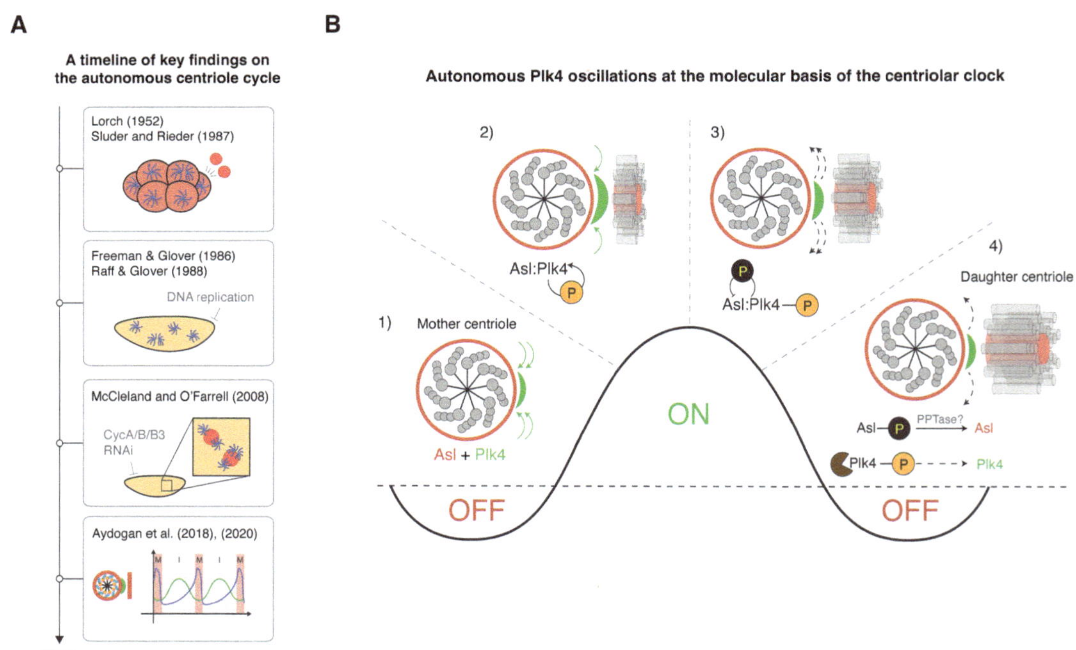

Fig. 2.2 Autonomy of centriole biogenesis cycle and its molecular underpinnings. (**a**) A chronicle of major advances leading to the discovery that centriole biogenesis cycle could run autonomously of nuclei, DNA replication, and CDK/cyclin activity. (**b**) Proposed molecular basis of the autonomous Plk4 oscillator driving centriole biogenesis cycles based on recent findings [16, 17]. Briefly, Plk4 (*green*) is recruited to centrioles by its centriolar receptor Asl/Cep152 (*red*) (step 1). Plk4 begins to activate itself by its phosphorylation in trans and trigger daughter centriole formation (step 2). Next, active Plk4 may phosphorylate its receptor to inhibit its own binding affinity, leading to Plk4's departure and/or potential degradation simultaneously to cease daughter centriole assembly (step 3). To reset the clock, the phosphorylated receptor may be dephosphorylated by a phosphatase to start the process over (step 4)

and the canonical cycles of DNA replication and segregation.

Subsequent studies in the 1980s further supported the mitotic autonomy of the centriole biogenesis cycle. For example, work from the laboratory of David Glover demonstrated in *Drosophila* embryos that centrioles can undergo repeated rounds of duplications in the absence of nuclear division, both in the presence and absence of ongoing DNA replication [35, 36] (Fig. 2.2a). Complementary studies by Doxsey [37], as well as by Sluder and Rieder [38], showed in *Xenopus* blastulae and sea urchin eggs that inhibition of protein synthesis—by then known to be required for multiple aspects of the nuclear division cycle [39–41]—does not prevent successive rounds of centriole duplication. Since these embryonic systems are characterized by minimal checkpoint control and the absence of canonical circadian clock machinery, these studies have collectively established a foundational body of evidence that the centriole biogenesis cycle can operate autonomously from the canonical cycles of DNA replication and segregation.

2.2.2 Molecular Underpinnings of the Autonomy: Plk4 Oscillations Independent from the CDKs

What could underlie the periodic nature of centriole biogenesis? For many years, the most obvious explanation centred on Cdk1 activity, which—unlike other canonical Cdks (e.g. Cdk2 or Cdk4/6)—is initially activated in the cytoplasm where centrioles are located [42–45]. This made Cdk1 an attractive candidate regulator for several reasons: first, its activity is inherently periodic [46], providing a plausible basis for the cyclic nature of centriole biogenesis; and second, its oscillations can be reconstituted in vitro in the absence of DNA replication and nuclei [47, 48]. At the same time, some reports had suggested that protein synthesis might be required for continued centriole biogenesis in certain experimental systems such as L929 mouse cells [49], seemingly contradicting findings from *Xenopus* blastulae and sea urchin eggs discussed above. These discrepancies raised the possibility that the key translational requirement might lie in the periodic synthesis of cyclins during each duplication cycle, rather than in the synthesis of centriole structural components themselves. In this view, continuous protein translation would be necessary primarily to sustain oscillatory Cdk1 activity.

As such, the temporal regulation of centriole biogenesis had long been thought to be metachronous with oscillatory Cdk1 activity in the cytoplasm, despite the absence of direct experimental evidence demonstrating that Cdk1 oscillations are sufficient to drive the intrinsic periodicity of the centriole biogenesis cycle. Nearly two decades after these debates began, McCleland and O'Farrell set out to formally test whether Cdk1 activity is indeed required for this process [50] (Fig. 2.2a). By treating early *Drosophila* embryos with a cocktail of double-stranded RNAs against all mitotic cyclins (A, B and B3), they demonstrated that cycles of centriole biogenesis can continue independently of Cdk1 activity. This result has challenged the long-held assumption that the oscillatory nature of centriole biogenesis is merely an epiphenomenon of Cdk1 activity. Yet, the mechanism that underlies the intrinsic periodicity of the centriole remained unknown.

Approaching centriole biogenesis from an orthogonal perspective, but also using early *Drosophila* embryos, we were interested in understanding how daughter centrioles consistently grow to the correct size during their duplication from their mothers. In this context, one key observation from early *Drosophila* embryos guided the way: when daughter centrioles grow more slowly, they do not simply end up shorter. Instead, they compensate by growing for a longer period [16]. Conversely, when growth proceeds more rapidly, daughter centrioles grow for a shorter duration yet still achieve the same final size. These findings suggested the existence of a molecular mechanism that enforces size homeostasis during centriole assembly by adjusting the period of centriole growth [16]. In other words, centriole size appeared to be intrinsically linked to the duration of its growth period. Screening for centriolar molecules that can elicit this homeostasis, we discovered that Polo-like kinase 4 can single-handedly account for this process [16]. For instance, when Polo-like kinase 4 levels are genetically halved, daughter centrioles by default grew slower, but for a longer period than their wild-type counterparts. But how could Plk4 achieve such homeostasis?

While attempting to address this question, we serendipitously discovered what proved to be a central feature of the molecular machinery: Plk4 levels oscillate at centrioles with a frequency and phase closely aligned with the cell cycle [17] (Fig. 2.2a). At the onset of interphase, when centrioles begin to grow, Plk4 levels start to rise, reaching their peak by mid-interphase. In contrast, during mitosis—a phase refractory to centriole growth—Plk4 levels fall to their lowest point (Fig. 2.2b). These observations suggested that oscillations in centriolar Plk4 levels might define the precise temporal window during which centriole biogenesis can occur within the cell cycle.

Indeed, when we monitored Plk4 dynamics and centriole growth simultaneously, we found that the same threshold level of Plk4 that triggered the initiation of centriole growth also helped terminate it. In other words, as Plk4 levels increased and crossed a critical accumulation threshold at the mother centriole, daughter centrioles began to grow; conversely, when Plk4 levels declined below this threshold, daughter centrioles ceased growing [17] (Fig. 2.2b). Together, these findings indicated that Plk4 oscillations not only initiate but also temporally delimit centriole biogenesis.

Importantly, our work further demonstrated Plk4 oscillations can persist and drive cycles of centriole biogenesis independently of CDK/cyclin oscillations, thereby formally establishing their autonomy from the canonical cycles of DNA replication and chromosome segregation. Under these autonomous conditions, Plk4 oscillations ran more slowly than under wild-type conditions, suggesting that the CDK system normally acts to accelerate Plk4 oscillations to match the pace of nuclear division cycle [17]. This interpretation is consistent with the historical observations described above, in which centriole biogenesis cycles also progressed more slowly in other systems like the enucleated sea urchin eggs [32].

Collectively, these studies in *Drosophila* provided the first experimental evidence for the molecular rhythms underlying the centriole biogenesis cycle—one that is typically entrained by CDK/cyclin activity to match the timing of DNA replication and chromosome segregation, yet retains an intrinsic capacity to run autonomously when CDK/cyclin function is silenced.

2.2.3 Emerging Mechanisms and Conserved Features of the Plk4 Clock for Centriole Biogenesis

Recent studies combining biochemistry, biophysics, genetics and mathematical modelling have begun to elucidate the mechanisms underlying Plk4 oscillations at the centriole (Fig. 2.2b). We have shown that Plk4 localizes to the centriole through binding to its centriolar docking molecule Asl/Cep152, marking the *rise phase* of the oscillation [17] (Fig. 2.2b, Step 1). This helps concentrate Plk4 at the centriole, promoting its self-dimerization and activation, thereby initiating centriole growth. Next, the fully active pool of centriolar Plk4 phosphorylates itself in trans (Fig. 2.2b, Step 2) and its docking molecule Asl/Cep152, corresponding to the *peak phase* of the oscillation (Fig. 2.2b, Step 3). As such, active Plk4 both promotes its own degradation via associated phosphodegrons and self-weakens the affinity toward its docking molecule Asl/Cep152 (Fig. 2.2b, Step 3), respectively, which defines the *peak-to-trough phase* of the oscillation. As a result, the effective pool of centriolar Plk4 begins to decline, likely through a combination of dissociation and degradation. Once Plk4 levels fall below a critical threshold, centriole growth ceases. To initiate another round of Plk4 oscillation and centriole duplication, the cycle is reset during *the trough phase*, likely when both the docking molecule Asl/Cep152 and the remaining cytoplasmic pool of phosphorylated Plk4 that escaped degradation are dephosphorylated [17] (Fig. 2.2b, Step 4).

Studies from other groups have confirmed that several mechanistic aspects of these findings are conserved across systems. For example, Kitagawa and colleagues have demonstrated in cultured human osteosarcoma cells that Plk4 localizes to centrioles in oscillatory cycles that occur in phase with the mammalian cell cycle [51]. From a more mechanistic perspective, the laboratory of Greg Rogers showed in cultured S2 cells that the negative feedback relationship between Plk4 and its docking protein Asl/Cep152 operates similarly to that observed in early fly embryogenesis: non-phosphorylated Asl binds Plk4 and promotes its kinase activity, but once Asl itself is phosphorylated by Plk4, this helps suppress Plk4 activity [52]. This negative feedback loop helps delimit Plk4 activity at the centriole to cyclic bursts, thereby defining repeated rounds of a temporal window when centriole duplication can occur. Furthermore, Bettencourt-Dias and colleagues demonstrated that elevating Plk4 levels in unfertilized eggs induces de novo centriole formation, followed by periodic rounds of centriole duplication from the same centrioles independently of cell cycle progression [6].

Combined with work from these studies, our work has revealed the existence of a molecular clock based on the oscillatory localization of Plk4 at the centriole that initiates and temporally patterns centriole biogenesis in repeated cycles independently of the CDK/cyclin oscillator.

2.3 Coupling of the Centriole Cycle to the Mitotic Cycle and Other Biological Clocks

Arguably, the most prominent role of centrioles is to enable the formation of centrosomes that nucleate astral microtubules, thereby ensuring the fidelity of spindle assembly during mitosis. To achieve bipolar spindle formation, cells must establish exactly one centrosome at each pole, which in turn requires that centriole number be reset to two at the start of each cell cycle following mitosis. This necessitates tight regulation to ensure that centrioles duplicate once—and only once—per cycle.

Indeed, the longstanding adage that '*centrioles duplicate once and only once during the cell cycle*' implies the existence of intrinsic mechanisms that constrain centriole biogenesis to occur in step with the DNA replication–segregation cycle. In other words, at least some regulatory mechanisms must link one or more steps of the mitotic cycle to the three principal stages of the centriole biogenesis cycle: first, the physical detachment of the newly formed daughter centriole from the old mother centriole to become a new mother (i.e. centriole disengagement); second, the duplication of centrioles from both the old and newly formed mother centrioles; and finally, the separation of the centrosome pair—each consisting of a mother centriole with its newly formed daughter—such that they can again undergo disengagement and initiate another round of duplication.

2.3.1 Entrainment of the Centriole Biogenesis Cycle by Mitotic Cues

Traditionally, the initiation of DNA replication during late G1/S phase has been viewed as the permissive condition for centriole duplication to begin [28, 53]. Classic studies from the laboratories of Greenfield Sluder and Tim Stearns demonstrated that centriole separation requires active Cdk2 together with its binding partners cyclin A and cyclin E [54, 55]. More recent studies have further revealed that, once DNA replication initiation is licensed by Cdc6 in the nucleus, DONSON—a component of the DNA helicase machinery—displaces Cdc6 and promotes its translocation to the cytoplasm [56] (Fig. 2.3). Cytoplasmic Cdc6 then localizes to the centrosome, where it prevents premature separation of newly forming daughter centrioles. Together with classic molecular processes that link mitotic exit to centriole duplication [57, 58], this mechanism ensures that DNA replication initiation not only provides a permissive window for the centriole biogenesis cycle to begin, but also helps restrict it to a single round per cell cycle.

In contrast to S phase, which acts as a permissive window for centriole biogenesis, the onset of mitosis creates a prohibitive state for centriole duplication. Toward the end of interphase and during early mitosis, centrosome separation is licensed by rising mitotic kinase activity. Plk1 counteracts PP1γ-mediated dephosphorylation of the Mst2–Nek2A complex [59], thereby dissolving the fibrous connections that tether the

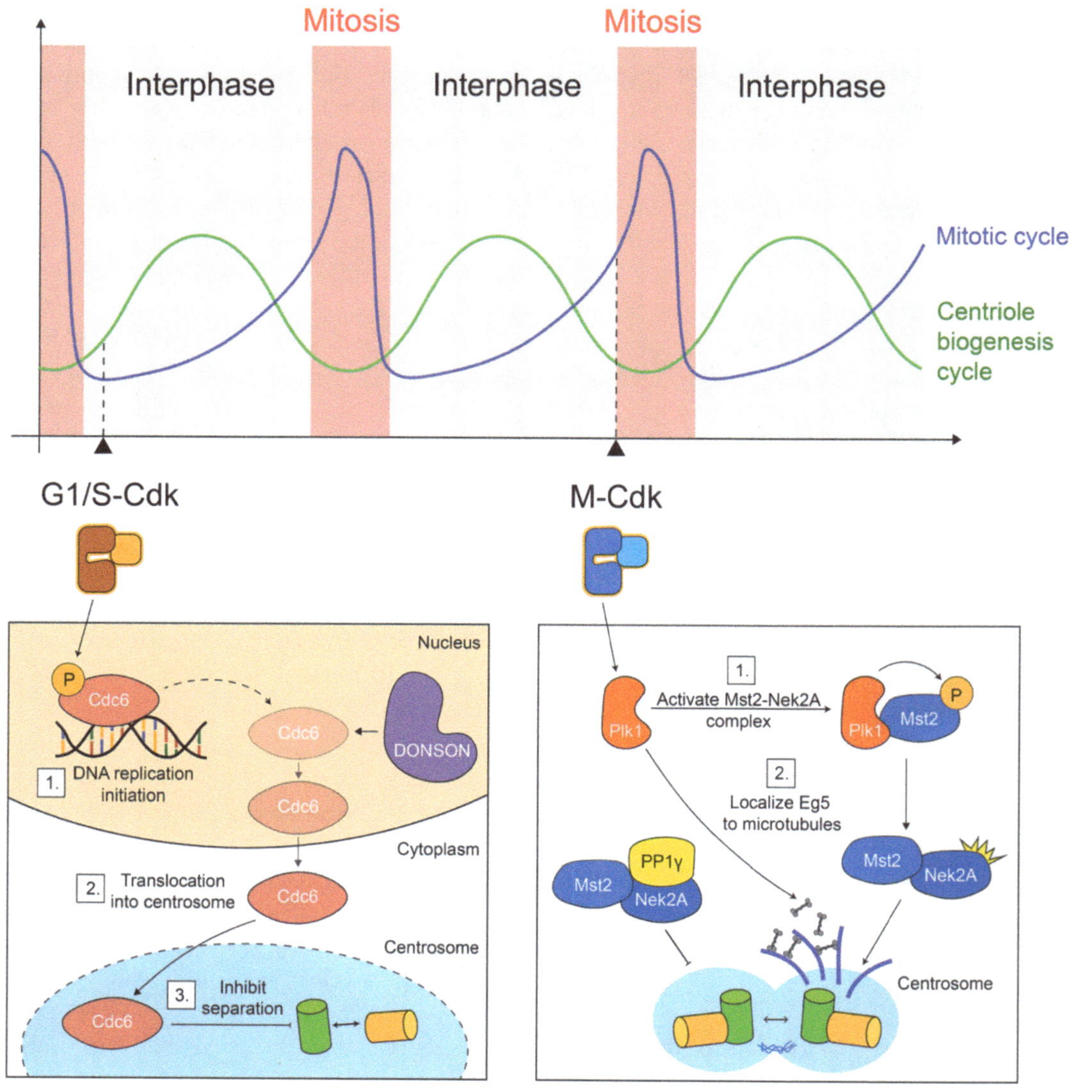

Fig. 2.3 Centriole biogenesis cycle can be *entrained* by cues from the mitotic cycle. In S phase, DNA replication initiation licensing by Cdc6 in the nucleus precedes DONSON displacing Cdc6 out of the nucleus into the cytoplasm. There, Cdc6 localizes to the centrosome, where it prevents premature separation of newly forming daughter centrioles (left panel). In contrast, during the onset of mitosis, rising mitotic kinase activity licenses centrosome separation by activating Plk1, which counteracts PP1γ-mediated dephosphorylation of Mst2-Nek2A complex. Concomitantly, Eg5 kinesin motors are phosphorylated to localize onto astral microtubules (right panel)

two centrosomes (Fig. 2.3). At the same time, Cdk1 phosphorylates Eg5 kinesin motors [60], promoting astral microtubule–dependent centrosome separation (Fig. 2.3). Although, in principle, separated centrosomes could initiate another round of centriole biogenesis during mitosis, recent studies have demonstrated that high Cdk1 activity imposes a biochemical restriction on centriole duplication. Specifically, Cdk1's inhibitory phosphorylation of STIL and Sas6—core cartwheel components—prevents the self-assembly of STIL and Sas6 into a new centriole [14, 61, 62]. As a result, centrioles can only initiate a new duplication cycle once Cdk1 activity declines at the onset of telophase.

Canonically, much of this regulation has been attributed to the activity of CDK–cyclin complexes [63]. Emerging evidence suggests that, although CDKs *entrain* key steps of the centriole biogenesis cycle to specific phases of the mitotic cycle, the centriole cycle itself can operate autonomously [17]. This notion of entrainment is supported by observations that, when cells are arrested for prolonged periods in interphase at the G0/G1 or G1/S boundary without increasing CDK2 activity, centrioles can nevertheless resume duplication [64, 65]. Thus, although rising CDK2 activity helps entrain the onset of centriole duplication to the initiation of DNA replication, CDK2 itself is not required for centriole duplication—a conclusion directly supported by studies showing that genetic deletion or biochemical inactivation of CDK2 does not prevent centriole duplication [66, 67]. Conversely, when cells are arrested in mitosis under high CDK1 activity, the autonomous Plk4 oscillatory machinery that normally drives centriole biogenesis can still continue to run, but its amplitude remains below threshold required to initiate centriole duplication [68] (Fig. 2.4a).

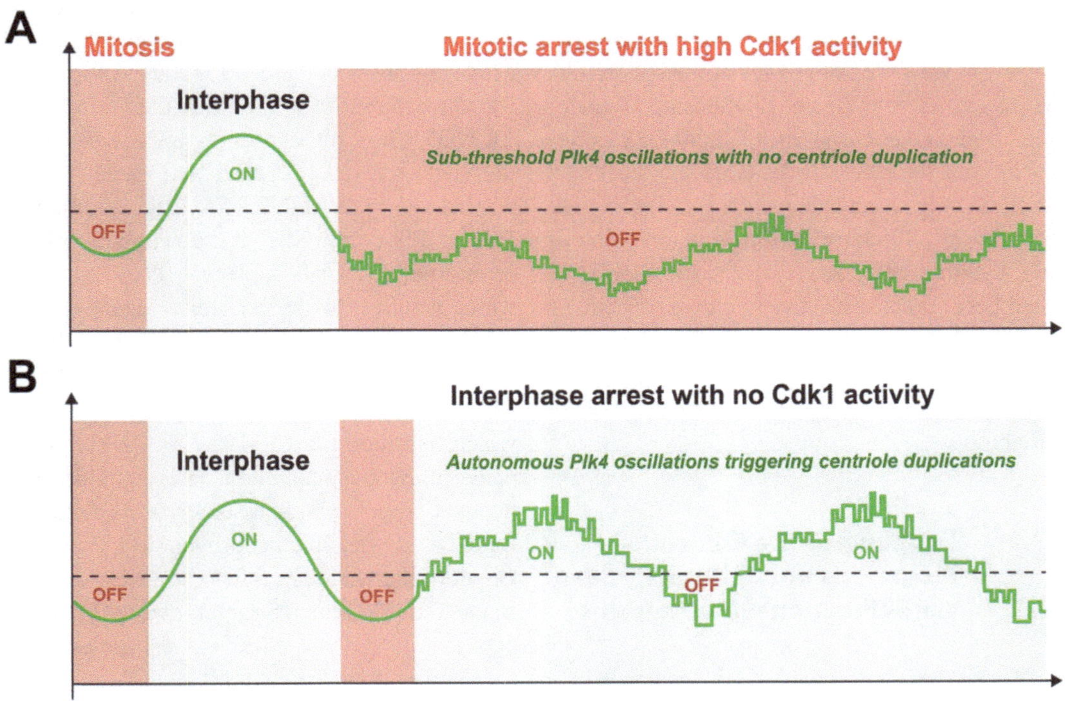

Fig. 2.4 Prolonged arrest in opposing phases of the cell cycle has contrasting effects on the centriole duplication cycle. (**a**) During prolonged mitotic arrest, high CDK1 activity permits autonomous Plk4 oscillations to continue, but their amplitude remains below the threshold required to trigger centriole duplication, thereby locking the centriole duplication cycle. (**b**) By contrast, during prolonged interphase arrest, when CDK1 activity is halted, Plk4 oscillations continue with sufficient amplitude to drive periodic bouts of centriole biogenesis

Duplication is therefore inhibited under these conditions (Fig. 2.4a), consistent with studies demonstrating an inhibitory role for CDK1 in centriole assembly [14, 61, 62]. Accordingly, genetically induced increases or decreases in CDK1 activity can shift the phase of Plk4 oscillation, causing it to peak prematurely or with a delay during interphase as Cdk1 activity rises towards mitosis [17]. Thus, while CDK1 helps entrain the termination of centriole duplication to the onset of mitosis, it is not required for centriole duplication itself. Indeed, as with CDK2, inhibition of CDK1 does not prevent continuous rounds of centriole duplication and its associated autonomous Plk4 oscillations [17, 50] (Fig. 2.4b).

Collectively, these findings indicate that CDK2-dependent control of centriole duplication initiation and CDK1-dependent control of its termination primarily serve to align the centriole biogenesis cycle with nuclear cycle progression, rather than to directly drive centriole duplication. This could also explain why different CDK/cyclin pairs both promote and prevent centriole biogenesis even within a single experimental system [69]. Such an entrainment relationship is analogous to circadian entrainment: light–dark cycles synchronize the phase of the circadian clock, yet the clock itself is capable of free-running in their absence [70]. Similarly, a complementary pair of CDK-cyclin complexes establishes permissive (interphase) and restrictive (mitosis) windows for centriole biogenesis to occur, while the underlying molecular clock—driven by the Plk4 oscillator—retains an intrinsic capacity to operate autonomously.

2.3.2 Coupling of the Centriole Biogenesis and Mitotic Cycles via Bidirectional Mechanisms

Such an entrainment model for the centriole biogenesis clock is further elaborated by other recent evidence demonstrating that the centriole biogenesis cycle might even be *coupled* with the CDK-cyclin oscillator. In other words, this would mean that, when active, CDK-cyclin oscillator can impose on the centriole biogenesis clock a rate limiting step that would lock the frequency of the latter's oscillations to its own oscillations. For instance, work in *Drosophila* embryos have demonstrated that Wee1 kinase—a key enzyme that antagonizes Cdk1 activity by its inhibitory phosphorylation—is required for coupling the peak activity time of the centriole biogenesis clock to the progression of interphase and rising Cdk1 activity levels [17] (Fig. 2.5; left). Since early embryos go through alternating S- and M-phases, this could indicate that a step of the centriole biogenesis cycle is coupled to the progression of DNA replication via a Wee1-dependent mechanism.

Importantly, such a coupling between the two cycles may not just be unidirectional from CDK-cyclins to the centriole cycle. Recent evidence has demonstrated that the centriole biogenesis cycle may in turn induce molecular breaks on the nuclear division cycle beyond traditional mechanisms of how under or over-replicated centrioles may lead to structural defects during mitosis, which in turn can activate broader cell death responses in the cell [71, 72]. Recent studies have identified the E3 ubiquitin ligase TRIM37 whose function may impose breaks on the nuclear division cycle based on Plk4 levels and activity in the cell [73, 74] (Fig. 2.5; middle). When Plk4 activity is downregulated by centrinone (a small molecule Plk4 inhibitor), which would slow the centriole biogenesis cycle [75], the nuclear division cycle appears to display a biphasic response based on the concomitant TRIM37 levels. If TRIM37 levels are high, which help degrade Cep192 [76, 77], cells prevent centrosome maturation, an essential step that licenses proper spindle assembly as pre-requisite for CDK-driven mitotic progression (Fig. 2.5). On the other hand, if TRIM37 levels are low, since TRIM37 cannot target excess Plk4 (the pool that is rendered inactive at centrioles due to centrinone) for degradation, cytoplasmic Plk4 mediated condensates facilitate acentrosomal microtubule organization and can help sustain the nuclear division cycles [76, 78]. As such, TRIM37 emerges as a molecule that can couple the activity of the centriole biogenesis cycle to the nuclear cycle via its ability to gauge centrosome

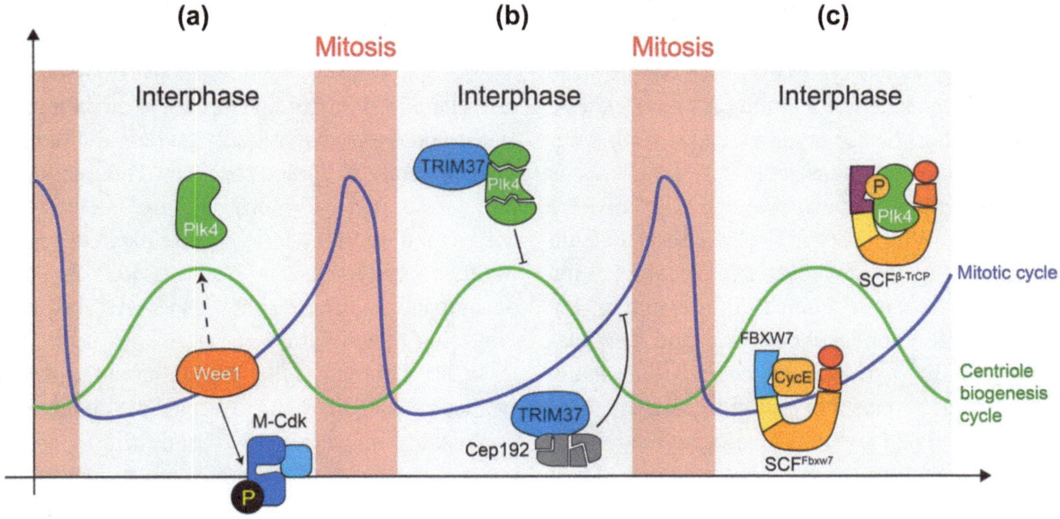

Fig. 2.5 Centriole biogenesis and mitotic cycles can be *coupled* via several mechanisms. There are three potential ways centriole biogenesis cycle can be coupled to mitotic cycles. (**a**) Model on the left: Wee1 kinase, a key antagonist of Cdk1 activity, is required for coupling the peak time of Plk4 accumulation at the centriole to the progression of interphase and rising Cdk1 activity levels. (**b**) Model in the middle: TRIM37, an E3 ubiquitin ligase, has been shown to target excess Plk4 for degradation, preventing centriole overduplication, while also targeting Cep192 for degradation to prevent early centrosome maturation, hence spindle assembly and subsequent CDK1-driven mitotic progression. As such, TRIM37 can gauge G2-M progression and mitosis can proceed normally under CDK1-Cyclin A/B control. (**c**) Model on the right: Both Plk4 and Cyclin E are targeted for degradation by Skp1-Cullin-F-box (SCF) ubiquitin ligases with respective F-box subunits

maturation and spindle assembly (Fig. 2.5), which indirectly feed into whether G2-M progression and mitosis respectively can proceed normally under CDK1-Cyclin A/B control.

Nevertheless, such potential coupling mechanisms between the two cycles are only impinging on each other indirectly: from the nuclear division cycle to the centriole biogenesis cycle, how exactly and whether Wee1 can directly control Plk4 oscillations remains unclear [17]. Similarly, although TRIM37 can couple centriole biogenesis cycle to the mitotic cycle via Plk4-mediated mechanisms, it is not clear how and whether a component of the CDK-cyclin machinery is a direct degradation target of TRIM37. Especially as to the latter, there are ongoing debates about the exact mechanisms by which TRIM37 transduces Plk4 activity to the regulation of the nuclear division cycle as investigated in different human cell types [74]. Likewise, although the TRIM ligases are an ancient group of enzymes, TRIM37 has been lost in invertebrates like *Drosophila* and *C. elegans* [79]. As such, the coupling between the centriole biogenesis and nuclear division cycles may be a modular trait of their oscillators with species or, even, cell-type specific variations as a function of animal evolution.

Finally, it is possible that the two cycles may be coupled via having a mutual re-setting mechanism (Fig. 2.5; right). Both Plk4 and cyclins, molecules at the core of oscillations in the two cycles, benefit from phosphatases or ubiquitin ligase systems to counteract their levels and activities. At the ligase level, Plk4 and cyclins share the Skp1-Cullin-F-box (SCF) surveillance: while Plk4 is targeted for degradation by SCF^{Slimb} (in *Drosophila*) and $SCF^{\beta\text{-TrCP}}$ (in mammals) via a phosphodegron created by Plk4 autophosphorylation [80–83], cyclin E (responsible for DNA replication step in the nuclear cycle) is targeted by SCF^{Fbw7} [84]. While the exact F-box subunits differ, Plk4 and cyclin E are both regulated by the SCF ligase.

As such, the 'off' phases for centriole biogenesis and DNA replication (overlapping in the mitotic phase of the nuclear cycle) may help couple these two cycles by virtue of a common ligase system.

At the phosphatase level, evidence from studies in *Drosophila* points to PP2A complexes as the shared point between the regulation of Plk4 and cyclins. While Plk4 is dephosphorylated and stabilized by PP2A complex containing the B-regulatory subunit Twins [85], preventing premature Plk4 autophosphorylation and degradation by SCF^{Slimb} mediated degradation, many CDK-cyclin phosphorylation targets are dephosphorylated by PP2A holoenzymes, helping CDK signalling during mitotic entry and exit [86, 87]. As such, the coupling of Plk4 oscillations and the mitotic cycle could potentially occur at the level of their resetting mechanisms involving SCF ubiquitin ligases and PP2A phosphatase complexes.

2.3.3 Synchronizing Centriole Biogenesis Cycle to Alternating Programmes of Ciliogenesis and Mitosis

Beyond their central mitotic role in forming centrosomes for spindle assembly, centrioles are also essential for templating primary cilia—antenna-like, axoneme-based organelles that are critical for cellular signalling in development and physiology. In proliferating cells, centrioles alternate between functioning as basal bodies to support ciliogenesis in G0 and as core components of centrosomes during G2/M [88]. Consequently, the centriole biogenesis cycle must be tightly coordinated in time with both the ciliogenesis programme and the nuclear division cycle through shared regulatory mechanisms, ensuring that centrosome formation and cilium assembly do not occur simultaneously. Here, I focus on emerging evidence for shared mechanisms that coordinate the timing and execution of these three processes across the cell cycle, rather than on pathways that are uniquely dedicated to coupling ciliogenesis with either centriole biogenesis (e.g. CCDC41) or with mitosis (e.g. Tctex-1, Nek2-Kif24 and Trichoplein-Aurora A-HDAC6 axes), which have been extensively reviewed elsewhere [89]. Instead, I emphasize on molecular frameworks that may broadly govern and temporally integrate all three processes. This section is also restricted to proliferating cells; the corresponding regulatory principles in post-mitotic systems are addressed in the accompanying Chap. 3 by Adel Al Jord and Gonzalo Ortiz Alvarez.

In proliferating cells, CP110 serves as a key molecular link between the centriole biogenesis cycle and ciliogenesis programme (Fig. 2.6). Its recruitment from the cytoplasm to centrioles yields distinct outcomes depending on both its spatial localization and the timing of its association. Early in the centriole biogenesis cycle, when nascent daughter centrioles just begin to assemble at the G1/S transition, CP110 localized proximal to the mother centriole to regulate Plk4 activity, thereby controlling centriole duplication—a phenomenon observed across multiple cell types and species [18, 90]. In contrast, at later stages, more mature centrioles are capped by CP110 at their distal ends, a configuration that suppresses cilium formation [91]. This inhibitory cap must be removed as cells enter quiescence to permit ciliogenesis prior to the next G1/S transition (Fig. 2.6). Thus, CP110 adopts two temporally and spatially distinct roles, coordinating whether centrioles engage in duplication or in cilium assembly.

Critical to understanding the relationship between the nuclear division cycle and the centriole-cilia biogenesis cycles, CP110 provides a unique point for integrating the three biological phenomena (Fig. 2.6). In the previous section, I highlighted the shared role of SCF ligases in regulating both the centriole biogenesis cycle (via Plk4) and the nuclear division cycle (via Cyclin E), suggesting a common molecular framework for coupling these processes. Consistent with this theme, CP110 localization to centrioles in cultured human cells is itself regulated by SCF ligases, as it is targeted for degradation by the SCF^{Fbxo1} complex during G2

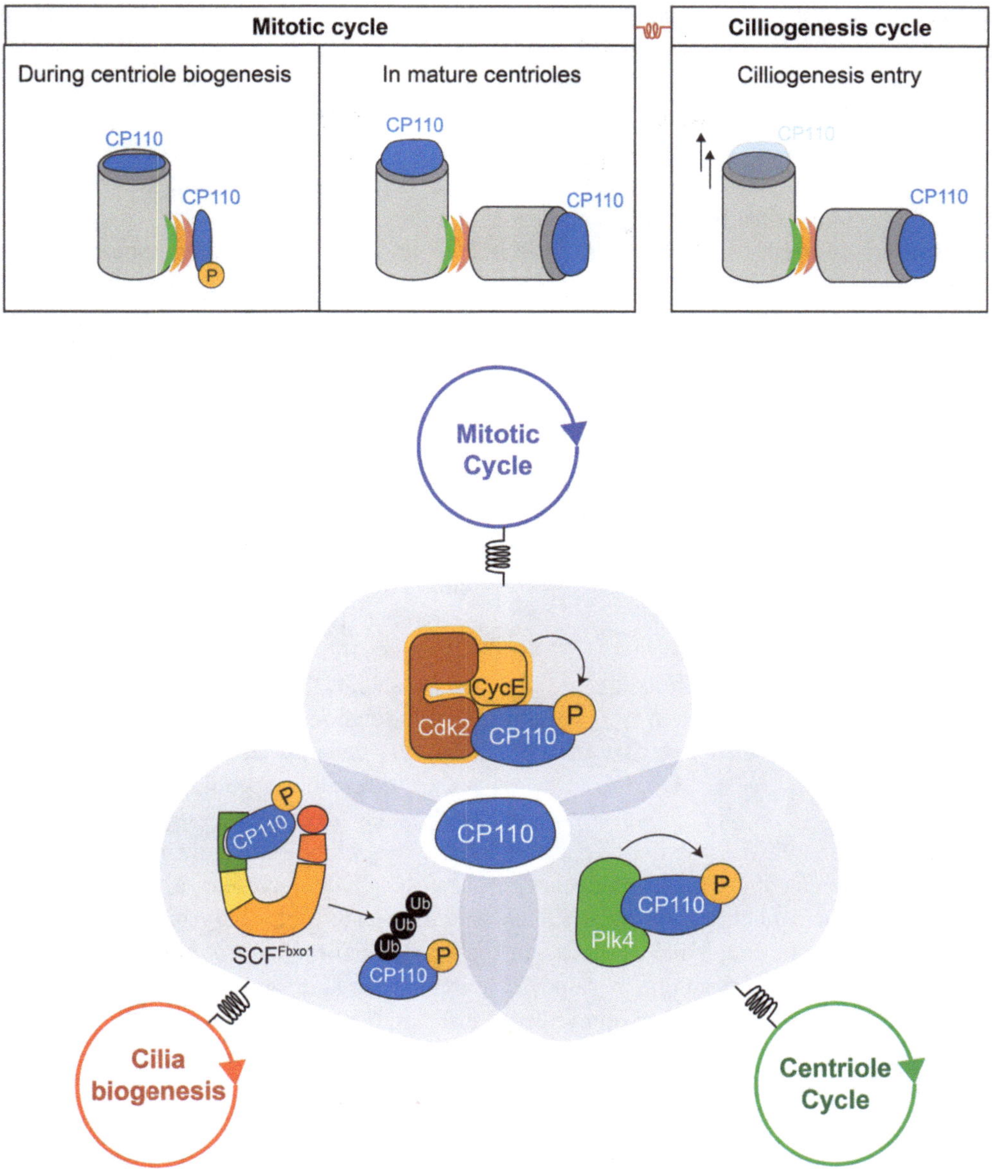

Fig. 2.6 Synchronizing centriole biogenesis cycle to alternating programmes of ciliogenesis and mitosis via CP110. (Top panels) CP110 has been observed to localize to the base of mother centrioles to regulate Plk4 activity before the daughter centriole growth starts. As centriole growth completes and they mature, the newly formed centrioles are capped at their distal ends to suppress cilium formation. This inhibitory cap is removed as cells enter quiescence to permit ciliogenesis prior to the next G1/S transition. (Bottom panel) CP110 lies at the centre of the tripartite relationship encapsulating nuclear division cycle, centriole biogenesis cycle, and ciliogenesis cycle. CP110 is both a Cdk2/cyclin E and Plk4 substrate, while being targeted by an SCF ubiquitin ligase complex for degradation

phase of the cell cycle [92]. Moreover, CP110 is among the few—if not the only—ciliary proteins that has been biochemically validated as a CDK substrate (specifically CDK2/cyclin E) [93, 94] while also playing a critical role in the centriole biogenesis cycle, further strengthening its candidacy as a coordinator of this tripartite relationship.

Beyond its established roles in controlling centriole duplication and suppressing ciliogenesis, CP110 has also been implicated in ensuring the fidelity of cell division [95]. This raises the possibility that CP110 functions as a temporal coordinator in proliferating cells, synchronizing the centriole biogenesis cycle with the alternating programmes of ciliogenesis and mitosis. More broadly, because all three programmes rely—at least in part—on SCF ligases to reset their key regulatory states, SCF ubiquitin ligases may serve as a common platform for coupling otherwise autonomous cellular cycles. Future studies will determine whether CP110 indeed plays such a central integrative role, and whether SCF ligases more generally act as core regulators in coordinating autonomous clocks across the cell cycle.

2.3.4 Coupling Beyond Mitosis: Centriole Biogenesis, Ciliogenesis and the Circadian Clock

Recent studies highlight the possibility that the mechanisms governing the centriole biogenesis cycle—and even their functional output, such as cilia formation—may be coupled to the circadian clock. Analysis of the post-mitotic hepatocyte proteome from circadian-entrained mice revealed that oscillations in Plk4, the master regulator of centriole biogenesis, are entrained to the circadian cycle [17], suggesting that these two oscillatory systems must somehow be coupled. Although the mechanisms underlying such coupling remain unknown, its functional significance is beginning to emerge in regenerative biology and neuroscience. The groups of Koji Ikegami and Hui-Yan Li have independently discovered, in mouse fibroblasts in culture [96] and neurons from the mouse brain [96, 97], a circadian dependence on the abundance and length of primary cilia (Fig. 2.7)—centriole-templated, microtubule-based cellular protrusions that enable cells to sense extracellular cues and communicate intrinsic signals to neighbouring cells, for example through extracellular vesicles.

Li and colleagues discovered that the abundance and length of primary cilia in neuromedin-S–producing neurons of the murine suprachiasmatic nucleus (SCN) oscillate in association with daily light–dark cycles [97] (Fig. 2.7). Beyond simple correlation, the relationship between ciliary dynamics and SCN-level circadian activity appears to be bidirectional. On one hand, rhythmic ciliary dynamics across the SCN promote interneuronal communication through ciliary Hedgehog signalling and thereby contribute to the synchronization and robustness of tissue-level circadian activity. On the other hand, abrogation of the core cell-intrinsic circadian clock in Bmal1$^{-/-}$ mice markedly reduces rhythmic ciliary behaviour [97], indicating that these processes are reciprocally reinforcing and coupled.

Parallel studies from Ikegami and colleagues independently identified rhythmic regulation of ciliary abundance and length in mature neurons from the mouse hippocampal dentate gyrus [96] (Fig. 2.7), demonstrating that this phenomenon is not restricted to a specific brain region. Importantly, they extended these findings to mouse embryonic fibroblasts, showing that rhythmic ciliary dynamics represent a fundamental cell-intrinsic feature of ciliary biology rather than a phenomenon limited to circadian behavioural circuits. Ikegami and colleagues further found that cilia resorption during the high-activity phase of the circadian cycle creates a permissive condition for fibroblast cortical activation and migration following wound induction [96]. Because previous work has shown that wound-triggered fibroblast migration occurs preferentially during the mouse's active circadian phase [98], these results suggest that coupling rhythmic ciliary dynamics to the circadian cycle—such that cilia disassemble during the active phase—may enhance wound responses. This likely occurs because cilium formation requires extensive cortical membrane and cytoskeletal remodelling [99, 100], processes that are also required for cell migration [101, 102]; preventing cilia formation during the active phase may thus allow fibroblasts to maximize migratory capacity.

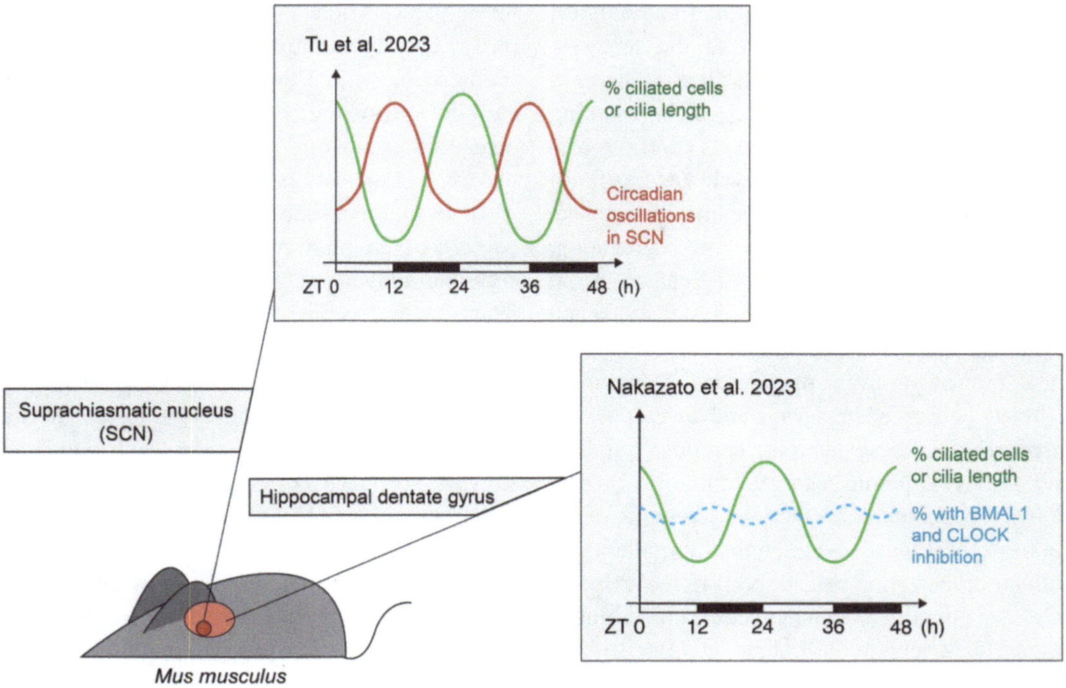

Fig. 2.7 Cycles of ciliogenesis coupled to the circadian clock. Tu et al. (2023) discovered that the percentage of ciliated neuromedin-S-producing neurons of the murine suprachiasmatic nucleus (SCN) and their cilia lengths oscillate in association with light–dark cycles. Independently, Nakazato et al. (2023) discovered in the mouse hippocampal dentate gyrus that their fraction of ciliated cells and cilia lengths also oscillate in association with light–dark cycles, and the oscillations were abrogated by inhibition of the circadian clock

Together, these findings raise two fundamental possibilities for further investigation. From the perspective of proliferating fibroblasts, because centriole-templated ciliogenesis occurs primarily during the G0/G1 phases of the cell cycle, a four-way coupling among the cycles of mitosis, centriole biogenesis, cilia assembly/resorption and the circadian clock is anticipated. This expectation is supported by prior discoveries of coupling between the mitotic cycle and centriole biogenesis, between the centriole biogenesis cycle and the circadian clock, between the mitotic cycle and ciliogenesis (all three directions as elucidated above) and between the mitotic cycle and circadian rhythms (as covered in detail elsewhere [103–105]).

By contrast, neurones are largely non-dividing, suggesting the existence of a more simplified coupling architecture in which the cycle of cilium assembly and resorption is tied directly to the circadian clock, independently of the cycles of centriole biogenesis and mitosis. Future studies aimed at elucidating the mechanistic basis of this four-way coupling in dividing cells—and understanding how terminally differentiated cells reduce this network to a two-way coupling between ciliary dynamics and circadian rhythms—promise to yield important insights into such modular integration and disintegration of cellular timing systems.

2.4 Pathophysiological Implications of the Autonomy of the Centriole Biogenesis Cycle

The emerging view that the centriole biogenesis cycle is entrained by, yet capable of operating autonomously from, the nuclear division cycle carries important physiological and pathological implications that extend beyond the textbook view based on mitosis in regularly dividing cells [106]. In physiological settings, tight coupling between an otherwise autonomous centriole biogenesis cycle and the mitotic cycle ensures faithful coordination between organelle inheritance, spindle assembly and chromosome segregation, thereby preserving the fidelity of cell division. However, when this coupling becomes weakened or uncoupled, the intrinsic autonomy of the centriole biogenesis cycle may instead recapitulate hallmarks observed in cancers and developmental disorders, including aberrant centriole numbers and defective spindle organization.

This section explores how the autonomy of the centriole biogenesis cycle confers modularity to cellular organization that supports tissue architecture, yet simultaneously creates vulnerabilities when its coupling to other cellular timing programmes fails. Moreover, the pace-making machinery of centriole biogenesis appears to be repurposed in post-mitotic contexts, particularly in multiciliated cell types that generate tens to hundreds of centrioles independently of cell division—a topic discussed in detail in Chap. 3 of this volume and in a prior review of autonomous clocks [106].

2.4.1 Erroneous Uncoupling of Centriole Biogenesis Cycle from the Mitotic Cycle

2.4.1.1 Part 1: Nuclear Vulnerabilities in Cancers and Other Developmental Disorders

Two well-established roles of centrosomes—the mature centriole together with its surrounding pericentriolar material that emanates astral microtubules—are to ensure fidelity of spindle assembly and chromosome segregation during mitosis, and to subsequently function as the principal microtubule-organizing centres (MTOCs) of the cell during interphase until the next mitosis. In Part 1, I discuss the former: the pathophysiological implications of nuclear aberrations that arise when the centriole biogenesis cycle is erroneously uncoupled from the mitotic cycle in proliferating cells. I limit this discussion specifically to mitosis, but I encourage readers to also consider analogous uncoupling phenotypes observed in meiosis as well [107].

In evaluating such uncoupling, we consider two critical endpoint phenotypes. First, when the centriole biogenesis cycle becomes uncoupled from cyclin-dependent kinase (CDK) activity—either because coupling mechanisms fail, or because the abundance of key precursors is reduced or the activity of regulatory enzymes is weakened on either end—the centriole cycle slows and produces fewer centrioles than required per mitotic cycle [17]. Conversely, the centriole biogenesis cycle can be pathologically accelerated, for example through overexpression of Plk4, allowing it to escape the frequency regime necessary for proper coupling with the cell cycle. Under these conditions, the cycle produces more than the ideal number of centrioles, leading to centriole amplification [108, 109]. As seen in either scenario, the resultant centriole numbers in the proliferating cell become aberrant.

Ever since the pioneering work of Theodor Boveri [110], it has been appreciated that defects in centriole number—often arising from uncoupling between the centriole biogenesis and mitotic cycles—are associated with tumour formation. Indeed, any deviation from the normal complement of two centrosomes is detrimental to proliferating cells [111], as both under- and over-duplication of centrosomes lead to aberrant spindle assembly and defective chromosome segregation—mechanisms that can drive chromosomal instability (CIN) and aneuploidy [112–115] (Fig. 2.8). In particular, supernumerary centrosomes arising from centriole over-duplication are commonly observed across aggressive cancers [111], such as invasive breast tumours, in which up to 80% of cells display excess centrosomes [116]. Although centrosome amplification can arise through multiple mechanisms, centriole over-duplication frequently stems from dysregulation of Plk4, the

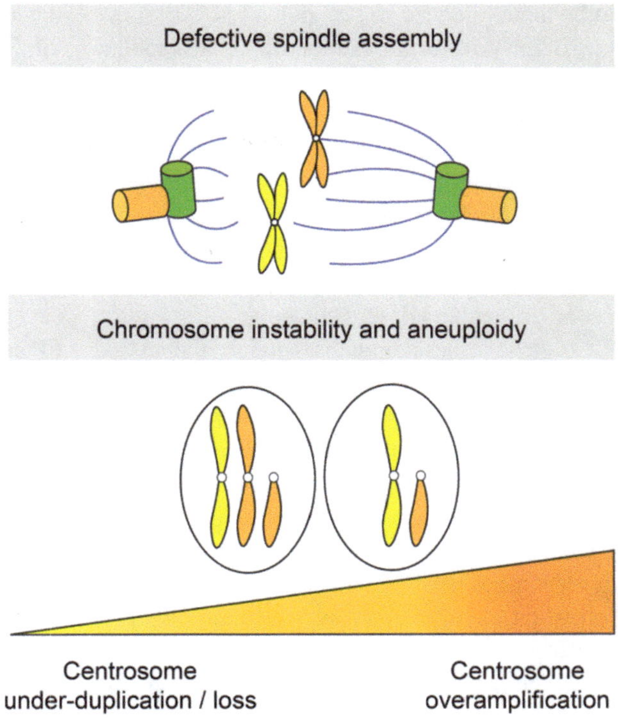

Fig. 2.8 Physiological implications of uncoupling the centriole biogenesis cycle from the mitotic cycle. Centrosome under- or over-amplification can both lead to aberrant spindle assembly and defective chromosome segregation, and subsequently drive chromosomal instability (CIN) and aneuploidy

key pacemaker of centriole biogenesis, as demonstrated in melanomas and gliomas [117–119]. Intriguingly, centriole over-elongation in cancer cells can further promote centriole over-amplification and chromosome missegregation [120], suggesting that both the rate and duration of centriole growth are tightly regulated, likely through Plk4-dependent mechanisms as has been discovered under wild-type conditions of other systems [17].

Furthermore, tumour cells are particularly susceptible to mitotic defects arising from abnormal centriole numbers, often owing to their impaired ability to activate cell death pathways that sense centrosome loss or amplification during mitosis. These surveillance mechanisms operate through the USP28–53BP1–p53–p21 and PIDD1–MDM2–p53 signalling axes [78, 121, 122], respectively. As these pathways and their implications for cancer biology are recently reviewed elsewhere [123, 124], I refer readers to these pieces for further insights. Collectively, these observations underscore the tumour vulnerability that arises when the intrinsic timing of the centriole biogenesis cycle escapes the regulatory constraints imposed by the mitotic cycle.

Beyond the formation and progression of tumours, maintaining tight coupling between centriole biogenesis and mitotic cycles appears critical for normal animal development. For instance, Malumbres and colleagues demonstrated that neural progenitor cells in mice lacking Cep135 (Cep135$^{-/-}$), a critical component of the centriole biogenesis machinery, exhibit centriole under-duplication that triggers p53-dependent apoptotic cell death [125]. Importantly, when Cep135$^{-/-}$ neural progenitors are combined with partial loss of p53 (p53$^{+/-}$), defective cells are no longer efficiently cleared, leading to cortical heterotopias—malformations also observed in congenital microcephaly (MCPH) patients carrying MCPH8 mutations (the Cep135-associated MCPH allele) [125]. While studies in neural progenitors illustrate how loss of centrioles is normally managed through p53-dependent cell clearance, centrosome overamplification appears to be more tolerated in certain tissues. For example, B-cell progenitors can transiently accumulate excess centrioles, likely due to aberrant duplication events, which are subsequently resolved during later stages of development [126]. In contrast, when B-cell progenitors lose centrioles following deletion of Plk4, they undergo Usp28–p53-dependent apoptosis [126], similar to responses observed in many other cell types, including the neural progenitors described above.

Together, these findings support the notion that uncoupling the centriole biogenesis cycle from the mitotic cycle has distinct cellular consequences depending on directionality. Evolutionarily, loss of centrioles may impose a stronger selective pressure because coupling to cyclin–CDK activity normally accelerates centriole biogenesis; when this coupling is disrupted, centriole formation slows, leaving proliferating cells particularly sensitive to insufficient centriole numbers. This asymmetry may also help explain why tumours more frequently exhibit centrosome overamplification rather than centriole loss [111]. Overamplification often arises from aberrant Plk4 overexpression—an abnormal condition for which cells may not have evolved equally robust surveillance or corrective mechanisms.

2.4.1.2 Part 2: Cytoplasmic Vulnerabilities in Cancers and Other Developmental Disorders

While Part 1 addressed the nuclear abnormalities that cells face upon the uncoupling of centriole biogenesis cycle from the mitotic cycle—an area that the field has traditionally focused on for over a century—in Part 2, I will focus on emerging evidence highlighting the cytoplasmic vulnerabilities that cells encounter following such uncoupling.

Pertinent to their role as the major MTOC in many cells, astral microtubules emanating from centrosomes help tether, organize and redistribute cytoplasmic organelles [127, 128]. For instance, Fuchs and colleagues demonstrated that peroxisomes—de novo forming organelles essential for fatty acid oxidation—in mammalian epidermal stem cells tether to astral microtubules via Pex11b [129]. Loss of Pex11b led to disorganized peroxisome positioning across the mitotic spindle, resulting in mitotic delay, asymmetric organelle inheritance and ultimately skewed daughter-cell fates in the epidermis. In terms of organelle organization, centrosomes also play a crucial role in shaping the distribution of the endoplasmic reticulum (ER) across the nuclear envelope [130–132] and in spreading the tubular ER network throughout the cell [133]. Like the spindle-associated tethering of peroxisomes, centrosome-mediated ER organization appears to be critical for proper mitotic spindle formation and homogeneous ER segregation [130, 134]. Furthermore, the distribution of numerous membrane-bound organelles and vesicles occurs along astral microtubules, facilitating their transport and positioning across the cell [128]. For example, mitochondria can tether directly to microtubules via proteins such as Miro and Milton, enabling their localization to regions of the cell with specific metabolic demands [135–138]. In neurons, for instance, microtubule networks bias mitochondrial localization into axons to meet their high energy requirements [139]. Together, these observations support the notion that deviations from the correct centriole number—and consequently the proper number of MTOCs—can disrupt cellular homeostasis, particularly at the level of organelle organization and coordination.

While all cellular organelles compartmentalize specific reactions and functions required by the cell, one major crossroads among them is redox metabolism, as oxidant-based signalling is integral to the function of many organelles [140–144]—far beyond mitochondria, where redox metabolism has historically been most intensively studied [145]. Indeed, while peroxisomes generate high levels of H_2O_2 during fatty acid oxidation, they also serve as key sites for spatially controlled H_2O_2 detoxification via catalase [142]. Similarly, although the ER utilizes H_2O_2 to facilitate oxidative protein folding, it also contains sophisticated isomerase systems that mitigate oxidant stress and preserve protein homeostasis [143]. Given that centrosomal microtubules influence the localization, distribution, organization and segregation of multiple organelles [128], redox metabolism emerges as a potential crossroad through which centriole number regulation may exert broad effects on cellular physiology.

An increasing number of studies in the field now provide concrete evidence supporting this possibility [146–149]. For instance, Godinho and colleagues demonstrated in human breast cancer cells that centrosome overamplification leads to NOX-mediated reactive oxygen species (ROS) production at the plasma membrane without inducing cell death [147]. This, in turn, facilitates the expression and secretion of IL-8, a well-known pro-inflammatory chemokine that promotes cell invasiveness in a paracrine manner. Although the precise mechanisms by which supernumerary centrosomes trigger ROS elevation in cancer cells remain incompletely understood, evidence from other systems suggests that the converse situation—a deficit in centriole numbers—can activate transcriptional programmes that promote antioxidant responses. In support of this idea, Peifer and colleagues showed in *Drosophila* wing imaginal discs that halting centriole biogenesis (via loss of Sas4 or Asl/Cep152) leads to the upregulation of numerous glutathione S-transferase genes [148], indicating activation of redox-buffering systems in cells lacking centrioles.

Together, these findings demonstrate that when the centriole biogenesis cycle becomes uncoupled from the mitotic cycle, cells respond asymmetrically depending on centriole number. Cells containing supernumerary centrioles tend to tolerate and even promote elevated oxidant levels, and vice versa [146]; whereas cells with insufficient centriole numbers activate antioxidant defences to limit cell death and preserve normal tissue development. Although this remains an emerging frontier in centriole biology, future studies are poised to uncover new insights into how erroneous uncoupling of the centriole biogenesis cycle renders cells susceptible to cytoplasmic vulnerabilities, such as oxidative stress, as described here.

2.5 Outstanding Questions and Emerging Opportunities

2.5.1 On Mechanisms Underlying the Autonomous Clock Governing Centriole Biogenesis Cycle

While Plk4 oscillations are emerging as the central pacemaker of centriole biogenesis, the core molecular circuitry required to sustain repeated cycles of its activity remains unresolved. A central outstanding question is therefore: what molecular components constitute the minimal self-sustaining oscillator that drives periodic centriole biogenesis?

2.5.1.1 The Paradox of Necessity and Sufficiency: The Case of CP110

Molecules such as CP110 and Cep97 have been shown to modulate Plk4 levels at the centriole and to influence the phase and amplitude of its oscillations [18]. However, whether these proteins form integral components of the core oscillator, or instead act as auxiliary regulators, remains unclear. This uncertainty arises in part from conflicting evidence regarding their essentiality for centriole growth. In somatic cells, CP110 and Cep97 have been reported to be required for proper centriole duplication [25, 90], whereas in the germline and early embryonic system, centriole growth can proceed in their absence [18, 25]. Such system-dependent requirements raise the possibility that these proteins do not constitute universal oscillator components, but instead participate in regulatory modules that adapt centriole biogenesis to specific cellular contexts.

As discussed in earlier sections, we have proposed that CP110 may function as part of a coupling mechanism linking the centriole biogenesis cycle to the mitotic cycle, particularly at the G1/S transition. In cellular systems that include extended gap phases—such as most somatic cells—the absence of CP110 may enable the mitotic cycle to impose a checkpoint-like restriction on centriole biogenesis unless the appropriate permissive signals are present. By contrast, in rapidly cycling reproductive or embryonic systems that largely lack canonical gap phases and alternate primarily between S and M phases, this coupling step may be reduced or bypassed altogether. This may also explain why altering CP110 levels can modulate parameters of Plk4 oscillations, yet its abrogation does not abolish the oscillator itself [18]. As such, future work aimed at defining the minimal set of core oscillator components, distinguishing them from context-dependent coupling modules, and determining how such coupling mechanisms may be bypassed in specialized cell-cycle architectures will be instrumental for establishing a mechanistic framework for the centriole biogenesis cycle.

2.5.1.2 What Resets the Centriole Clock: A Potential Role for Phosphatases

Mechanisms that reset the Plk4 oscillator—whose circuit appears to depend on phosphorylation of both Plk4 itself and its centriolar docking molecule Asl/Cep152—remain poorly defined. We previously proposed that at least two phosphatase activities could be involved [17]: one that dephosphorylates Asl/Cep152 to permit renewed recruitment of Plk4, and another that dephosphorylates unbound but phosphorylated Plk4 (the residual pool that escapes complete degradation by SCF^{Slimb}). However, the molecular identities of the phosphatases responsible for resetting the Plk4 oscillator remain unknown.

Studies identifying phosphatases such as PP2A and PP5 as regulators of Plk4 suggest that these enzymes represent promising candidates [85, 150]. However, these candidates present a conceptual challenge similar to that posed by CP110: both PP2A and PP5 also participate in resetting events within the mitotic cycle [151, 152]. As such, they may function primarily as coupling molecules that coordinate the centriole biogenesis and mitotic cycles, rather than as dedicated resetting factors specific to the Plk4 oscillator. Alternatively, evolution may have favoured shared molecular reset mechanisms that simultaneously regulate both oscillators, thereby providing efficient interfaces for coupling their temporal dynamics. These possibilities are not mutually exclusive, and future work aimed at distinguishing dedicated oscillator-resetting modules from shared coupling mechanisms will be essential for defining the core molecular circuitry that drives centriole biogenesis timing.

2.5.1.3 How Do Biophysical Features of Centriole Growth Cooperate with the Plk4 Clock?

While biochemical investigations will continue to refine the molecular composition of centriole biogenesis clocks, understanding the systems-level relationship between Plk4 oscillations and centriole growth will require quantitative biophysical investigation. A striking feature of centriole assembly is its switch-like behaviour: once Plk4 levels rise above a critical threshold, centriole growth initiates abruptly and proceeds linearly [17]. This observation implies that the cartwheel subunits required to build the centriole core must be highly concentrated at that moment, allowing growth to proceed in a sustained linear manner rather than through a gradually accelerating (i.e. exponential) process. Likewise, linear centriole growth terminates abruptly when Plk4 levels later fall below the same threshold, suggesting that growth cessation does not occur simply because structural components are depleted [17]. Instead, this behaviour indicates that cartwheel building blocks remain locally abundant even at the point of termination. Together, these features suggest that oscillatory Plk4 activity must be coupled to biophysical mechanisms that transiently concentrate centriole assembly factors—such as Sas-6 and Ana2/STIL—within the centriolar environment.

One attractive and mechanistically plausible explanation for this coupling lies in Plk4's ability to undergo phase separation. Recent studies have demonstrated that Plk4 can form condensates through self-organisation both in the cytosol and at centrioles, indicating that condensate formation is an intrinsic biophysical property of the enzyme [12, 13]. Importantly, these condensates have been shown to recruit Sas-6 and Ana2/STIL as client proteins [12], suggesting that Plk4 phase separation may serve not only as the biochemical engine of oscillatory timing, but also as the physical scaffold that enables cooperative centriole assembly.

To address these mechanistic questions and extend beyond current models, reconstitution of a synthetic Plk4 oscillator in vitro would represent a major advance—and potentially the ultimate proof of principle—analogous to the landmark reconstitutions of the cyanobacterial circadian clock in cyanobacteria [153] and the core CDK–cyclin oscillator in frog egg extracts [47, 48].

2.5.2 Towards a Unified Machinery that Couples the Cycles of Centriole Biogenesis, Mitosis and the Circadian Clock

While it is increasingly apparent that centriole biogenesis, mitotic and circadian cycles can couple in pairs—namely between centriole biogenesis and mitosis, centriole biogenesis and the circadian clock and mitosis and the circadian clock—there is also the intriguing possibility that all three cycles may be coupled through shared molecular machinery. For example, the protein kinase Wee1—canonically known for restraining Cdk1 activity—has been shown to couple both centriole biogenesis to the mitotic cycle [17] and the mitotic cycle to circadian clock [154, 155]. Although it remains unclear whether Wee1 also directly links centriole biogenesis to the circadian clock, an important future direction will be to determine whether Wee1—or related regulatory mechanisms—can function as a master coupling node across eukaryotic timing systems.

Such a discovery would represent a major conceptual advance, not only for centriole biology but for the broader field of biological timekeeping. It would shift our understanding of cellular timing from a framework centred on individual clocks to one focused on their coordinated interactions within an interconnected regulatory network.

2.5.3 Single-Cell Dissection and Therapeutic Opportunities of the Centriole Clock in Cancers

We have described at length in this chapter the cancer vulnerabilities that arise when the centriole biogenesis cycle becomes uncoupled from the mitotic cycle. Despite the recognized association between centriole defects and cancer for over a century, surprisingly few therapeutic strategies have been developed that specifically target centriole loss or overamplification in tumour tissues (e.g. the Plk4 inhibitor RP-1664 whose Phase 1 trials were recently terminated as of its last update on 21 November 2025 [156], or the Plk4 inhibitor CFI-400945 whose Phase 2 trial is underway as of its last update on 18 February 2026 [157]). Indeed, no FDA-approved centriole-targeting drug is yet available. This challenge is relevant not only in primary tissues, where alterations in the pace of the centriole biogenesis cycle have been implicated in tumour initiation, but also in later stages of disease, particularly in cells undergoing epithelial-to-mesenchymal transition (EMT) during metastasis.

However, targeting cells solely based on centriole number is not straightforward, largely because tumours exhibit substantial heterogeneity in this feature. Although many cancers are associated with centrosome amplification, not all cells within the same tumour display this phenotype. This has also been a key roadblock in the development of therapeutic strategies targeting centrosome overamplification in cancer. As such, future approaches that focus on single-cell-level dissection and manipulation of the centriole biogenesis cycle and its clock may prove especially fruitful.

For example, at present there is no single-cell reporter capable of tracking the dynamics of the Plk4 oscillator in real time. The development of such a tool would represent a major technological advance, as it would enable direct visualization of the temporal uncoupling between the Plk4 oscillator and the mitotic cycle, which through Plk4 overexpression, underlies centrosome amplification defects observed in cancer pathophysiology. Moreover, it would open the door to functional assays designed to identify molecular strategies for tuning the period of Plk4 oscillations in controlled ways. For instance, if centrosome overamplification—such as that observed under conditions of Plk4 overexpression—results from acceleration of the centriole biogenesis cycle, it is conceivable that specific missense variants of Plk4 could slow the oscillator and thereby restore more normal centriole numbers. To this end, a real-time single-cell Plk4 reporter would make it possible to perform saturation mutagenesis of Plk4, enabling systematic dissection of the genetic landscape that governs oscillator tuning.

Indeed, analogous strategies have been transformative in the circadian clock field. Random mutagenesis approaches pioneered by Takao Kondo were instrumental in identifying mutations that define the period of the cyanobacterial circadian clock [158], while more recent work by Eric Zhang and colleagues has used saturation mutagenesis of RUVBL2 to probe the regulatory landscape of eukaryotic circadian timing at large [159]. Applying similar strategies to the Plk4 oscillator holds great promise for uncovering principles of temporal regulation that may ultimately be leveraged for therapeutic intervention in diseases marked by centriole dysregulation.

2.5.4 Evolution and Evolvability of the Centriole Biogenesis Clock

An important frontier in understanding the centriole biogenesis clock lies in determining its evolutionary origins and diversification across animal lineages. While core regulators such as Plk4 are broadly conserved among organisms that possess canonical centrioles such as holozoans [160–162], the degree to which their oscillatory behaviour represents an ancient or universal feature of animals remains unclear.

To this end, our initial discovery of the Plk4 oscillator in early fly embryos [16, 17] bears conceptual similarity to how Tim Hunt and colleagues identified the cyclin oscillator in early sea urchin embryos [163]. The absence of gap phases, weak cell-cycle checkpoints and the abundance of maternally deposited components make early embryonic systems particularly conducive to uncovering the core oscillatory elements of biological cycles [164]. Nonetheless, much like the oscillations in CDK activity, Plk4 oscillations are now beginning to be detected in single-cell studies of human cells [51], suggesting that principles first revealed in embryonic contexts extend broadly across biological systems.

Furthermore, the discovery of Plk4 oscillations also did not arise from an explicit search for a molecular clock governing centriole duplication, but rather from our curiosity to understand how centriole size is regulated [16]. In this regard, it is striking to consider that Paul Nurse's landmark discovery of kinases controlling the cell cycle similarly emerged from his initial pursuit of regulators of cell size—exemplified by the identification of Wee1 through its characteristic 'wee', or small-cell, mutant phenotype [165, 166]. These historical parallels underscore a recurring theme in the evolution of such autonomous cellular clocks [106]: fundamental timing mechanisms are often uncovered indirectly through investigations of their direct roles in the size, growth, or structural regulation of cellular organelles and other sub-cellular entities.

Taken together, these observations suggest that—like the core CDK oscillator that governs cell growth and division—the Plk4 oscillator most plausibly originated as an autonomous timing mechanism that initiates and regulates cycles of centriole biogenesis. Only later, through evolutionary pressures favouring coordination with nuclear events, was this intrinsic oscillator likely entrained to DNA replication and chromosome segregation via the emergence of coupling modules, potentially including its regulation by CDK-dependent phosphorylation networks.

While the above attempts to invigorate future work to dissect the evolution of centriole biogenesis cycle, it remains unclear if the centriole biogenesis cycle itself evolvable. If so, at what level of its organization does this evolvability reside?

Comparative studies across species provide a powerful framework to address this question. While Plk4 and its functional orthologues are broadly conserved across animals, several molecules I postulated above that may couple the Plk4 oscillator to the mitotic cycle—such as TRIM37—are present in vertebrate systems but absent in invertebrates. This observation suggests that the molecular interfaces linking centriole biogenesis clock to other cellular cycles may have diversified over evolutionary time. Such differences raise the possibility that, while the core oscillator architecture remains conserved, the coupling mechanisms that synchronize centriole biogenesis to mitotic and circadian cycles could be modular and subject to lineage- or cell-type–specific variation. This may provide the centriole biogenesis clock with a biological platform for evolvability and adaptability in its resident system.

In sum, future studies aimed at mapping the conservation of core oscillator components, identifying species- or context-specific coupling molecules, and determining how shared regulatory systems—such as ubiquitin ligases and phosphatases—have been repurposed across distinct biological clocks will be instrumental in defining the evolutionary logic that governs the temporal regulation of the centriole biogenesis cycle.

Acknowledgements This work was funded by NIH DP2GM154328 (M.G.A.), and Howard Hughes Medical Institute, Freeman Hrabowski Program (M.G.A.).

Disclosures The authors declare no conflict of interest.

References

1. Fernandes-Mariano C, Bugalhão JN, Santos D, Bettencourt-Dias M (2025) Centrosome biogenesis and maintenance in homeostasis and disease. Curr Opin Cell Biol 94:102485
2. Gönczy P (2026) Critical constituents and assembly principles of centriole biogenesis in human cells. Nat Rev Mol Cell Biol 27:260–277
3. Sir J-H, Pütz M, Daly O, Morrison CG, Dunning M, Kilmartin JV, Gergely F (2013) Loss of centrioles causes chromosomal instability in vertebrate somatic cells. J Cell Biol 203:747–756
4. Nabais C, Pereira SG, Bettencourt-Dias M (2017) Noncanonical biogenesis of centrioles and basal bodies. Cold Spring Harb Symp Quant Biol 82:123–135
5. Wang W-J, Acehan D, Kao C-H, Jane W-N, Uryu K, Tsou M-FB (2015) De novo centriole formation in human cells is error-prone and does not require SAS-6 self-assembly. elife 4:e10586
6. Nabais C, Pessoa D, de-Carvalho J, van Zanten T, Duarte P, Mayor S, Carneiro J, Telley IA, Bettencourt-Dias M (2021) Plk4 triggers autonomous de novo centriole biogenesis and maturation. J Cell Biol 220:e202008090
7. Hilbert M, Noga A, Frey D et al (2016) SAS-6 engineering reveals interdependence between cartwheel and microtubules in determining centriole architecture. Nat Cell Biol 18:393–403
8. Vasquez-Limeta A, Loncarek J (2021) Human centrosome organization and function in interphase and mitosis. Semin Cell Dev Biol 117:30–41
9. Kim T-S, Park J-E, Shukla A et al (2013) Hierarchical recruitment of Plk4 and regulation of centriole biogenesis by two centrosomal scaffolds, Cep192 and Cep152. Proc Natl Acad Sci 110:E4849–E4857
10. Sonnen KF, Gabryjonczyk A-M, Anselm E, Stierhof Y-D, Nigg EA (2013) Human Cep192 and Cep152 cooperate in Plk4 recruitment and centriole duplication. J Cell Sci 126:3223–3233
11. Ohta M, Ashikawa T, Nozaki Y, Kozuka-Hata H, Goto H, Inagaki M, Oyama M, Kitagawa D (2014) Direct interaction of Plk4 with STIL ensures formation of a single procentriole per parental centriole. Nat Commun 5:5267
12. Park J-E, Zhang L, Bang JK, Andresson T, DiMaio F, Lee KS (2019) Phase separation of Polo-like kinase 4 by autoactivation and clustering drives centriole biogenesis. Nat Commun 10:4959
13. Ryniawec JM, Buster DW, Slevin LK, Boese CJ, Amoiroglou A, Dean SM, Slep KC, Rogers GC, Wignall S (2023) Polo-like kinase 4 homodimerization and condensate formation regulate its own protein levels but are not required for centriole assembly. Mol Biol Cell 34:ar80
14. Qi F, Yin S, Yang X et al (2025) Dynamic SAS-6 phosphorylation aids centrosome duplication and elimination in C. elegans oogenesis. EMBO Rep 26: 3411–3444
15. Nievergelt AP, Banterle N, Andany SH, Gönczy P, Fantner GE (2018) High-speed photothermal off-resonance atomic force microscopy reveals assembly routes of centriolar scaffold protein SAS-6. Nat Nanotechnol 13:696–701
16. Aydogan MG, Wainman A, Saurya S, Steinacker TL, Caballe A, Novak ZA, Baumbach J, Muschalik N, Raff JW (2018) A homeostatic clock sets daughter centriole size in flies. J Cell Biol 217:1233–1248

17. Aydogan MG, Steinacker TL, Mofatteh M et al (2020) An autonomous oscillation times and executes centriole biogenesis. Cell 181:1566–1581.e27
18. Aydogan MG, Hankins LE, Steinacker TL, Mofatteh M, Saurya S, Wainman A, Wong S-S, Lu X, Zhou FY, Raff JW (2022) Centriole distal-end proteins CP110 and Cep97 influence centriole cartwheel growth at the proximal end. J Cell Sci 135:jcs260015
19. Lin Y, Chang C, Hsu W, Tang CC, Lin Y, Chou E, Wu C, Tang TK (2013) Human microcephaly protein CEP135 binds to hSAS-6 and CPAP, and is required for centriole assembly. EMBO J 32:1141–1154
20. Lin Y-N, Wu C-T, Lin Y-C, Hsu W-B, Tang C-JC, Chang C-W, Tang TK (2013) CEP120 interacts with CPAP and positively regulates centriole elongation. J Cell Biol 202:211–219
21. Comartin D, Gupta GD, Fussner E et al (2013) CEP120 and SPICE1 cooperate with CPAP in centriole elongation. Curr Biol 23:1360–1366
22. Sharma A, Aher A, Dynes NJ et al (2016) Centriolar CPAP/SAS-4 imparts slow processive microtubule growth. Dev Cell 37:362–376
23. Schmidt TI, Kleylein-Sohn J, Westendorf J, Le Clech M, Lavoie SB, Stierhof Y-D, Nigg EA (2009) Control of centriole length by CPAP and CP110. Curr Biol 19:1005–1011
24. Delgehyr N, Rangone H, Fu J, Mao G, Tom B, Riparbelli MG, Callaini G, Glover DM (2012) Klp10A, a microtubule-depolymerizing kinesin-13, cooperates with CP110 to control *Drosophila* centriole length. Curr Biol 22:502–509
25. Dobbelaere J, Schmidt Cernohorska M, Huranova M, Slade D, Dammermann A (2020) Cep97 is required for centriole structural integrity and cilia formation in *Drosophila*. Curr Biol 30:3045–3056.e7
26. Ryniawec JM, Hannaford MR, Zibrat ME, Fagerstrom CJ, Galletta BJ, Aguirre SE, Guice BA, Dean SM, Rusan NM, Rogers GC (2023) Cep104 is a component of the centriole distal tip complex that regulates centriole growth and contributes to *Drosophila* spermiogenesis. Curr Biol 33:4202–4216.e9
27. Iyer SS, Chen F, Ogunmolu FE et al (2025) Centriolar cap proteins CP110 and CPAP control slow elongation of microtubule plus ends. J Cell Biol 224:e202406061
28. Nigg EA, Stearns T (2011) The centrosome cycle: centriole biogenesis, duplication and inherent asymmetries. Nat Cell Biol 13:1154–1160
29. Sluder G (2014) One to only two: a short history of the centrosome and its duplication. Philos Trans R Soc Lond Ser B Biol Sci 369:20130455
30. Rattner JB, Phillips SG (1973) Independence of centriole formation and DNA synthesis. J Cell Biol 57: 359–372
31. Kuriyama R, Borisy GG (1981) Centriole cycle in Chinese hamster ovary cells as determined by whole-mount electron microscopy. J Cell Biol 91:814–821
32. Sluder G, Miller FJ, Rieder CL (1986) The reproduction of centrosomes: nuclear versus cytoplasmic controls. J Cell Biol 103:1873–1881
33. Sluder G, Lewis K (1987) Relationship between nuclear DNA synthesis and centrosome reproduction in sea urchin eggs. J Exp Zool 244:89–100
34. Lorch IJ (1952) Enucleation of sea-urchin blastomeres with or without removal of asters. J Cell Sci s3-93:475–486
35. Freeman M, Nüsslein-Volhard C, Glover DM (1986) The dissociation of nuclear and centrosomal division in gnu, a mutation causing giant nuclei in Drosophila. Cell 46:457–468
36. Raff JW, Glover DM (1988) Nuclear and cytoplasmic mitotic cycles continue in Drosophila embryos in which DNA synthesis is inhibited with aphidicolin. J Cell Biol 107:2009–2019
37. Gard DL, Hafezi S, Zhang T, Doxsey SJ (1990) Centrosome duplication continues in cycloheximide-treated Xenopus blastulae in the absence of a detectable cell cycle. J Cell Biol 110: 2033–2042
38. Sluder G, Miller FJ, Cole R, Rieder CL (1990) Protein synthesis and the cell cycle: centrosome reproduction in sea urchin eggs is not under translational control. J Cell Biol 110:2025–2032
39. Gerhart J, Wu M, Kirschner M (1984) Cell cycle dynamics of an M-phase-specific cytoplasmic factor in Xenopus laevis oocytes and eggs. J Cell Biol 98: 1247–1255
40. Kimelman D, Kirschner M, Scherson T (1987) The events of the midblastula transition in Xenopus are regulated by changes in the cell cycle. Cell 48:399–407
41. Dabauvalle MC, Doree M, Bravo R, Karsenti E (1988) Role of nuclear material in the early cell cycle of xenopus embryos. Cell 52:525–533
42. Pines J, Hunter T (1991) Human cyclins A and B1 are differentially located in the cell and undergo cell cycle-dependent nuclear transport. J Cell Biol 115: 1–17
43. Bailly E, Pines J, Hunter T, Bornens M (1992) Cytoplasmic accumulation of cyclin B1 in human cells: association with a detergent-resistant compartment and with the centrosome. J Cell Sci 101:529–545
44. Jackman M, Lindon C, Nigg EA, Pines J (2003) Active cyclin B1–Cdk1 first appears on centrosomes in prophase. Nat Cell Biol 5:143–148
45. Gavet O, Pines J (2010) Activation of cyclin B1–Cdk1 synchronizes events in the nucleus and the cytoplasm at mitosis. J Cell Biol 189:247–259
46. Guan Y, Li Z, Wang S, Barnes PM, Liu X, Xu H, Jin M, Liu AP, Yang Q (2018) A robust and tunable mitotic oscillator in artificial cells. elife 7:e33549
47. Murray AW, Kirschner MW (1989) Cyclin synthesis drives the early embryonic cell cycle. Nature 339: 275–280

48. Murray AW, Solomon MJ, Kirschner MW (1989) The role of cyclin synthesis and degradation in the control of maturation promoting factor activity. Nature 339:280–286
49. Phillips SG, Rattner JB (1976) Dependence of centriole formation on protein synthesis. J Cell Biol 70:9–19
50. McCleland ML, O'Farrell PH (2008) RNAi of mitotic cyclins in *Drosophila* uncouples the nuclear and centrosome cycle. Curr Biol 18:245–254
51. Takao D, Yamamoto S, Kitagawa D (2019) A theory of centriole duplication based on self-organized spatial pattern formation. J Cell Biol 218:3537–3547
52. Boese CJ, Nye J, Buster DW, McLamarrah TA, Byrnes AE, Slep KC, Rusan NM, Rogers GC, Bloom KS (2018) Asterless is a Polo-like kinase 4 substrate that both activates and inhibits kinase activity depending on its phosphorylation state. Mol Biol Cell 29:2874–2886
53. Breslow DK, Holland AJ (2019) Mechanism and regulation of centriole and cilium biogenesis. Annu Rev Biochem 88:691–724
54. Hinchcliffe EH, Li C, Thompson EA, Maller JL, Sluder G (1999) Requirement of Cdk2-cyclin E activity for repeated centrosome reproduction in xenopus egg extracts. Science 283:851–854
55. Lacey KR, Jackson PK, Stearns T (1999) Cyclin-dependent kinase control of centrosome duplication. Proc Natl Acad Sci 96:2817–2822
56. Matsuhashi K, Ito KK, Nagai K et al (2025) The DNA replication machinery transmits dual signals to prevent unscheduled licensing and execution of centrosome duplication. Nat Commun 16:7799
57. Tsou M-FB, Stearns T (2006) Mechanism limiting centrosome duplication to once per cell cycle. Nature 442:947–951
58. Tsou M-FB, Wang W-J, George KA, Uryu K, Stearns T, Jallepalli PV (2009) Polo kinase and separase regulate the mitotic licensing of centriole duplication in human cells. Dev Cell 17:344–354
59. Mardin BR, Agircan FG, Lange C, Schiebel E (2011) Plk1 controls the Nek2A-PP1γ antagonism in centrosome disjunction. Curr Biol 21:1145–1151
60. Smith E, Hégarat N, Vesely C et al (2011) Differential control of Eg5-dependent centrosome separation by Plk1 and Cdk1. EMBO J 30:2233–2245
61. Zitouni S, Francia ME, Leal F et al (2016) CDK1 prevents unscheduled PLK4-STIL complex assembly in centriole biogenesis. Curr Biol 26:1127–1137
62. Huang F, Xu X, Xin G, Zhang B, Jiang Q, Zhang C (2022) Cartwheel disassembly regulated by CDK1-cyclin B kinase allows human centriole disengagement and licensing. J Biol Chem 298:102658
63. Nigg EA, Holland AJ (2018) Once and only once: mechanisms of centriole duplication and their deregulation in disease. Nat Rev Mol Cell Biol 19:297–312
64. Balczon R, Bao L, Zimmer WE, Brown K, Zinkowski RP, Brinkley BR (1995) Dissociation of centrosome replication events from cycles of DNA synthesis and mitotic division in hydroxyurea-arrested Chinese hamster ovary cells. J Cell Biol 130:105–115
65. Durcan TM, Halpin ES, Casaletti L, Vaughan KT, Pierson MR, Woods S, Hinchcliffe EH (2008) Centrosome duplication proceeds during mimosine-induced G1 cell cycle arrest. J Cell Physiol 215: 182–191
66. Matsumoto Y, Maller JL (2002) Calcium, calmodulin, and CaMKII requirement for initiation of centrosome duplication in Xenopus egg extracts. Science 295:499–502
67. Duensing A, Liu Y, Tseng M, Malumbres M, Barbacid M, Duensing S (2006) Cyclin-dependent kinase 2 is dispensable for normal centrosome duplication but required for oncogene-induced centrosome overduplication. Oncogene 25:2943–2949
68. Aydogan MG, Steinacker TL, Mofatteh M, Gartenmann L, Wainman A, Saurya S, Conduit PT, Zhou FY, Boemo MA, Raff JW (2019) A free-running oscillator times and executes centriole biogenesis:510875
69. Haase SB, Winey M, Reed SI (2001) Multi-step control of spindle pole body duplication by cyclin-dependent kinase. Nat Cell Biol 3:38–42
70. Golombek DA, Rosenstein RE (2010) Physiology of circadian entrainment. Physiol Rev 90:1063–1102
71. Cuomo ME, Knebel A, Morrice N, Paterson H, Cohen P, Mittnacht S (2008) p53-Driven apoptosis limits centrosome amplification and genomic instability downstream of NPM1 phosphorylation. Nat Cell Biol 10:723–730
72. Bazzi H, Anderson KV (2014) Acentriolar mitosis activates a p53-dependent apoptosis pathway in the mouse embryo. Proc Natl Acad Sci 111:E1491–E1500
73. Domínguez-Calvo A, Gönczy P, Holland AJ, Balestra FR (2021) TRIM37: a critical orchestrator of centrosome function. Cell Cycle 20:2443–2451
74. Tkach JM, Philip R, Sharma A, Strecker J, Durocher D, Pelletier L (2022) Global cellular response to chemical perturbation of PLK4 activity and abnormal centrosome number. eLife 11:e73944
75. Wong YL, Anzola JV, Davis RL et al (2015) Reversible centriole depletion with an inhibitor of Polo-like kinase 4. Science 348:1155–1160
76. Meitinger F, Ohta M, Lee K-Y et al (2020) TRIM37 controls cancer-specific vulnerability to PLK4 inhibition. Nature 585:440–446
77. Yeow ZY, Lambrus BG, Marlow R et al (2020) Targeting TRIM37-driven centrosome dysfunction in 17q23-amplified breast cancer. Nature 585:447–452
78. Meitinger F, Anzola JV, Kaulich M et al (2016) 53BP1 and USP28 mediate p53 activation and G1 arrest after centrosome loss or extended mitotic duration. J Cell Biol 214:155–166
79. Marín I (2012) Origin and diversification of TRIM ubiquitin ligases. PLoS One 7:e50030

80. Rogers GC, Rusan NM, Roberts DM, Peifer M, Rogers SL (2009) The SCF Slimb ubiquitin ligase regulates Plk4/Sak levels to block centriole reduplication. J Cell Biol 184:225–239
81. Cunha-Ferreira I, Rodrigues-Martins A, Bento I, Riparbelli M, Zhang W, Laue E, Callaini G, Glover DM, Bettencourt-Dias M (2009) The SCF/Slimb ubiquitin ligase limits centrosome amplification through degradation of SAK/PLK4. Curr Biol 19: 43–49
82. Guderian G, Westendorf J, Uldschmid A, Nigg EA (2010) Plk4 trans-autophosphorylation regulates centriole number by controlling βTrCP-mediated degradation. J Cell Sci 123:2163–2169
83. Holland AJ, Lan W, Niessen S, Hoover H, Cleveland DW (2010) Polo-like kinase 4 kinase activity limits centrosome overduplication by autoregulating its own stability. J Cell Biol 188:191–198
84. Koepp DM, Schaefer LK, Ye X, Keyomarsi K, Chu C, Harper JW, Elledge SJ (2001) Phosphorylation-dependent ubiquitination of cyclin E by the SCFFbw7 ubiquitin ligase. Science 294: 173–177
85. Brownlee CW, Klebba JE, Buster DW, Rogers GC (2011) The Protein Phosphatase 2A regulatory subunit Twins stabilizes Plk4 to induce centriole amplification. J Cell Biol 195:231–243
86. Mochida S, Hunt T (2012) Protein phosphatases and their regulation in the control of mitosis. EMBO Rep 13:197–203
87. Nilsson J (2018) Protein phosphatases in the regulation of mitosis. J Cell Biol 218:395–409
88. Plotnikova OV, Pugacheva EN, Golemis EA (2009) Primary cilia and the cell cycle. In: Methods in cell biology. Academic Press, pp 137–160
89. Izawa I, Goto H, Kasahara K, Inagaki M (2015) Current topics of functional links between primary cilia and cell cycle. Cilia 4:12
90. Kleylein-Sohn J, Westendorf J, Le Clech M, Habedanck R, Stierhof Y-D, Nigg EA (2007) Plk4-induced centriole biogenesis in human cells. Dev Cell 13:190–202
91. Spektor A, Tsang WY, Khoo D, Dynlacht BD (2007) Cep97 and CP110 suppress a cilia assembly program. Cell 130:678–690
92. D'Angiolella V, Donato V, Vijayakumar S, Saraf A, Florens L, Washburn MP, Dynlacht B, Pagano M (2010) SCF (Cyclin F) controls centrosome homeostasis and mitotic fidelity through CP110 degradation. Nature 466:138–142
93. Chen Z, Indjeian VB, McManus M, Wang L, Dynlacht BD (2002) CP110, a cell cycle-dependent CDK substrate, regulates centrosome duplication in human cells. Dev Cell 3:339–350
94. Hu S, Lu Y, Orr B, Godek K, Mustachio LM, Kawakami M, Sekula D, Compton DA, Freemantle S, Dmitrovsky E (2015) Specific CP110 phosphorylation sites mediate anaphase catastrophe after CDK2 inhibition: evidence for cooperation with USP33 knockdown. Mol Cancer Ther 14:2576–2585
95. Tsang WY, Spektor A, Luciano DJ, Indjeian VB, Chen Z, Salisbury JL, Sánchez I, Dynlacht BD, Davis T (2006) CP110 cooperates with two calcium-binding proteins to regulate cytokinesis and genome stability. Mol Biol Cell 17:3423–3434
96. Nakazato R, Matsuda Y, Ijaz F, Ikegami K (2023) Circadian oscillation in primary cilium length by clock genes regulates fibroblast cell migration. EMBO Rep 24:EMBR202356870
97. Tu H-Q, Li S, Xu Y-L et al (2023) Rhythmic cilia changes support SCN neuron coherence in circadian clock. Science 380:972–979
98. Hoyle NP, Seinkmane E, Putker M et al (2017) Circadian actin dynamics drive rhythmic fibroblast mobilization during wound healing. Sci Transl Med 9:eaal2774
99. Kim J, Jo H, Hong H, Kim MH, Kim JM, Lee J-K, Heo WD, Kim J (2015) Actin remodelling factors control ciliogenesis by regulating YAP/TAZ activity and vesicle trafficking. Nat Commun 6:6781
100. Phua SC, Chiba S, Suzuki M et al (2017) Dynamic remodeling of membrane composition drives cell cycle through primary cilia excision. Cell 168:264–279.e15
101. Ridley AJ (2011) Life at the leading edge. Cell 145: 1012–1022
102. Seetharaman S, Etienne-Manneville S (2020) Cytoskeletal crosstalk in cell migration. Trends Cell Biol 30:720–735
103. Masri S, Cervantes M, Sassone-Corsi P (2013) The circadian clock and cell cycle: interconnected biological circuits. Curr Opin Cell Biol 25:730–734
104. Feillet C, van der Horst GTJ, Levi F, Rand DA, Delaunay F (2015) Coupling between the circadian clock and cell cycle oscillators: implication for healthy cells and malignant growth. Front Neurol. https://doi.org/10.3389/fneur.2015.00096
105. Farshadi E, van der Horst GTJ, Chaves I (2020) Molecular links between the circadian clock and the cell cycle. J Mol Biol 432:3515–3524
106. Mofatteh M, Echegaray-Iturra F, Alamban A, Dalla Ricca F, Bakshi A, Aydogan MG (2021) Autonomous clocks that regulate organelle biogenesis, cytoskeletal organization, and intracellular dynamics. elife 10:e72104
107. Schvarzstein M, Pattabiraman D, Bembenek JN, Villeneuve AM (2013) Meiotic HORMA domain proteins prevent untimely centriole disengagement during Caenorhabditis elegans spermatocyte meiosis. Proc Natl Acad Sci 110:E898–E907
108. Coelho PA, Bury L, Shahbazi MN et al (2015) Overexpression of Plk4 induces centrosome amplification, loss of primary cilia and associated tissue hyperplasia in the mouse. Open Biol 5:150209
109. Levine MS, Bakker B, Boeckx B et al (2017) Centrosome amplification is sufficient to promote

spontaneous tumorigenesis in mammals. Dev Cell 40:313–322.e5
110. Boveri T (2008) Concerning the origin of malignant tumours by Theodor Boveri. Translated and annotated by Henry Harris. J Cell Sci 121:1–84
111. Godinho SA, Basto R (2025) Centrosomes and cancer: balancing tumor-promoting and inhibitory roles. Trends Cell Biol 35:515–526
112. Basto R, Brunk K, Vinadogrova T, Peel N, Franz A, Khodjakov A, Raff JW (2008) Centrosome amplification can initiate tumorigenesis in flies. Cell 133: 1032–1042
113. Ganem NJ, Godinho SA, Pellman D (2009) A mechanism linking extra centrosomes to chromosomal instability. Nature 460:278–282
114. Wang M, Nagle RB, Knudsen BS, Cress AE, Rogers GC (2020) Centrosome loss results in an unstable genome and malignant prostate tumors. Oncogene 39:399–413
115. Morretton J, Simon A, Herbette A et al (2022) A catalog of numerical centrosome defects in epithelial ovarian cancers. EMBO Mol Med 14: EMMM202215670
116. Lingle WL, Barrett SL, Negron VC, D'Assoro AB, Boeneman K, Liu W, Whitehead CM, Reynolds C, Salisbury JL (2002) Centrosome amplification drives chromosomal instability in breast tumor development. Proc Natl Acad Sci 99:1978–1983
117. Denu RA, Shabbir M, Nihal M, Singh CK, Longley BJ, Burkard ME, Ahmad N (2018) Centriole overduplication is the predominant mechanism leading to centrosome amplification in melanoma. Mol Cancer Res 16:517–527
118. Wang B, Yu R-Z, Zhang X-Y, Ren Y, Zhen Y-W, Han L (2024) Polo-like kinase 4 accelerates glioma malignant progression and vasculogenic mimicry by phosphorylating EphA2. Cancer Lett 611:217397
119. Wang B, Zhang X, Li Z, Wei C, Yu R, Du X, He Y, Ren Y, Zhen Y, Han L (2024) Polo-like kinase 4 promotes tumorigenesis and glucose metabolism in glioma by activating AKT1 signaling. Cancer Lett 585:216665
120. Marteil G, Guerrero A, Vieira AF et al (2018) Over-elongation of centrioles in cancer promotes centriole amplification and chromosome missegregation. Nat Commun 9:1258
121. Lambrus BG, Daggubati V, Uetake Y, Scott PM, Clutario KM, Sluder G, Holland AJ (2016) A USP28–53BP1–p53–p21 signaling axis arrests growth after centrosome loss or prolonged mitosis. J Cell Biol 214:143–153
122. Fava LL, Schuler F, Sladky V et al (2017) The PIDDosome activates p53 in response to supernumerary centrosomes. Genes Dev 31:34–45
123. Burigotto M, Fava LL (2021) The PIDDosome: centrosome guardian and backup on the DNA damage response. Mol Cell Oncol 8:1893625
124. Stracker TH (2024) Regulation of p53 by the mitotic surveillance/stopwatch pathway: implications in neurodevelopment and cancer. Front Cell Dev Biol. https://doi.org/10.3389/fcell.2024.1451274
125. González-Martínez J, Cwetsch AW, Martínez-Alonso D et al (2021) Deficient adaptation to centrosome duplication defects in neural progenitors causes microcephaly and subcortical heterotopias. JCI Insight. https://doi.org/10.1172/jci.insight.146364
126. Schapfl MA, LoMastro GM, Braun VZ, Hirai M, Levine MS, Kiermaier E, Labi V, Holland AJ, Villunger A (2024) Centrioles are frequently amplified in early B cell development but dispensable for humoral immunity. Nat Commun 15:8890
127. Waterman-Storer CM, Sanger JW, Sanger JM (1993) Dynamics of organelles in the mitotic spindles of living cells: membrane and microtubule interactions. Cell Motil Cytoskeleton 26:19–39
128. Barlan K, Gelfand VI (2017) Microtubule-based transport and the distribution, tethering, and organization of organelles. Cold Spring Harb Perspect Biol 9:a025817
129. Asare A, Levorse J, Fuchs E (2017) Coupling organelle inheritance with mitosis to balance growth and differentiation. Science 355:eaah4701
130. Karabasheva D, Smyth JT (2019) A novel, dynein-independent mechanism focuses the endoplasmic reticulum around spindle poles in dividing Drosophila spermatocytes. Sci Rep 9:12456
131. Diaz U, Bergman ZJ, Johnson BM, Edington AR, de Cruz MA, Marshall WF, Riggs B (2019) Microtubules are necessary for proper Reticulon localization during mitosis. PLoS One 14:e0226327
132. Maheshwari R, Rahman MM, Drey S, Onyundo M, Fabig G, Martinez MAQ, Matus DQ, Müller-Reichert T, Cohen-Fix O (2023) A membrane reticulum, the centriculum, affects centrosome size and function in *Caenorhabditis elegans*. Curr Biol 33: 791–806.e7
133. Friedman JR, Webster BM, Mastronarde DN, Verhey KJ, Voeltz GK (2010) ER sliding dynamics and ER–mitochondrial contacts occur on acetylated microtubules. J Cell Biol 190:363–375
134. Araújo M, Tavares A, Vieira DV, Telley IA, Oliveira RA (2023) Endoplasmic reticulum membranes are continuously required to maintain mitotic spindle size and forces. Life Sci Alliance. https://doi.org/10.26508/lsa.202201540
135. Morris RL, Hollenbeck PJ (1995) Axonal transport of mitochondria along microtubules and F-actin in living vertebrate neurons. J Cell Biol 131:1315–1326
136. Stowers RS, Megeath LJ, Górska-Andrzejak J, Meinertzhagen IA, Schwarz TL (2002) Axonal transport of mitochondria to synapses depends on Milton, a novel *Drosophila* protein. Neuron 36:1063–1077
137. Guo X, Macleod GT, Wellington A, Hu F, Panchumarthi S, Schoenfield M, Marin L, Charlton MP, Atwood HL, Zinsmaier KE (2005) The GTPase dMiro is required for axonal transport of mitochondria to Drosophila synapses. Neuron 47: 379–393

138. Glater EE, Megeath LJ, Stowers RS, Schwarz TL (2006) Axonal transport of mitochondria requires Milton to recruit kinesin heavy chain and is light chain independent. J Cell Biol 173:545–557
139. Sheng Z-H, Cai Q (2012) Mitochondrial transport in neurons: impact on synaptic homeostasis and neurodegeneration. Nat Rev Neurosci 13:77–93
140. Kurz T, Eaton JW, Brunk UT (2010) Redox activity within the lysosomal compartment: implications for aging and apoptosis. Antioxid Redox Signal 13:511–523
141. Go Y-M, Jones DP (2010) Redox control systems in the nucleus: mechanisms and functions. Antioxid Redox Signal 13:489–509
142. Walker CL, Pomatto LCD, Tripathi DN, Davies KJA (2018) Redox regulation of homeostasis and proteostasis in peroxisomes. Physiol Rev 98:89–115
143. Ushioda R, Nagata K (2019) Redox-mediated regulatory mechanisms of endoplasmic reticulum homeostasis. Cold Spring Harb Perspect Biol 11:a033910
144. Kellokumpu S (2019) Golgi pH, ion and redox homeostasis: how much do they really matter? Front Cell Dev Biol. https://doi.org/10.3389/fcell.2019.00093
145. Lennicke C, Cochemé HM (2021) Redox metabolism: ROS as specific molecular regulators of cell signaling and function. Mol Cell 81:3691–3707
146. Chae S, Yun C, Um H, Lee JH, Cho H (2005) Centrosome amplification and multinuclear phenotypes are induced by hydrogen peroxide. Exp Mol Med 37:482–487
147. Arnandis T, Monteiro P, Adams SD et al (2018) Oxidative stress in cells with extra centrosomes drives non-cell-autonomous invasion. Dev Cell 47:409–424.e9
148. Poulton JS, McKay DJ, Peifer M (2019) Centrosome loss triggers a transcriptional program to counter apoptosis-induced oxidative stress. Genetics 212:187–211
149. Na H-J, Sung M-J, Park J-S (2024) Age- and oxidative stress-induced centrosome amplification and renal stones in Drosophila Malpighian tubules. Biol Open 13:bio061743
150. Ábrahám E, Réthi-Nagy Z, Vilmos P, Sinka R, Lipinszki Z (2023) Plk4 is a novel substrate of protein phosphatase 5. Int J Mol Sci 24:2033
151. Hinds TD, Sánchez ER (2008) Protein phosphatase 5. Int J Biochem Cell Biol 40:2358–2362
152. Wlodarchak N, Xing Y (2016) PP2A as a master regulator of the cell cycle. Crit Rev Biochem Mol Biol 51:162–184
153. Nakajima M, Imai K, Ito H, Nishiwaki T, Murayama Y, Iwasaki H, Oyama T, Kondo T (2005) Reconstitution of circadian oscillation of cyanobacterial KaiC phosphorylation in vitro. Science 308:414–415
154. Matsuo T, Yamaguchi S, Mitsui S, Emi A, Shimoda F, Okamura H (2003) Control mechanism of the circadian clock for timing of cell division in vivo. Science 302:255–259
155. Gérard C, Goldbeter A (2012) Entrainment of the mammalian cell cycle by the circadian clock: modeling two coupled cellular rhythms. PLoS Comput Biol 8:e1002516
156. Repare Therapeutics (2025) Phase 1 trial of the safety, pharmacokinetics, pharmacodynamics, and preliminary clinical activity of RP-1664 in participants with advanced solid tumors. clinicaltrials.gov
157. Canadian Cancer Trials Group (2026) A phase II study of CFI-400945 and durvalumab in patients with advanced/metastatic triple negative breast cancer (TNBC). clinicaltrials.gov
158. Ito-Miwa K, Furuike Y, Akiyama S, Kondo T (2020) Tuning the circadian period of cyanobacteria up to 6.6 days by the single amino acid substitutions in KaiC. Proc Natl Acad Sci 117:20926–20931
159. Liao M, Liu Y, Xu Z et al (2025) The P-loop NTPase RUVBL2 is a conserved clock component across eukaryotes. Nature 642:165–173
160. Carvalho-Santos Z, Machado P, Branco P, Tavares-Cadete F, Rodrigues-Martins A, Pereira-Leal JB, Bettencourt-Dias M (2010) Stepwise evolution of the centriole-assembly pathway. J Cell Sci 123:1414–1426
161. Hodges ME, Scheumann N, Wickstead B, Langdale JA, Gull K (2010) Reconstructing the evolutionary history of the centriole from protein components. J Cell Sci 123:1407–1413
162. Nabais C, Peneda C, Bettencourt-Dias M (2020) Evolution of centriole assembly. Curr Biol 30:R494–R502
163. Evans T, Rosenthal ET, Youngblom J, Distel D, Hunt T (1983) Cyclin: a protein specified by maternal mRNA in sea urchin eggs that is destroyed at each cleavage division. Cell 33:389–396
164. Yuan K, Seller CA, Shermoen AW, O'Farrell PH (2016) Timing the *Drosophila* mid-blastula transition: a cell cycle-centered view. Trends Genet 32:496–507
165. Nurse P (1975) Genetic control of cell size at cell division in yeast. Nature 256:547–551
166. Thuriaux P, Nurse P, Carter B (1978) Mutants altered in the control co-ordinating cell division with cell growth in the fission yeast Schizosaccharomyces pombe. Mol Gen Genet 161:215–220

Repurposing Cell Cycle Oscillators for Multiciliated Cell Differentiation

3

Gonzalo Ortiz-Álvarez and Adel Al Jord

Abstract

Repurposing extant resources is at the basis of functional innovation in evolution. In cell biology, conserved regulatory circuits, with minor adjustments to a subset of components, can be recruited for functional novelty. A prominent example concerns multiciliated cells that generate cilia-based fluid flows, vital for organism development and homeostasis. During their post-mitotic differentiation, these cells repurpose the autonomous oscillators that normally govern the cell division cycle to drive massive production of cilia-nucleating organelles, named centrioles, in the cytoplasm without committing to cell division. In this chapter, we review recent advances in understanding how a repurposed cell cycle machinery regulates these cytoplasmic dynamics essential for motile ciliogenesis.

Keywords

Cell cycle · Cell differentiation · Centriole amplification · Multiciliated cells · Motile ciliogenesis · Organelle remodeling · Oscillators

G. Ortiz-Álvarez
Department of Medicine and Life Sciences, Universitat Pompeu Fabra (UPF), Barcelona, Spain

A. Al Jord (✉)
Centre for Genomic Regulation (CRG), The Barcelona Institute of Science and Technology, Barcelona, Spain

Universitat Pompeu Fabra (UPF), Barcelona, Spain
e-mail: adel.aljord@crg.eu

3.1 Introduction

3.1.1 Repurposing for Biological Novelty

Repurposing extant structures is central to sustaining life on Earth. François Jacob once stated, "*Evolution does not produce novelties from scratch. It works on what already exists, either transforming a system to give it new functions or combining several systems to produce a more elaborate one*" [1]. Indeed, resource repurposing has sustained radical evolutionary innovation on countless occasions and across spatial scales [1–9]: from the feather, deviated from its primordial role of heat insulation to subsequently support flight [2, 3], down to the skeletal gene Osteocrin, co-opted by the primate brain to presumably support augmented cognitive complexity [8]. Complex networks of regulatory proteins can also be redeployed and tweaked for novelty. For instance, the developmental Hedgehog signaling pathway [10, 11], through variations in the regulation of a subgroup of components, is recruited in butterfly eyespot evolution [12]. Another salient example of regulatory circuit repurposing was more recently discovered in multiciliated cells [13–20] (MCCs), whose differentiation is marked by the genesis of up to several hundred hair-like

M. G. Aydogan (ed.), *Biological Time Control*, Advances in Experimental Medicine and Biology 1521,
https://doi.org/10.1007/978-3-032-36189-9_3

protrusions known as motile cilia. During post-mitotic differentiation, these cells repurpose cell cycle oscillators from their typical task in regulating cell proliferation to drive, instead, cytoplasmic dynamics leading to motile ciliogenesis [21–26]. Below, we review mammalian MCCs and discuss the current progress on how the autonomous oscillators that normally govern the cell cycle machinery could also orchestrate MCC differentiation.

3.1.2 Multiciliated Cells in Health and Disease

MCCs are specialized epithelial cells with a bundle of microscopic motile cilia at their apical surface [14]. The coordinated beating of these cilia propels physiological fluids and cells along lumens of certain organs, including brain ventricles, airways, and both male and female reproductive tracts (Fig. 3.1a) [14]. Motile cilia-powered flows have diverse critical contributions to organism development and homeostasis. For example, brain MCCs known as ependymal cells (Fig. 3.1b, c) aid the circulation of the cerebrospinal fluid in ventricles. These flows are essential for nutrient delivery as well as waste clearance in the central nervous system [27–29], and even guide neuronal migration in the developing brain [30]. Airway MCCs (Fig. 3.1b, c) drive retrograde transport of inhaled particles and mucus, produced by goblet and secretory cells, as a first line of defense against pathogens and chemicals, thereby contributing to immunity [31]. In the male reproductive system, MCCs line the epithelium of efferent ducts in the testes where they help transport and prevent aggregation of sperm [32]. In females, MCCs lining oviducts (Fig. 3.1b, c) assist the transport of sperm toward the ova and, post-fertilization, that of the embryo toward the uterus [33].

These key functions of MCCs rationalize why defects in their differentiation and motile ciliogenesis, which interrupt vital cilia-based flows, can result in distinct deadly pathologies and organ failures [14, 34]. In the brain for instance, hindered cerebrospinal fluid flows cause ventricular enlargement and constriction of the brain parenchyma. This can lead to hydrocephalus, a condition that can be fatal if untreated or cause serious symptoms such as motor, cognitive and sensory impairment, epilepsy, endocrine dysfunction, vomiting, depression and chronic headaches [35, 36]. MCC malfunction in airways impedes mucociliary clearance resulting in chronic infections and compromised immunity [37], whereas reproductive MCC dysfunction increases risks of extra-tubal embryo implantation and infertility [14, 32, 33].

3.2 Main Text

3.2.1 Multiciliated Cell Differentiation: Centriole Amplification for Motile Ciliogenesis

Motile cilia stem from cylindrical root-like organelles named centrioles. As is the case in most animal cells, progenitors of MCCs contain two of these microtubule-based membrane-less organelles which, in the context of cell division, duplicate to provide each daughter cell with the microtubule organizing center known as the centrosome [38] (Fig. 3.2a). In dividing progenitors, each centrosomal centriole serves as scaffold for the duplication of a new procentriole as of the G_1/S transition of the cell cycle. These newly formed centrioles grow and mature during the G_2 phase. At mitotic entry, the two centrosomes travel to opposing cell poles to form the mitotic spindle, prior to segregating the chromosomes and disengaging the centrioles from one another. Each daughter cell inherits one centrosome, in a process that—like DNA inheritance—is semiconservative: each cell incorporates a centrosome composed of an older (mother) and a newly formed (daughter) centriole [38, 39] (Fig. 3.2a).

As the post-mitotic progenitor cell, with a single centrosome, initiates MCC differentiation, the centriole population massively expands up to several hundred for the nucleation of an identical number of motile cilia [14]. Electron microscopy-based studies attempted to

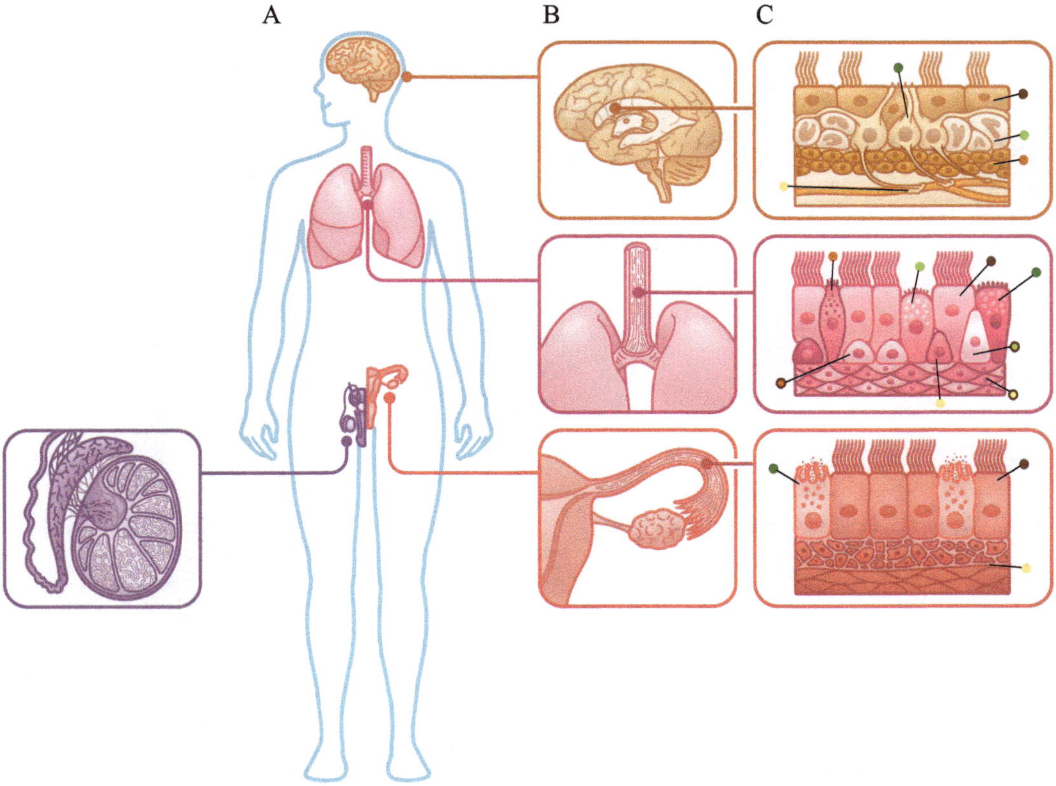

Fig. 3.1 Mammalian multiciliated cells across tissues. (**a**) Mammalian multiciliated cells (MCCs) are found in the central nervous system, including brain and spinal cord, the respiratory and reproductive systems of males and females [14]. (**b**) MCCs form a continuous epithelium lining the ventricular system of the central nervous system, formed by the brain ventricles (lateral ventricles, third and fourth ventricles) and the central canal in the spinal cord. They form a communicating lumen where near-wall cerebrospinal fluid propulsion promotes nutrient and signal delivery, and clearing of waste and neurotoxins. The multiciliated epithelium also lines the cavities of the trachea and bronchi, as well as the nasal cavities, paranasal sinuses and parts of the auditory canal. In airways, MCCs engage in defense against outer insults via their role in mucociliary clearance, to keep external agents out of the respiratory canal. Finally, MCCs line the lumen of the efferent ducts of the testes, where they perform a sperm-concentrating function, to supposedly ensure that high-quality sperm reaches the epididymis. In females, motile cilia play a role in transport of the ova from the ovaries into the uterus, as well as the sperm in the opposite direction to help it reach and fertilize the egg [13, 14]. (**c**) The lateral ventricle epithelium is the most studied niche for ependymal cells (brown dot) in the nervous system. There, they form, along with adult neural stem (B1) cells (dark green dot), pinwheel or rosette-like structures with B1 cells displaying a small apical cell surface where their primary cilium protrudes into the ventricular lumen. Ependymal cells, with a much larger apical surface, arrange themselves around B1 cells and protrude their motile cilia into the ventricular lumen [105]. B1 cells engage in adult neurogenesis, i.e., the generation of new neurons during adulthood, by dividing and giving rise to C cells (light green dot), neuronal-committed progenitors that in turn generate A cells (red dot). These can divide further and generate the immature neurons that migrate out of the niche [106, 107]. B1 cells also possess a basal projection and an endfoot that often contacts underlying blood vessels (yellow dot) [105]. In airways, MCCs (brown dot) are interspersed with many other cell types. Secretory cells, like goblet (dark green dot) and club cells (light green dot) secrete mucus and factors with protective and immunomodulatory functions, respectively. Rare chemosensory cell types like tuft cells (red dot) and pulmonary endocrine cells (yellow dot) coordinate interactions with the external environment. Ionocytes (light green with brown outline dot), also rare, play a role in mucus production regulation. Basal cells (red dot with brown outline dot) are multipotent stem cells that serve as progenitors of all these ciliated, secretory and regulatory cell types. Finally, stromal cells like smooth muscle cells and fibroblasts (yellow with brown outline dot) control airway epithelial cell turnover and can constrict airways [31, 108]. In the female reproductive system, MCCs (brown dot) are also intercalated with transcriptomically heterogeneous secretory cells (dark green dot) [109]. Together, they support fertilization and preimplantation development. Connective tissue and muscle cells (yellow dot) underlie the oviduct epithelium [110]. In males (not shown), resorptive cells increase sperm concentration and form an epithelium together with MCCs [111]. Smooth muscle cells contract to push sperm toward the epididymis and provide structural rigidity [32]

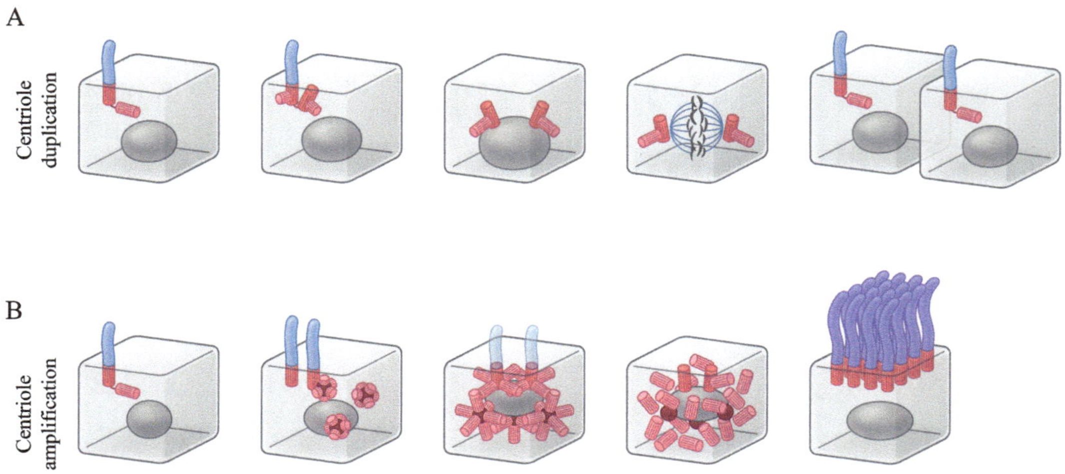

Fig. 3.2 Centriole duplication in dividing cells and centriole amplification in differentiating multiciliated cells. (**a**) Centriole duplication. Animal cells have a centrosome, composed of an older mother and a younger daughter (recently synthesized) centriole, associated via a linker (cell #1). The parent centriole nucleates the primary cilium, an antenna-like microtubule-based protrusion of the cell membrane, which integrates signaling for cell development and homeostasis [38, 112]. At the G_1|S transition, one procentriole, the new daughter centriole, starts to be synthesized at the proximal end of each centriole, now the two new mother centrioles (cell #2) [38, 112]. Upon reaching G_2, the new daughter centrioles are still immature, but by the G_2|M transition, they reach maturity, the link tethering mother centrioles is lost and each new centrosome migrates to one pole of the cell. At this point, the primary cilium is also resorbed (cell #3) [38, 112]. During mitosis, each centrosome, as microtubule organizing centers, contributes to the formation of the mitotic spindle that enables sister chromatid separation, with each daughter cell receiving an exact replica of the genome (cell #4) [113]. At the end of mitosis, the mother and daughter centrioles separate from each other, a process known as "centriole disengagement" [113, 114]. In the end, two cells are formed, each with one centrosome formed by an older mother centriole and a younger daughter centriole (generated in the previous cell cycle). A new primary cilium grows at G_0|G_1 (cells #5) [38, 112]. (**b**) Centriole amplification. MCC progenitors start with a centrosome and a non-motile primary cilium (cell #1), as most other animal cells [44]. When centriole amplification starts, the proximal end of the daughter centriole starts to nucleate deuterosomes. Procentrioles start budding from these and successively pass into the cytoplasm, until sufficient deuterosome/procentrioles complexes are formed [44]. The nucleus of ependymal cells appears to increase in size at the onset of centriole amplification [115] (amplification phase, cell #2). Procentrioles on deuterosome platforms and on walls of the centrosomal centrioles grow simultaneously, forming flower-like structures [44]. Around 50% of ependymal cells grow a second primary cilium in vivo and in vitro [49] (growth phase, cell #3). All newly formed centrioles disengage simultaneously to migrate to the apical surface (disengagement phase, cell #4) [44]. Eventually, the centrioles, now basal bodies, nucleate motile cilia. The nucleus of fully mature ependymal cells has an increased volume despite being diploid [21, 69], and is flatter and closer to the centriolar patch when compared to progenitor cells (cell#5) [115]

comprehend this cellular transformation since the late 60s [40–42]. For decades, mammalian MCCs were assumed to undergo centriole amplification using a cell type-specific assembly line [40–43]. However, these assumptions were challenged by advanced microscopy approaches that precisely documented the spatiotemporal dynamics of centriole multiplication primarily in ependymal MCC progenitors [44, 45]. They revealed that centriole amplification in MCCs acts out a more structured play than expected (Fig. 3.2b). As such, three distinct and successive phases of centriole production characterize the differentiation of MCCs:

- *Amplification Phase*: The production of procentrioles initiates in the vicinity of the centrosome where a "pericentrosomal cloud" forms in a microtubule-dependent manner [44, 46]. The formation of procentrioles is

accompanied by the recruitment of regulatory and scaffold proteins of centriole biogenesis used by dividing cells during centriole duplication, which include Cep152, Plk4, and Sas6 [38]. However, these proteins are asymmetrically distributed in favor of the daughter centriole during ependymal MCC differentiation [44], in stark contrast to their equal recruitment to centrosomal centrioles during centriole duplication [44, 47, 48]. The majority of procentrioles are nucleated around MCC-specific spherical platforms known as deuterosomes [40], which seem to form specifically on the daughter centriole. Procentrioles, while still associated to deuterosome platforms, are sequentially released into the cytoplasm, but remain latent at an early stage of biogenesis until the last procentriole is formed [44] (Fig. 3.2b).
- *Growth Phase*: When the final amount of procentrioles is produced, they all simultaneously grow, forming 3-dimensional flower-like structures, while still attached to deuterosomes [44]. During this phase, some ependymal MCCs grow a second cilium on the daughter centriole in addition to the preexisting primary cilium nucleated by the mother centriole, before they both collapse during the transition into the subsequent phase [49] (Fig. 3.2b).
- *Disengagement Phase*: Grown centrioles synchronously disengage from their centriolar and deuterosomal platforms around the nucleus before migrating apically, docking, and nucleating motile cilia [44, 46] (Fig. 3.2b).

Centriole amplification is a highly robust process because neither the depletion of the deuterosomes, the parental centrosome, nor both at the same time impairs centriole amplification [49–52]. Notably, when deuterosomes are not present, centriole biogenesis from the walls of parental centrioles is greatly upregulated to cope with the high demand of centriole amplification [49]. Removing both the centrosome and deuterosomes leads to centriolar amplification from the "pericentrosomal cloud" [49]. At the single organelle level, a potential key regulator of early steps of centriole amplification may be Plk4: a protein described long ago to be necessary for centriole duplication [53–55] as well as shown to induce centriole amplification from parental centrioles in proliferating cells [56], and, more recently, found to regulate centriole growth in the syncytial *Drosophila* embryo [57, 58]. In this single multinucleated cell with hundreds of centrioles, local Plk4 oscillations establish a permissive window for centriole biogenesis: above a certain Plk4 threshold, centrioles grow, duplicating from mother centrioles, and below it, they do not grow or stop growing, thereby achieving a constant centriolar size in all centriolar pairs. Furthermore, these oscillations, although entrained to the cell cycle, can happen autonomously, since disrupting the mitotic oscillator Cyclin-Dependent Kinase (CDK)/Cyclin leads to spatially stochastic yet periodic—when looking at individual organelles—centriole duplication events. Also in *Drosophila*, unfertilized eggs that do not engage in the cell cycle can duplicate their centrioles autonomously when overexpressing Plk4 [59]. In MCCs, Plk4 is indeed recruited asymmetrically to daughter centrioles [44] and deuterosomes [60], raising the question of whether it could be regulating the amplification of hundreds of procentrioles and their growth to a constant size in these non-dividing cells. The answer is not trivial due to conflicting results about the role of Plk4 in MCC centriole amplification. On one hand, inhibiting Plk4 activity or knocking down its expression does not affect the final outcome of MCC differentiation, despite some delays in centriole amplification dynamics [49, 50, 52]. On the other hand, genetic ablation of the Plk4 gene (knock-out) prevents MCC differentiation altogether [60], suggesting that Plk4 may be necessary as a scaffold component at the base of the nascent procentrioles. Whether its activity could locally modulate procentriole growth dynamics in MCCs, in an analogous manner to centriole duplications in the syncytial *Drosophila* embryo [57, 58], remains to be probed.

The process of centriole amplification also seems conserved across multiciliated tissues and even among diverse species. Indeed, parallel examinations of developing mouse ciliated ependyma and airways show similar phases of centriole biogenesis, implying that mammalian MCCs share this *en masse* organelle production pipeline [44]. This notion is also supported by the convergence of transcriptomes of mice and humans across multiciliated tissues, with distinct embryological origins (ependyma, airways and oviducts), for most of the ciliogenesis and ciliary function genes [61]. Furthermore, this logic may even be extended to other vertebrates, since deuterosomes and their associated proteins are also present in *Xenopus* [62, 63].

Overall, centriole amplification in differentiating progenitors of MCCs somewhat mirrors the centriole duplication cycle of dividing cells, which also originates from the centrosome and, similarly, follows three phases (i.e., formation, growth, disengagement) [38]. In turn, the findings establish the centrosome's unpredicted capacity to support either the duplication of centrioles in dividing progenitors or their amplification in differentiating ones. Such remarkable phase-based progression of centriole production can thus unify the cellular basis of centriole biogenesis in dividing and differentiating cells. Differentiating cells, however, repurpose the centrosome's ability to duplicate to meet the needs of motile ciliogenesis, instead of their canonical role implicated in cell division.

3.2.2 Repurposed Cell Cycle Drives Differentiation Instead of Proliferation

How do MCCs regulate the clock-like phase-based progression of massive centriole production? An analogy between centriole duplication and amplification was key in addressing this issue and in uncovering the molecular drivers steering centrioles toward motile ciliogenesis [21]. Dividing cells undergoing centriole duplication coordinate centriole copy number, growth, and dynamics using key promoters of cell cycle progression [38], mainly the conserved mitotic oscillator. However, this clock-like regulatory circuit, centered on the CDK1-APC/C axis, is a universal trigger for cell division that drives G_2|M-phase and metaphase|anaphase transitions [64] in dividing cells. This introduced the counterintuitive question of whether differentiating MCCs coordinate centriole copy number, growth, and dynamics using a conserved cell division trigger in absence of division. The first answers to this question were delivered by immunocytochemistry experiments that showed timely and transient expression of CDK1, Cyclin B1, and other key mitotic proteins in differentiating progenitors of MCCs undergoing centriole amplification. The mitotic machinery was not only expressed, but also active, as judged by the sequential appearance of mitotic phosphorylations in the cytoplasm and nucleus, albeit at lower intensities than in mitosis [21]. This is the case in differentiating mouse ependymal MCCs in vivo and in vitro, with some comparable evidence in differentiating MCCs of mouse tracheas in vivo [21], further suggesting the conservation of the multiciliation program across tissues. The chemical modulation of pivotal elements of the mitotic oscillator coupled to live-imaging of centriole amplification revealed the oscillator's impact on MCC differentiation. Restraining the oscillator's activity by inhibition of CDK1 activity prolongs the duration of the procentriole amplification phase and increases procentriole numbers by impeding centriole amplification phase transitions. Inversely, hyper-activating CDK1 accelerates centriole amplification phase transitions, thereby decreasing final centriole number as a result of a shortened amplification phase [21]. A striking discovery was that this heightened CDK1 activity, due to pharmacological inhibition of CDK1 inhibitors (e.g. Myt/Wee1 and APC/C [64–66]), drives these post-mitotic progenitors into mitosis (Table 3.1). This means that differentiating MCC progenitors redeploy the conserved mitotic circuit, in a calibrated form with a dampened oscillatory activity, to orchestrate cytoplasmic organelle remodeling associated with motile ciliogenesis while preventing nuclear division [20, 21].

Table 3.1 Summary of the main DNA synthesis and chromosome condensation events occurring in multiciliated cells upon tampering with cell cycle-associated proteins

Phenotype	Target	Method	Model	References
Partial DNA Synthesis	Cyclin E2	Gene over-expression	Mouse ependymal cells in vitro	Serizay et al. [25]
	E2F7	Gene knock out	Mouse tracheal epitelial cells in vitro	Choksi et al. [24]
	Emi2	Gene over-expression	*Xenopus laevis* larval skin in vivo	Kim et al. [83]
	Emi1	Gene over-expression	Mouse ependymal cells in vitro	Serizay et al. [25]
Partial Chromosome Condensation	Cyclin A2	Gene over-expression	Mouse ependymal cells in vitro	Serizay et al. [25]
	Wee1/ Myt1	Pharmacological inhibition (drug: PD 166285)	Mouse ependymal cells in vitro	Al Jord et al. [21]
	APC/ CCdc20	Pharmacological inhibition (drug: proTAME)	Mouse ependymal cells in vitro	Al Jord et al. [21]

The central Emi1 row has both events: partial DNA synthesis and partial chromosome condensation.

Other key players of the cell cycle were subsequently found to play a role in MCC differentiation. For instance, CDK2, which promotes the G_1|S and S|G_2 transitions in complex with cyclins E and A2 [67, 68], is expressed in mouse tracheal epithelial cell culture (mTECs, tracheal MCCs in vitro). Together with cyclin A1, CDK2 is necessary to drive motile ciliogenesis at the cellular and transcriptomic levels [22]. Also, hyperphosphorylated retinoblastoma (Rb) protein, whose phosphorylation allows the G_1|S transition [67], is present during centriole amplification in ependymal cells, but not before or once MCC differentiation terminates [69].

The true magnitude of the redeployment of cell cycle players during the multiciliation program was recently revealed by single-cell RNA sequencing across mammals (human and mouse) and tissues (ependyma and MCCs in the trachea), both in vivo and in vitro [24, 25]. Two independent studies have reached similar conclusions, claiming a massive repurposing of cell cycle genes during MCC differentiation. This has been labeled as a new or alternative cell cycle variant, which operates during centriole amplification. The successive steps of MCC differentiation are characterized by a cyclical trajectory of their transcriptome that mirrors that of the cell cycle. Early MCCs express G_0|G_1-associated genes, including G_0|G_1 transition cyclins (cyclins D1 and D2) and CDKs (CDK4, CDK6). As they differentiate, they express S-phase associated genes like cyclin E1 and CDK2, though at lower levels to possibly evade the DNA replication threshold, followed by G_2|M-associated genes like cyclin B1 and CDK1. Finally, mature MCCs return to a G_0|G_1-like state, thereby completing the cycle [24, 25] (Fig. 3.3).

As during the cell cycle, waves of cyclins, both canonical and non-canonical, mark the progression through differentiation. Early in differentiation, a surge of Cyclins D1 [24] and D2 [25] is detected (Fig. 3.3). These cyclins classically bind to CDK4/6 in dividing cells to promote G1 progression in the presence of mitogenic cues [70, 71]. Blocking CDK4/6 activity inhibits not only differentiation, but commitment to the MCC lineage, as seen by the absence of early fate determination marker expression like *Gmnc* (GemC1), *Foxj1* or *Myb* [72–76]. CDK4/6 activity is then necessary to promote the G_0|G_1-like (MCC progenitors) to S-like (MCCs in the amplification phase) transition during MCC differentiation [24] (Fig. 3.3). This step is followed by a surge and then sustained expression of CDK2 during the entire differentiation process [22, 24]

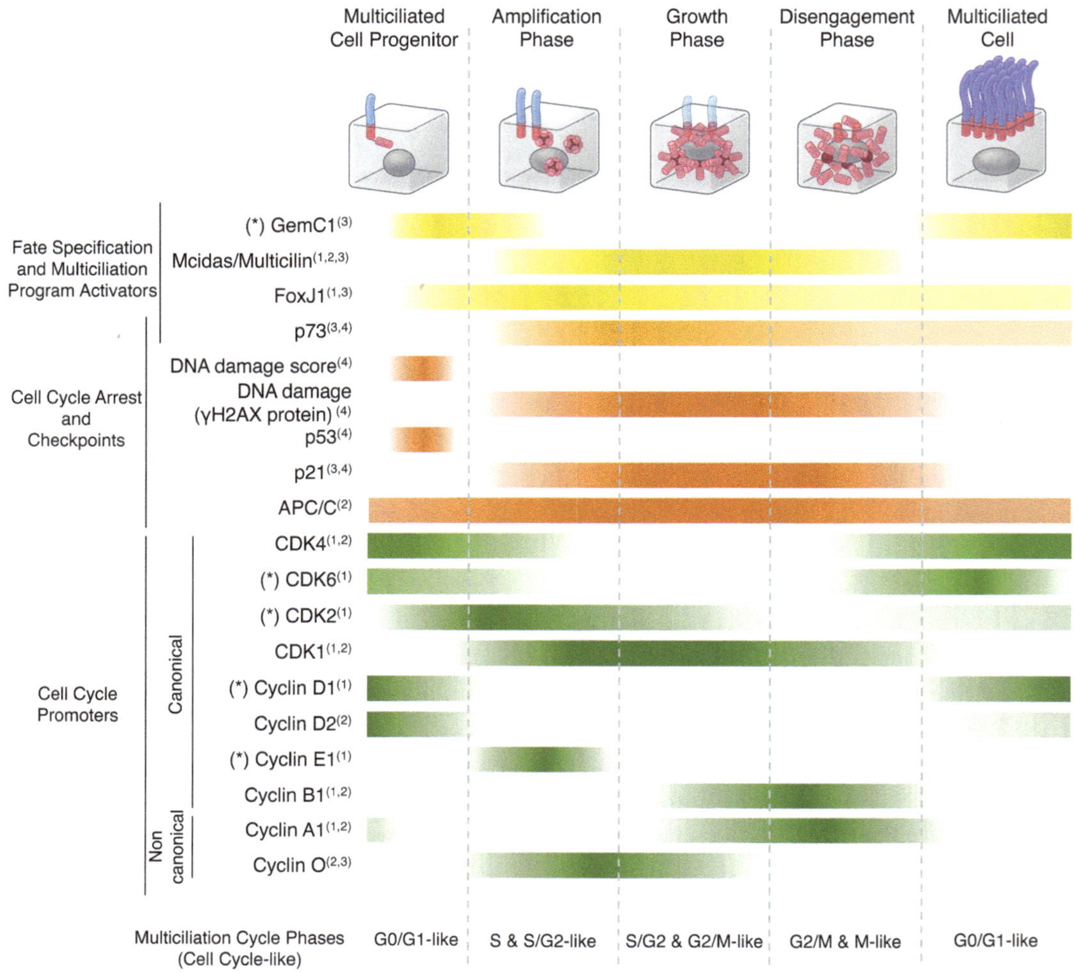

Fig. 3.3 Repurposing of cell cycle oscillators during centriole amplification in multiciliated cells. Integration of single-cell RNA sequencing studies to depict the expression level of fate specification and multiciliation gene regulatory network activators (in yellow), cell cycle arrest inducers and checkpoints (in red) and cell cycle promoters (in green), relative to distinct phases of centriole amplification—from an MCC progenitor (two centrosomal centrioles) to a mature MCC. Note that p73 expression is depicted in orange as it was described as a multiciliation transcriptional activator and a cell cycle brake inducer [89, 90]. Intensity of bar colors indicates expression levels. The studies from which the single-cell RNA sequencing data was assessed are indicated with numbers next to each protein name: (1) Choksi et al. [24], (2) Serizay et al. [25], (3) Khoury Damaa et al. [26], (4) Ortiz-Álvarez et al. [69]. The cell cycle-like phase of each of the steps of centriole amplification is indicated in the lowermost panel. This is based on the high percentage of cell cycle phase-specific genes that are differentially expressed at each one of the centriole amplification phases. Minor discrepancies (indicated with an asterisk) arise among the cited studies: GemC1 is barely detected in only a few progenitors and some differentiating cells in [69]. However, [26] find sustained higher expression in progenitors and mature MCCs. CDK6 presents a low, barely oscillatory behavior according to [25], but it is highly transcribed with strong expression peaks and valleys, as deduced from [24]. CDK2 is absent or very low in [25]. In [22] it is expressed, albeit low and non-oscillatory, as seen via quantitative real-time PCR and it is necessary. In [24], a peak of expression is suggested at the beginning of the differentiation process, peaking just after CDK4 and before CDK1. Cyclin D1 and E1 are at very low levels in [25], although they do show a mild peak at the progenitor and the amplification stages, respectively. High surges for the expression of both, at the same phases, are detected in [24]. Note that the single-cell RNA sequencing dataset from [25, 26] was generated from mouse ependymal (brain) MCCs in vitro, whereas that of [24] are derived from mTECs in vitro. Data from [69] were produced from ependymal mouse cells in vivo. The discrepancies might derive from differences in the data's origin (ependyma/trachea) or experimental approach (in vivo/in vitro). The expression levels for the DNA damage protein, γH2AX are depicted, since they greatly differ from the DNA damage score, based on several genes. This discrepancy is possibly due to post-translational modifications that cannot be accounted for with single-cell RNA sequencing data [69]

and an increase of non-canonical Cyclin O [25, 77] and Cyclin E1 [24] during the amplification and growth phases (Fig. 3.3), which correspond to an S|G_2|M-like transcriptomic state [24, 25]. In cycling cells, E type cyclins in association with CDK2 promote the phosphorylation of Rb to enable the G_1|S transition and S progression via the phosphorylation of substrates involved in centrosome duplication or DNA synthesis and repair [78, 79]. As mentioned above, CDK2 inhibition depletes the MCC differentiation program [22]. Cyclin O is necessary to enter the cell cycle-like MCC differentiation, since mice lacking the *Ccno* gene do not acquire an S-like transcriptomic signature, and are depleted of brain ependymal cells and most MCCs in the trachea. However, Cyclin O is not essential for cell fate acquisition as judged by expression of *Gmnc*, *Foxj1* and *Mcidas* [26]. Finally, these cells reach a wave of Cyclin B1 and the non-canonical Cyclin A1 in mid and late differentiating MCCs, which display a G_2|M-like transcriptomic state (Fig. 3.3). The former was previously shown to work with CDK1 [21], and the latter, with CDK2 [22]. Consistently, the deletion of the Cyclin A1 gene, *Ccna1*, also leads to a decrease of MCCs in murine tracheas [22].

The expression of so many cell cycle-related genes during differentiation makes MCCs poised to develop real cell cycle events, such as DNA replication and chromosome condensation, under certain conditions (Table 3.1). For instance, the overexpression of canonical Cyclin E2 or the inhibition of Emi1 (an inhibitor of the APC/C [80]) leads to an increase of DNA synthesis alone or with chromosome condensation, respectively [25]. In a similar manner, ectopic Cyclin A2 expression (absent or little expressed in normal MCC differentiation [22, 25], but essential in normal cell cycle progression [81]) causes mitotic events (chromosome condensation) [25]. Depletion of *E2f7*, a regulator of the G1|S transition [82], also increases DNA synthesis in mTECs [24]. Finally, inhibiting APC/C by targeting Emi2 in other vertebrates, namely *Xenopus*, also triggers abnormal partial DNA replication in the MCCs of this amphibian's embryonic epithelium [83].

Since MCC differentiation is coupled to the activity of proteins that promote cell cycle progression, it seems plausible that proteins that hinder it also have a role in the process. CDK inhibitors (CKIs) function as brakes to halt cell cycle progression under unfavorable conditions by restraining the activity of various CDKs and cyclins [84, 85]. Other proteins can also arrest the cell cycle due to internal and external stresses, including DNA damage, hypoxia, nutrient deprivation, and cancer cell risk, to end in cell cycle exit, senescence or apoptosis, like the vastly studied tumor suppressor gene *Trp53* (p53) [86]. Stabilization of p53 and consequent cycle arrest are, for instance, one outcome of the loss or over-amplification of the centrosome during cell division [38]. The gene *Trp73* (p73 protein) belongs to the p53 family and, like p53, can induce cell cycle arrest and apoptosis [87], as a result of genotoxic stress [88]. The protein p73 also binds and promotes the expression of target genes in the multiciliogenesis program, including *Foxj1* [89]. By now, it comes without surprise that p73, p53, and its downstream effector, the CKI p21, are expressed during multiciliogenesis [69, 89, 90]. *Trp73−/−* mice display hydrocephalus due to a lack of normal ependyma in the brain, as well as infertility and chronic infections of the airways, due to the absence of MCCs in the reproductive and respiratory tracts. However, *Trp53−/−* and *Cdkn1a−/−* (p21 KO) mice do not develop hydrocephalus [90, 91], despite the proteins p53 and p21 being expressed in differentiating ependymal cells [69]. Their absence only causes delays in differentiation that are visible at early stages of ependymal development, although later timepoints remain to be probed in more detail [69].

3.3 Concluding Remarks

This last decade of research convincingly implicated a cell cycle variant in orchestrating the massive remodeling of cilia-nucleating organelles in the cytoplasm of differentiating MCCs. Some cell cycle regulators, including CDK4/6, CDK2, Cyclin A1 or Cyclin O are necessary for multiciliogenesis. Other cell cycle players seem to be important to avoid cell cycle events, like partial DNA replication or pseudo mitosis, both probably deleterious for multiciliation. Finally, CKIs like p21 and p53 family members (p53 and p73) are also recruited during centriole amplification. At present, the role of the cell cycle machinery in regulating these MCC cytoplasmic dynamics appears quasi elucidated. However, the machinery's impact on the dynamics of other major cell compartments, including the nucleus, remains poorly explored during MCC differentiation. In dividing cells, cell cycle oscillations and phosphorylations regulate the dynamics of nuclear architecture along with its internal functional components like chromatin, biomolecular condensates, and RNA molecules [92–102]. Thus, a promising yet challenging research path in the MCC field is to venture into unexplored grounds of how repurposed cell cycle oscillators impact the MCC nucleus across spatiotemporal scales—from the dynamics of nuclear architecture down to that of chromatin, condensates, and RNA processing.

The MCC findings collectively establish the capacity of epithelial cells to physiologically repurpose the cellular and molecular machineries of the cell cycle for multiciliation instead of proliferation. Conceptually, the MCC cell cycle variant is analogous to other cell cycle variants that evolved to sustain endocycles, like endoreplication and endomitosis, in various differentiated cell types of numerous species [103]. A future challenge consists in comparing and capturing a more complete picture of these diverse cell cycle variants, to better understand how cells alter molecular switches regulating the dynamics of cellular events leading to functional novelty. Beyond cell differentiation, identifying sets of either unique or unifying strategies that cells deploy to bypass canonical cell cycle molecular switches is critical from a disease perspective, as it may open new therapeutic doors to prevent cancer cell attempts to tweak cell cycle circuits for enhanced proliferative drives [104].

Acknowledgments and Grants The Al Jord "Mechanics of Organelle Remodeling" Laboratory is supported by the Spanish Ministry of Science and Innovation through the Agencia Estatal de Investigación grant PID2024-157329NA-I00, the Fondation Jérôme Lejeune grant 2523_GRT-2025B, the Centro de Excelencia Severo Ochoa (CEX2020-001049-S, MCIN/AEI /10.13039/501100011033 / FEDER, UE), the Generalitat de Catalunya CERCA programme, and the CRG Core Technologies programme. We thank M. Ducom for designing the illustrations following discussions with authors. G.O-A. received the support of a post-doctoral fellowship from "la Caixa" Foundation (ID 100010434). The fellowship code is "LCF/BQ/PR23/11980036."

Author Contributions A.A.J. and G.O-A. discussed and co-wrote the manuscript.

Disclosures The authors declare no conflict of interest.

References

1. Jacob F (1977) Evolution and tinkering. Science (1979) 196:1161–1166
2. Prum RO (1999) Development and evolutionary origin of feathers. J Exp Zool 285:291–306
3. Heers AM, Dial KP (2012) From extant to extinct: locomotor ontogeny and the evolution of avian flight. Trends Ecol Evol 27:296–305
4. Glasauer SMK, Neuhauss SCF (2014) Whole-genome duplication in teleost fishes and its evolutionary consequences. Mol Gen Genomics 289:1045–1060
5. Carelli FN, Liechti A, Halbert J, Warnefors M, Kaessmann H (2018) Repurposing of promoters and enhancers during mammalian evolution. Nat Commun 9:4066
6. Kellenberger RT et al (2023) Multiple gene co-options underlie the rapid evolution of sexually deceptive flowers in Gorteria diffusa. Curr Biol 33: 1502–1512.e8
7. True JR, Carroll SB (2002) Gene co-option in physiological and morphological evolution. Annu Rev Cell Dev Biol 18:53–80
8. Ataman B et al (2016) Evolution of Osteocrin as an activity-regulated factor in the primate brain. Nature 539:242–247
9. Barve A, Wagner A (2013) A latent capacity for evolutionary innovation through exaptation in metabolic systems. Nature 500:203–206
10. Ingham PW, Nakano Y, Seger C (2011) Mechanisms and functions of Hedgehog signalling across the metazoa. Nat Rev Genet 12:393–406

11. Lee RTH, Zhao Z, Ingham PW (2016) Hedgehog signalling. Development (Cambridge) 143:367–372
12. Keys DN et al (1999) Recruitment of a hedgehog regulatory circuit in butterfly eyespot evolution. Science 283:532–534
13. Brooks ER, Wallingford JB (2014) Multiciliated cells. Curr Biol 24:R973–R982
14. Spassky N, Meunier A (2017) The development and functions of multiciliated epithelia. Nat Rev Mol Cell Biol 18:423–436
15. Mahjoub MR, Nanjundappa R, Harvey MN (2022) Development of a multiciliated cell. Curr Opin Cell Biol 77:102105
16. Boutin C, Kodjabachian L (2019) Biology of multiciliated cells. Curr Opin Genet Dev 56:1–7
17. Collins C, Ventrella R, Mitchell BJ (2021) Building a ciliated epithelium: transcriptional regulation and radial intercalation of multiciliated cells. Curr Top Dev Biol 145:3–39
18. Lewis M, Stracker TH (2021) Transcriptional regulation of multiciliated cell differentiation. Semin Cell Dev Biol 110:51–60
19. Meunier A, Azimzadeh J (2016) Multiciliated cells in animals. Cold Spring Harb Perspect Biol 8:a028233
20. Al Jord A, Spassky N, Meunier A (2019) Motile ciliogenesis and the mitotic prism. Biol Cell 111: 199–212
21. Al Jord A et al (2017) Calibrated mitotic oscillator drives motile ciliogenesis. Science 358:803–806
22. Vladar EK et al (2018) Cyclin-dependent kinase control of motile ciliogenesis. elife 7:1–23
23. Revinski DR et al (2018) CDC20B is required for deuterosome-mediated centriole production in multiciliated cells. Nat Commun 9:1–15
24. Choksi SP et al (2024) An alternative cell cycle coordinates multiciliated cell differentiation. Nature 630:214–221
25. Serizay J et al (2025) Cyclin switch tailors a cell cycle variant to orchestrate multiciliogenesis. Cell Rep 44: 115103
26. Khoury Damaa M et al (2025) Cyclin O controls entry into the cell-cycle variant required for multiciliated cell differentiation. Cell Rep 44:115117
27. Jiménez AJ, Domínguez-Pinos M-D, Guerra MM, Fernández-Llebrez P, Pérez-Fígares J-M (2014) Structure and function of the ependymal barrier and diseases associated with ependyma disruption. Tissue Barriers 2:e28426
28. Siyahhan B et al (2014) Flow induced by ependymal cilia dominates near-wall cerebrospinal fluid dynamics in the lateral ventricles. J R Soc Interface 11: 20131189
29. Faubel R, Westendorf C, Bodenschatz E, Eichele G (2016) Cilia-based flow network in the brain ventricles. Science (1979) 353:176–178
30. Sawamoto K et al (2006) New neurons follow the flow of cerebrospinal fluid in the adult brain. Science (1979) 311:629–632
31. Zepp JA, Morrisey EE (2019) Cellular crosstalk in the development and regeneration of the respiratory system. Nat Rev Mol Cell Biol 20:551–566
32. Hoque M, Kim EN, Chen D, Li FQ, Takemaru KI (2022) Essential roles of efferent duct multicilia in male fertility. Cells 11:341
33. Lyons RA, Saridogan E, Djahanbakhch O (2006) The reproductive significance of human Fallopian tube cilia. Hum Reprod Update 12:363–372
34. Reiter JF, Leroux MR (2017) Genes and molecular pathways underpinning ciliopathies. Nat Rev Mol Cell Biol 18:533–547
35. Kahle KT et al (2024) Paediatric hydrocephalus. Nat Rev Dis Primers 10:35
36. Vinchon M, Rekate H, Kulkarni AV (2012) Pediatric hydrocephalus outcomes: a review. Fluids Barriers CNS 9:18
37. Tilley AE, Walters MS, Shaykhiev R, Crystal RG (2015) Cilia dysfunction in lung disease. Annu Rev Physiol 77:379–406
38. Nigg EA, Holland AJ (2018) Once and only once: mechanisms of centriole duplication and their deregulation in disease. Nat Rev Mol Cell Biol. https://doi.org/10.1038/nrm.2017.127
39. Stearns T (2001) Centrosome duplication: a centriolar pas de deux. Cell 105:417–420
40. Sorokin SP (1968) Reconstructions of centriole formation and ciliogenesis in mammalian lungs. J Cell Sci 3:207–230
41. Dirksen ER (1971) Centriole morphogenesis in developing ciliated epithelium of the mouse oviduct. J Cell Biol 51:286–302
42. Anderson RGW, Brenner RM (1971) The formation of basal bodies (centrioles) in the Rhesus monkey oviduct. J Cell Biol 50:10–34
43. Zhao H et al (2013) The cep63 paralogue deup1 enables massive de novo centriole biogenesis for vertebrate multiciliogenesis. Nat Cell Biol 15:1434–1444
44. Al Jord A et al (2014) Centriole amplification by mother and daughter centrioles differs in multiciliated cells. Nature 516:104–107
45. Boudjema A-R et al (2024) Live-imaging centriole amplification in mouse brain multiciliated cells. In: Mennella V (ed) Cilia: methods and protocols. Springer US, New York, pp 167–180. https://doi.org/10.1007/978-1-0716-3507-0_10
46. Boudjema A-R et al (2024) The cell cycle variant in multiciliated cells incorporates 2 centriole biogenesis cycles. eLife 13:RP96584 https://doi.org/10.7554/eLife.96584.2
47. Arquint C, Nigg EA (2016) The PLK4–STIL–SAS-6 module at the core of centriole duplication. Biochem Soc Trans 44:1253
48. Hatch EM, Kulukian A, Holland AJ, Cleveland DW, Stearns T (2010) Cep152 interacts with Plk4 and is required for centriole duplication. J Cell Biol 191: 721–729

49. Mercey O et al (2019) Dynamics of centriole amplification in centrosome-depleted brain multiciliated progenitors. Sci Rep 9:13060
50. Zhao H et al (2019) Parental centrioles are dispensable for deuterosome formation and function during basal body amplification. EMBO Rep 20:e46735
51. Mercey O et al (2019) Massive centriole production can occur in the absence of deuterosomes in multiciliated cells. Nat Cell Biol 21:1544–1552
52. Nanjundappa R et al (2019) Regulation of cilia abundance in multiciliated cells. elife 8:e44039
53. Bettencourt-Dias M et al (2005) SAK/PLK4 is required for centriole duplication and flagella development. Curr Biol 15:2199–2207
54. Holland AJ et al (2012) Polo-like kinase 4 controls centriole duplication but does not directly regulate cytokinesis. Mol Biol Cell 23:1838
55. Habedanck R, Stierhof Y-D, Wilkinson CJ, Nigg EA (2005) The Polo kinase Plk4 functions in centriole duplication. Nat Cell Biol 7:1140–1146
56. Kleylein-Sohn J et al (2007) Plk4-induced centriole biogenesis in human cells. Dev Cell 13:190–202
57. Aydogan MG et al (2018) A homeostatic clock sets daughter centriole size in flies. J Cell Biol 217:1233–1248
58. Aydogan MG et al (2020) An autonomous oscillation times and executes centriole biogenesis. Cell 181:1566–1581.e27
59. Nabais C et al (2021) Plk4 triggers autonomous de novo centriole biogenesis and maturation. J Cell Biol 220:e202008090
60. Lomastro GM et al (2022) PLK4 drives centriole amplification and apical surface area expansion in multiciliated cells. elife 11:e80643
61. Chae S, Park TJ, Kwon T (2023) Convergent differentiation of multiciliated cells. Sci Rep 13(1):23028
62. Kim SK et al (2021) A role for Cep70 in centriole amplification in multiciliated cells. Dev Biol 471:10–17
63. KlosDehring DA et al (2013) Deuterosome mediated centriole biogenesis. Dev Cell 27:103–112
64. Wieser S, Pines J (2015) The biochemistry of mitosis. Cold Spring Harb Perspect Biol 7:a015776
65. Pines J, Hagan I (2011) The Renaissance or the cuckoo clock. Philos Trans R Soc Lond B Biol Sci 366:3625–3634
66. Pines J (2011) Cubism and the cell cycle: the many faces of the APC/C. Nat Rev Mol Cell Biol 12:427–438
67. Henley SA, Dick FA (2012) The retinoblastoma family of proteins and their regulatory functions in the mammalian cell division cycle. Cell Div 7:10
68. Gopinathan L et al (2014) Loss of Cdk2 and cyclin A2 impairs cell proliferation and tumorigenesis. Cancer Res 74:3870–3879
69. Ortiz-Álvarez G et al (2022) p53/p21 pathway activation contributes to the ependymal fate decision downstream of GemC1. Cell Rep 41:111810
70. Poon RYC (2016) Cell cycle control: a system of interlinking oscillators. Methods Mol Biol 1342:3–19
71. Hume S, Dianov GL, Ramadan K (2020) A unified model for the G1/S cell cycle transition. Nucleic Acids Res 48:12483
72. Kyrousi C et al (2015) Mcidas and GemC1/Lynkeas are key regulators for the generation of multiciliated ependymal cells in the adult neurogenic niche. Development 142:3661–3674
73. Terré B et al (2016) GEMC1 is a critical regulator of multiciliated cell differentiation. EMBO J 35:942–960
74. Arbi M et al (2016) GemC1 controls multiciliogenesis in the airway epithelium. EMBO Rep 17:400–413
75. Yu X, Ng CP, Habacher H, Roy S (2008) Foxj1 transcription factors are master regulators of the motile ciliogenic program. Nat Genet 40:1445–1453
76. Tan FE et al (2013) Myb promotes centriole amplification and later steps of the multiciliogenesis program. Development 140:4277–4286
77. Funk MC et al (2015) Cyclin O (Ccno) functions during deuterosome-mediated centriole amplification of multiciliated cells. EMBO Rep 34:1078–1089
78. Fagundes R, Teixeira LK (2021) Cyclin E/CDK2: DNA replication, replication stress and genomic instability. Front Cell Dev Biol 9:774845
79. Caldon CE, Musgrove EA (2010) Distinct and redundant functions of cyclin E1 and cyclin E2 in development and cancer. Cell Div 5:2
80. Machida YJ, Dutta A (2007) The APC/C inhibitor, Emi1, is essential for prevention of rereplication. Genes Dev 21:184
81. Bendris N et al (2012) Cyclin A2: a genuine cell cycle regulator? Biomol Concepts 3:535–543
82. Westendorp B et al (2012) E2F7 represses a network of oscillating cell cycle genes to control S-phase progression. Nucleic Acids Res 40:3511
83. Kim S, Chien Y-H, Ryan A, Kintner C (2022) Emi2 enables centriole amplification during multiciliated cell differentiation. Sci Adv 8(13):eabm7538
84. Pack LR, Daigh LH, Meyer T (2019) Putting the brakes on the cell cycle: mechanisms of cellular growth arrest. Curr Opin Cell Biol 60:106
85. Lim S, Kaldis P (2013) Cdks, cyclins and CKIs: roles beyond cell cycle regulation. Development 140:3079–3093
86. Wang H, Guo M, Wei H, Chen Y (2023) Targeting p53 pathways: mechanisms, structures and advances in therapy. Signal Transduct Target Ther 8(1):92
87. Nemajerova A, Moll UM (2019) Tissue-specific roles of p73 in development and homeostasis. J Cell Sci 132:jcs233338
88. Yoon MK, Ha JH, Lee MS, Chi SW (2015) Structure and apoptotic function of p73. BMB Rep 48:81–90
89. Marshall CB et al (2016) P73 is required for multiciliogenesis and regulates the Foxj1-associated gene network. Cell Rep 14:2289–2300

90. Gonzalez-Cano L et al (2016) p73 is required for ependymal cell maturation and neurogenic SVZ cytoarchitecture. Dev Neurobiol 76:730–747
91. Benson EK, Zhao B, Sassoon DA, Lee SW, Aaronson SA (2009) Effects of p21 deletion in mouse models of premature aging. Cell Cycle 8:2002
92. Champion L, Linder MI, Kutay U (2017) Cellular reorganization during mitotic entry. Trends Cell Biol 27:26–41
93. Carlton JG, Jones H, Eggert US (2020) Membrane and organelle dynamics during cell division. Nat Rev Mol Cell Biol 21:151–166
94. Lima JT, Ferreira JG (2024) Mechanobiology of the nucleus during the G2-M transition. Nucleus 15: 2330947
95. Dantas M, Lima JT, Ferreira JG (2021) Nucleus-cytoskeleton crosstalk during mitotic entry. Front Cell Dev Biol 9:1–9
96. Walczak CE, Cai S, Khodjakov A (2010) Mechanisms of chromosome behaviour during mitosis. Nat Rev Mol Cell Biol 11:91–102
97. Nagano T et al (2017) Cell-cycle dynamics of chromosomal organization at single-cell resolution. Nature 547:61–67
98. Hirose T, Ninomiya K, Nakagawa S, Yamazaki T (2023) A guide to membraneless organelles and their various roles in gene regulation. Nat Rev Mol Cell Biol 24:288–304
99. Giudice J, Jiang H (2024) Splicing regulation through biomolecular condensates and membraneless organelles. Nat Rev Mol Cell Biol 25:683–700
100. Petasny M et al (2021) Splicing to keep cycling: the importance of pre-mRNA splicing during the cell cycle. Trends Genet 37:266–278
101. Barutcu AR et al (2022) Systematic mapping of nuclear domain-associated transcripts reveals speckles and lamina as hubs of functionally distinct retained introns. Mol Cell 82:1035–1052.e9
102. Soggia G, ElMaghloob Y, Boromangnaeva A-K, Al Jord A (2025) Mechanical remodeling of nuclear biomolecular condensates. Physiology 40:102–113
103. Edgar BA, Zielke N, Gutierrez C (2014) Endocycles: a recurrent evolutionary innovation for post-mitotic cell growth. Nat Rev Mol Cell Biol 15:197–210
104. Matthews HK, Bertoli C, de Bruin RAM (2022) Cell cycle control in cancer. Nat Rev Mol Cell Biol 23: 74–88
105. Mirzadeh Z, Merkle FT, Soriano-Navarro M, Garcia-Verdugo JM, Alvarez-Buylla A (2008) Neural stem cells confer unique pinwheel architecture to the ventricular surface in neurogenic regions of the adult brain. Cell Stem Cell 3:265–278
106. Doetsch F, Caillé I, Lim DA, García-Verdugo JM, Alvarez-Buylla A (1999) Subventricular zone astrocytes are neural stem cells in the adult mammalian brain. Cell 97:703–716
107. Lim DA, Alvarez-Buylla A (2016) The adult ventricular–subventricular zone (V-SVZ) and olfactory bulb (OB) neurogenesis. Cold Spring Harb Perspect Biol 8:a018820
108. Hewitt RJ, Lloyd CM (2021) Regulation of immune responses by the airway epithelial cell landscape. Nat Rev Immunol 21(6):347–362
109. Ford MJ et al (2021) Oviduct epithelial cells constitute two developmentally distinct lineages that are spatially separated along the distal-proximal axis. Cell Rep 36:109677
110. Barton BE et al (2020) Roles of steroid hormones in oviductal function. Reproduction 159:R125–R137
111. Vinay L, Hess RA, Belleannée C (2024) Human efferent ductules and epididymis display unique cell lineages with motile and primary cilia. Andrology. https://doi.org/10.1111/ANDR.13732
112. Plotnikova OV, Pugacheva EN, Golemis EA (2009) Chapter 7 – Primary cilia and the cell cycle. In: Sloboda RD (ed) Methods in cell biology, vol 94. Academic Press, pp 137–160
113. Hoffmann I (2021) Centrosomes in mitotic spindle assembly and orientation. Curr Opin Struct Biol 66: 193–198
114. Tsou M-FB, Stearns T (2006) Mechanism limiting centrosome duplication to once per cell cycle. Nature 442:947–951
115. Basso M et al (2025) Actin-based deformations of the nucleus control mouse multiciliated ependymal cell differentiation. Dev Cell 60:749–761.e5

Autonomous Redox Rhythms and Their Coupling to the Circadian Clock

4

Sargis Karapetyan and Xinnian Dong

Abstract

Circadian rhythms have traditionally been considered to be outputs of genetic transcription-translation feedback loop (TTFL) oscillators. However, the discovery of circadian peroxiredoxin oxidation rhythms in anucleate red blood cells suggests that cells also possess an autonomous circadian redox rhythm. The conservation of this rhythm across all lineages of life, as well as in TTFL-defective backgrounds in multiple organisms, further indicates its importance. Rather than functioning as an isolated system, the redox rhythm is bidirectionally coupled to the genetic circadian clock. Perturbation of one oscillation may alter amplitude or shift period of the other, depending on biological context. In animals, the interplay between the redox rhythm and the genetic clock is complex, and the biological significance of the redox oscillation has yet to be established. In plants, immune-related redox perturbation by salicylic acid can reinforce the genetic clock, while the redox rhythm itself gates immune-induced programmed cell death toward morning as a circadian output. We propose that the redox rhythm reflects an intrinsic metabolic cycling in which cells alternate between the high-metabolic states that generate energy, reducing power, and reactive oxygen species (ROS), and the lower-metabolic states that favor detoxification, repair, and restoration of redox homeostasis. In this model, the redox rhythm is driven by the feedback between glycolysis, pentose phosphate pathway, ROS production via electron transport chains, antioxidant capacity, and cellular repair, and becomes entrained to the external cues via diurnal energy generation such as photosynthesis in green lineages.

Keywords

Circadian clock · Redox · Redox rhythm · Peroxiredoxin · Glycolysis · Pentose phosphate pathway (PPP) · Reactive oxygen species (ROS) · Salicylic acid · Plant immunity · Programmed cell death (PCD) · Electron transport chain · Photosynthesis

4.1 What Constitutes a Circadian Rhythm?

Daily rhythms are perhaps the most conspicuous of the biological oscillations with the sleep/wake cycles being a good example. However, the question whether these oscillations are driven purely by the environment or are intrinsic in nature was not considered until recently. French natural philosopher de Mairan, in 1729, observed that the

S. Karapetyan (✉) · X. Dong
Howard Hughes Medical Institute, Department of Biology, Duke University, Durham, NC, USA
e-mail: sargis.karapetyan@duke.edu

M. G. Aydogan (ed.), *Biological Time Control*, Advances in Experimental Medicine and Biology 1521,
https://doi.org/10.1007/978-3-032-36189-9_4

leaf movement of the *Mimosa Pudica* persisted in a sealed cave without any discernible environmental cues [1]. However, limited by his experimental design, he could only hypothesize that the plants could "feel the sun without seeing it at all." Although he proposed several experiments, such as controlling the temperature, he conceded that "the progress of true [science], which is experimental, can only be very slow." Indeed, a century had passed before, in 1832, the Swiss botanist de Candolle repeated the experiment in a controlled environment to measure the leaf movement oscillations under the free running (i.e., constant) conditions and found it to be close to, but not exactly, Earth's 24 h day/night cycle, indicating their endogenous nature [2]. This is the first property of circadian rhythms, the origin of their modern name: from Latin circa (about) and *diem* (day). The term "circadian" has to be distinguished from "diurnal": whereas the first term refers to a free running oscillator under constant conditions, the second describes an oscillation that is entrained by the 24 h day/night cycle. This entrainment encompasses the second property and the biological function of the circadian rhythms: to synchronize the organism to the environment via cues, such as light or temperature. Circadian rhythms are also distinctive by the ability to maintain relatively constant period across a wide range of temperatures [3], despite being biochemical oscillators subject to temperature-dependent reaction rates predicted by the Arrhenius equation. This last hallmark of the circadian rhythms, the "temperature compensation," ensures that the intrinsic period remains ~24 h even with changes in the environment.

The innate nature of the circadian rhythms indicates that they are driven by a self-sustaining oscillator. The mechanism of the rhythm generation for this oscillator was, and still is a subject of numerous studies. A major breakthrough came in 1971 with the isolation of three Drosophila mutants exhibiting either arrhythmia or altered period, and all mutations being mapped to the same section of the X chromosome [4]. The gene responsible for this behavior, fittingly named *period* (*per*), was finally identified in 1984 [5, 6], a landmark discovery for which the 2017 Nobel Prize was awarded. In the subsequent years, clock genes have been discovered in most lineages of life, first in fungi [7], then plants [8], mammals [9], and finally cyanobacteria [10], with archaea being a notable exception. These findings seemingly solved the mystery of the circadian rhythms by elucidating the underlying molecular oscillators consisting of transcription-translation feedback loops (TTFLs), which drive the ~24 h rhythmic behavior. However, it was soon discovered that TTFLs were not always necessary for generating the circadian rhythms. In 2005, cyanobacterial circadian clock proteins, KaiA, KaiB, and KaiC, assembled in vitro with correct stoichiometry, were found to reconstitute a phosphorylation-based circadian oscillator in the presence of ATP [11]. Interestingly, it was later discovered that the KaiABC oscillator senses not only ATP/ADP [12] reflecting the energy status of the cell, but also the redox state as well [13], suggesting a role for redox signals in circadian regulation. The remarkable discovery of a bona fide non-TTFL circadian redox oscillation in red blood cells (RBCs) [14] provided the first compelling evidence that the cellular redox does not merely influence the genetic clock, but may constitute an autonomous circadian rhythm distinct from the canonical genetic clock.

4.2 The Discovery of an Autonomous Redox Rhythm

In 2011, O'Neill and Reddy reported the oscillations of over/hyperoxidized 2-Cys peroxiredoxins (PRX-$SO_{2/3}$) in anucleate human RBCs lacking the TTFL circadian oscillator [14]. These redox oscillations satisfied all three criteria for the circadian rhythms: (1) having a free-running period of ~24 h, (2) being entrainable by temperature cycles, and (3) exhibiting temperature compensation. To account for possible contamination by nucleated cells, residual transcription and translation were inhibited by α-amanitin and cycloheximide (CHX), respectively, and the PRX-$SO_{2/3}$ oscillations still persisted. Besides peroxiredoxins, primary electron carriers,

NADH and NADPH, also exhibited 24 h oscillations in RBCs, indicating that a global redox rhythm is not restricted to a single class of antioxidant proteins. A second manuscript, a collaborative effort between Reddy and Millar labs, was published concurrently, exploring the PRX-$SO_{2/3}$ oscillations in the unicellular alga *Ostreococcus tauri* [15]. It possesses a chloroplast-targeted PRX that shares the sequence identity with human PRX2 at the catalytically active site. Under prolonged darkness (DD), *O. tauri* shuts down mRNA transcription, however PRX-$SO_{2/3}$ oscillations were still detected under those conditions, as well as in the presence of transcriptional and translational inhibitors cordycepin and CHX, respectively, just as in RBCs. This was an important observation as it confirmed the existence of PRX-$SO_{2/3}$ oscillations in nucleated cells. Overall, these pieces of evidence suggested an existence of an alternative, non-TTFL, autonomous metabolic oscillation mechanism driving the redox rhythms.

Even with the immense evolutionary distance between humans and unicellular algae, the same redox rhythm appeared to persist. Thus, the next step was to investigate whether other organisms also possess the redox rhythm. A collaborative effort showed that, remarkably, the catalytic sites of 2-Cys PRXs were conserved, and the PRX-$SO_{2/3}$ oscillations persisted across all major lineages of life [16], even in archaea, which have no known TTFL circadian clock components. This was in stark contrast to TTFL clocks, which evolved independently in major lineages. The extension of the redox rhythm to new model systems allowed further investigations of its relation to the genetic basis of the canonical circadian clock, particularly by probing clock mutants/overexpression lines lacking the rhythmicity of canonical TTFL components and outputs in cyanobacteria (*Synechococcus elongatus*), fungi (*Neurospora crassa*), unicellular algae (*O. tauri*), plants (*Arabidopsis thaliana*), and flies (*Drosophila melanogaster*). For example, overexpression of *TIMING OF CAB EXPRESSION 1* (*TOC1-OX*), causing arrhythmic behavior in the genetic clock of green lineages, had little effect on the PRX-$SO_{2/3}$ oscillation. However, in other cases, the impairment of TTFL caused changes in period and/or phase of the redox oscillations (Fig. 4.1), indicating that the otherwise autonomous redox cycle is normally coupled to the canonical circadian clock machinery. Interestingly, circadian PRX-$SO_{2/3}$ oscillations were also reported in the nematode *Caenorhabditis elegans* lacking known genes for the animal TTFL clock, accompanied by rhythmic olfactory-cued behavior as well as RNA and protein level changes [17]. Nevertheless, in all these systems, the presence of a vestigial or non-canonical TTFL circadian oscillator could not be ruled out. Therefore, RBCs were used further to study the redox rhythm in isolation from the genetic clock.

4.3 What Drives the Autonomous Redox Rhythms?

The RBCs provide a suitable system to reveal mechanisms underlying the redox rhythm for two primary reasons. First, they lack the genetic clock altogether thus relying purely on non-TTFL mechanisms for generating a circadian rhythm. Second, RBCs are highly sensitive to cellular redox changes because of their oxygen-transporting activity and, therefore, are expected to rely on redox-based mechanisms for regulation. Soon after the discovery of PRX-$SO_{2/3}$ oscillations, they were found to be directly linked to the oxygen-carrying function of RBCs, as the oscillations were largely abolished when the mouse RBCs were treated by carbon monoxide, preventing hemoglobin autoxidation [18]. In the same study, the selective proteasomal degradation of hyperoxidized peroxiredoxins was also found to play a key role in RBC PRX-$SO_{2/3}$ oscillations since pharmaceutical inhibition of 20S proteosome by either MG132 or lactacystin right before the peak of PRX hyperoxidation abolished the decreasing phase of the PRX-$SO_{2/3}$ oscillation, suggesting that degradation was responsible for resetting the oscillatory cycle. However, since the redox rhythm is not limited to PRX-$SO_{2/3}$ oscillations [14], a more global approach was required to better understand the metabolic reactions behind the redox oscillations.

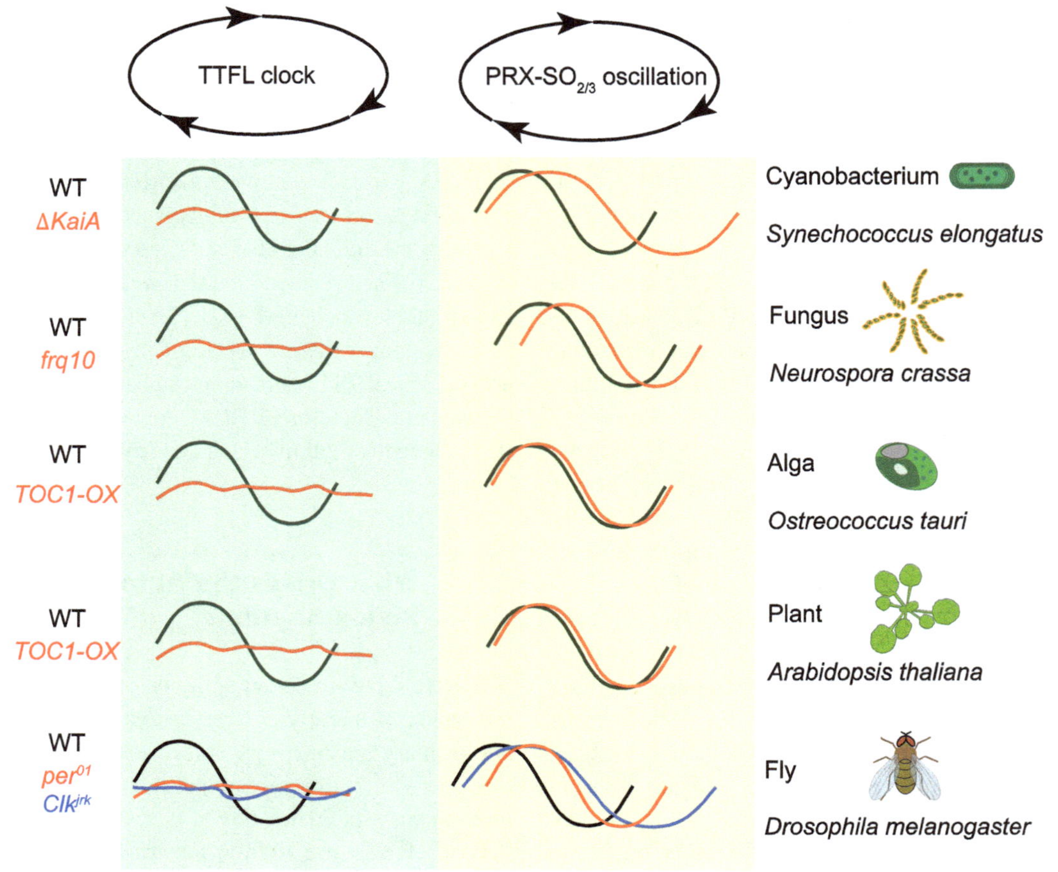

Fig. 4.1 PRX-$SO_{2/3}$ oscillations persist in TTFL-defective backgrounds across different lineages with various alterations to their phase and period. PRX-$SO_{2/3}$ oscillations are largely unchanged in *TOC1*-overexpression lines (*TOC1-OX*) in *A. thaliana* and *O. tauri*, have altered phase in mutants of *period* (*Clk*^{jrk}) in *D. melanogaster* and *frequency* (*frq10*) in *N. crassa* and markedly longer period in the mutants of *KaiA* (*DKaiA*) in *S. elongatus* and *clock* (*per*^{01}) in *D. melanogaster* as reported by Edgar et al. [16]

More recently, to reveal parts of the metabolic network that are connected to the redox rhythm, an unbiased metabolomics approach was applied to RBCs, revealing that ~20% of the detected metabolites exhibited circadian oscillations, primarily consisting of carbohydrate and redox metabolites enriched in glycolysis and the pentose phosphate pathway (PPP) [19]. These two pathways are the primary glucose metabolism pathways with their respective fluxes (i.e., the rate of glucose conversion through each pathway) determined by the cellular needs [20]: When ATP demand increases, glycolysis flux rises, whereas flux through the PPP generates NADPH and precursors for nucleotide production. Thus, these two pathways are central regulators for the cellular redox state (Fig. 4.2). ^{13}C-labeled glucose was used to estimate the metabolic flux of these two pathways, and both fluxes oscillated, with opposite phases, with glycolysis flux peaking in the morning and PPP flux in the evening (Fig. 4.2). Inhibition of key glycolysis enzyme glyceraldehyde 3-phosphate dehydrogenase (GAPDH) by heptelidic acid (HA) abolished circadian oscillations of both glycolytic flux and PRX-$SO_{2/3}$. Similarly, the glycolytic flux did not oscillate when PPP was inhibited using 6-aminonicotinamide (6AN), a competitive inhibitor of glucose-6-phosphate dehydrogenase (G6PD), though a longer period (~36 h)

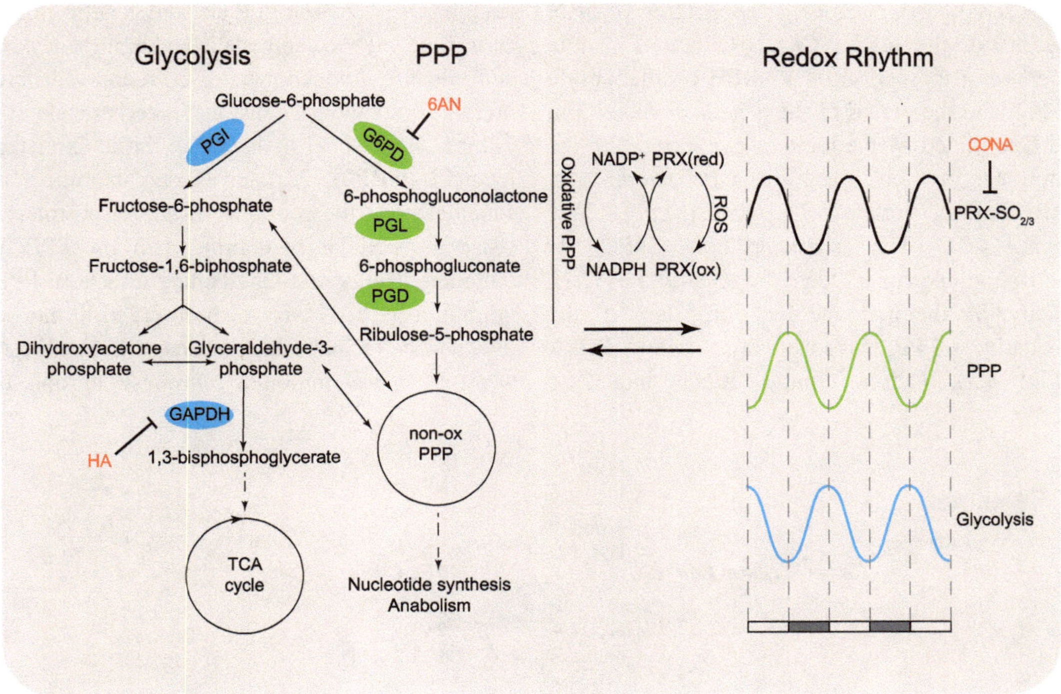

Fig. 4.2 Redox rhythm is tied to glucose metabolism through glycolysis and pentose phosphate pathway (PPP) in red blood cells. Glucose-6-phosphate can be metabolized either by glycolysis or PPP which run in parallel and are connected through shared intermediates. Glycolysis and PPP eventually either terminate at tricarboxylic acid (TCA) cycle for energy production or contribute to anabolic processes such as nucleotide synthesis, respectively. Key enzymes shown are phosphoglucose isomerase (PGI), glyceraldehyde-3-phosphate dehydrogenase (GAPDH), glucose-6-phosphate dehydrogenase (G6PD), 6-phosphogluconolactonase (PGL) and phosphogluconate dehydrogenase (PGD). NADPH produced from oxidative PPP supports the reduction of oxidized (ox) PRX generated after ROS detoxification to its reduced (red) form. Glycolytic and PPP fluxes peak at dawn and dusk, respectively, with PRX-$SO_{2/3}$ having an intermediate phase, peaking mid-day. Inhibition of either oscillation by conoidin A (CONA, for PRX), heptelidic acid (HA, for GAPDH) or 6-aminonicotinamide (6AN, for G6PD) largely abolishes the redox rhythm

oscillation of PRX-$SO_{2/3}$ persisted. Outputs of glycolysis (ATP) and PPP (GSSG) also lost circadian rhythmicity in RBCs treated by HA and 6AN, respectively. Inhibition of PRX-$SO_{2/3}$ oxidation by conoidin A or degradation by MG132 abolished the rhythmic glycolytic flux as well. Thus, the redox rhythm in RBCs appears to be driven by the rhythmic glucose metabolism through glycolysis and PPP, which are subject to feedback via ROS production, PRX-mediated ROS scavenging, and continuous reduction of PRXs from the sulfenic (PRX-SOH) form utilizing NADPH produced from PPP (Fig. 4.2). However, despite the progress made in studying the redox rhythm in RBCs, there are still numerous challenges in understanding the regulation and function of an autonomous redox rhythm in nucleated cells with a genetic clock.

4.4 The Interplay Between the Redox Rhythm and the Genetic Clock

Even before the findings from the RBCs, it had been hypothesized that the PPP, as the central redox pathway generating NADPH required for the reduction of PRXs, would have a profound impact on the redox rhythms, and, by extension, on the genetic clock as well [21]. In human osteosarcoma (U2OS) cells, application of 6AN perturbed the PRX-$SO_{2/3}$ oscillations, while increasing the period of NADPH oscillations to ~30 h and the period of the genetic clock by 3 h, as determined by the luciferase reporter with the promoter of clock gene Basic helix-loop-helix ARNT-like protein 1 (*Bmal1p:luc*) [21] (Fig. 4.3). The silencing of G6PD or phosphogluconate dehydrogenase (PGD), catalyzing the first and the third steps of the oxidative PPP, respectively, via siRNA (Fig. 4.2), had a similar effect, indicating coupling between the PPP and the genetic clock. Based on ChIP-Seq epigenetic and motif analysis, and siRNA knockdown, two redox-sensitive proteins, archetypal histone acetyltransferase P300 [22] and nuclear factor erythroid 2-related factor 2 (NRF2) [23], have been identified as candidates for the coupling. Both these proteins sense the cellular oxidation when the PPP is inhibited. The period-lengthening effects of PPP inhibition by 6AN were also observed in mouse liver and suprachiasmatic nucleus cells. However, when PPP was inhibited in mouse fibroblasts

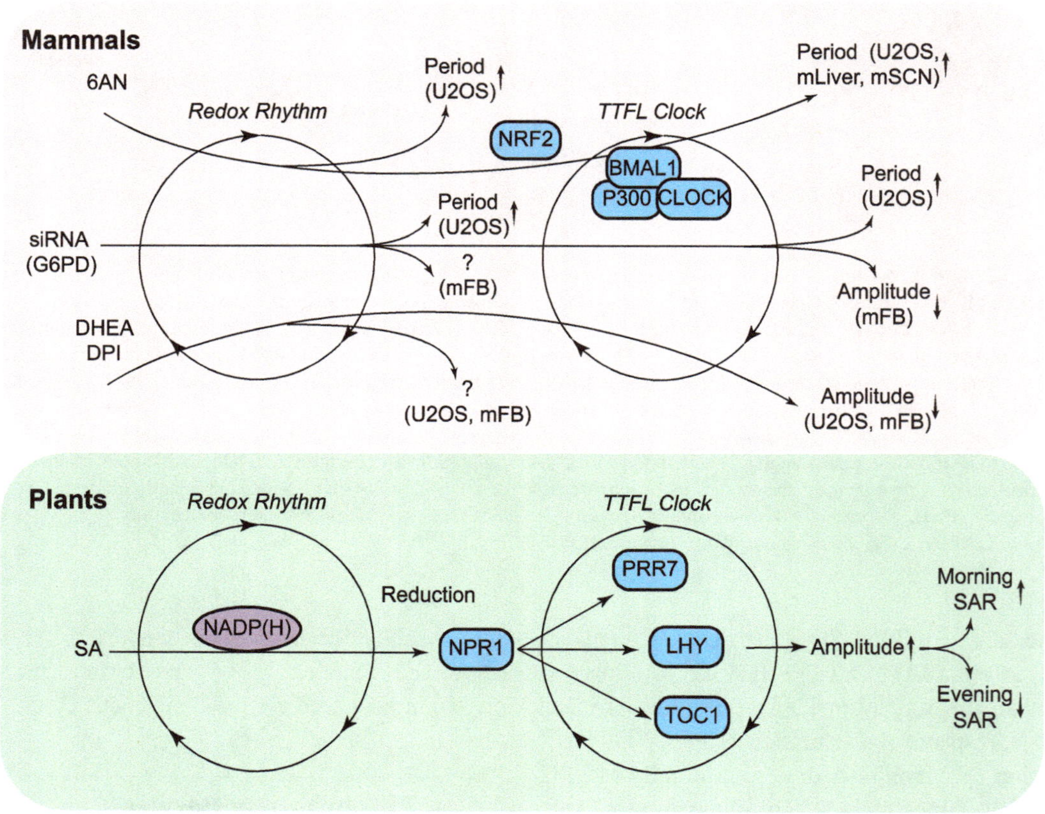

Fig. 4.3 Interplay between the redox rhythm and the TTFL clock in mammals and plants. In osteosarcoma (U2OS) cells, inhibition of G6PD by 6AN or siRNA leads to a longer redox rhythm. 6AN lengthens the TTFL clock period via BMAL1/CLOCK complex associated with P300 histone acetyltransferase or via redox-sensitive transcription factor NRF2. Knockdown of G6PD via siRNA lengthens TTFL clock period in U2OS, but only lowers the amplitude in mouse fibroblasts (mFBs). Inhibition of G6PD via dehydroepiandrosterone (DHEA) or diphenyleneiodonium (DPI) dampens the genetic clock without affecting the period. In plants, application of defense hormone salicylic acid (SA) induces the systemic acquired resistance (SAR) by perturbing the redox rhythm, shifting the balance towards reduction (NADPH) and activating the master immune regulator NPR1 to translocate into the nucleus. NPR1 promotes the expression of morning (*LHY*, *PRR7*) and evening (*TOC1*) TTFL components, resulting in an increase in the amplitude of the genetic clock and gating of SAR towards the morning

(mFB) via siRNA knockdown of G6PD, the period lengthening did not occur [24]. Furthermore, inhibiting G6PD by dehydroepiandrosterone (DHEA) or diphenyleneiodonium (DPI) only dampened the amplitude of the genetic clock oscillations in both U2OS and mFB cells, without a significant effect on the period, although their effect on the redox rhythm was not fully explored (Fig. 4.3). There was also little consistency in how glycolysis inhibition by HA affected different cell lines [24]. These discrepancies indicate that the crosstalk between the redox rhythm and the genetic clock in animals is not straightforward and might be specific to the nature of the redox perturbation, as well as to the cell type and/or organism.

The model plant *Arabidopsis* represents an attractive alternative to animals for studying the redox rhythm due to being a genetically uniform (as a self-pollinating species) and tractable higher eukaryote. Indeed, another mechanistic link between the redox rhythm and the genetic clock was soon discovered in *Arabidopsis* during systemic acquired resistance (SAR) [25] (Fig. 4.3). SAR is a broad-spectrum, whole-plant immune response, which is induced after local infection to protect the plant from secondary infections. SAR is mediated by the defense hormone salicylic acid (SA) through the master regulator NONEXPRESSER OF PR GENES 1 (NPR1) [26]. Upon SA treatment, NPR1 is reduced by thioredoxins H3 and H5 (TRX-H3 and TRX-H5) [27], releasing it from the cytoplasmic oligomer form [28]. NPR1 then translocates into the nucleus to activate defense gene expression in a complex with TGA transcription factors [29]. It was discovered that application of SA perturbed the circadian NADP (H) oscillation, shifting the redox balance toward reduction [25]. In contrast to U2OS cells, however, this perturbation of the redox rhythm did not affect the period or the phase of the clock bioluminescent reporter *TOC1p:LUC*, but rather reinforced it by increasing its amplitude and expression level via direct enhancement of *TOC1* gene expression by NPR1. Furthermore, both *npr1* and *trx-h3 trx-h5* mutants expressed *TOC1p:LUC* at a lower amplitude (with the same period), and with insensitivity to SA. However, the unchanged period, which is usually directly correlated with *TOC1* levels, could not be explained [8, 30]. Therefore, the P2012 *Arabidopsis* clock model [31] was used to systematically couple NPR1 to each clock gene to recapitulate the observed phenotype. Combining modeling and experimental approaches identified two morning-phased genes, *PSEUDO-RESPONSE REGULATOR 7 (PRR7)* and *LATE ELONGATED HYPOCOTYL (LHY)* as NPR1 targets, in addition to evening-phased *TOC1*. This balanced network architecture relays the redox rhythm perturbation to both morning and evening nodes of the genetic clock, hence reinforcing it. Since LHY and its partner CIRCADIAN CLOCK ASSOCIATED 1 (CCA1) are positive regulators of the defense [32, 33], and TOC1 suppresses the immunity at night [25], the increased amplitude results in clock-mediated gating of the immune response toward the morning, when the infection is more likely [33]. Thus, SA-induced redox changes result in a circadian immune response through a functional interplay between the redox rhythm and the genetic clock.

4.5 Biological Significance of the Autonomous Redox Rhythms

While it became increasingly apparent that perturbation of the redox rhythm has significant physiological consequences, the biological function of the oscillation itself remained elusive. Which physiological processes could benefit from being circadian gated by the autonomous redox rhythm, instead of the genetic clock? Attempts have been made to find the distinct redox rhythm transcriptional targets in both fly and human mutant cells lacking the rhythmicity of the TTFL oscillator (Fig. 4.4). The *Drosophila* S2 cells are known to lack several central TTFL components [34, 35] and were chosen as a TTFL-arrhythmic system. RNA-seq of S2 cells revealed 482 rhythmic transcripts enriched in metabolism and protein biosynthesis [36]. The total protein content and 342 individual proteins, enriched in metabolic enzymes, also exhibited ~24 h

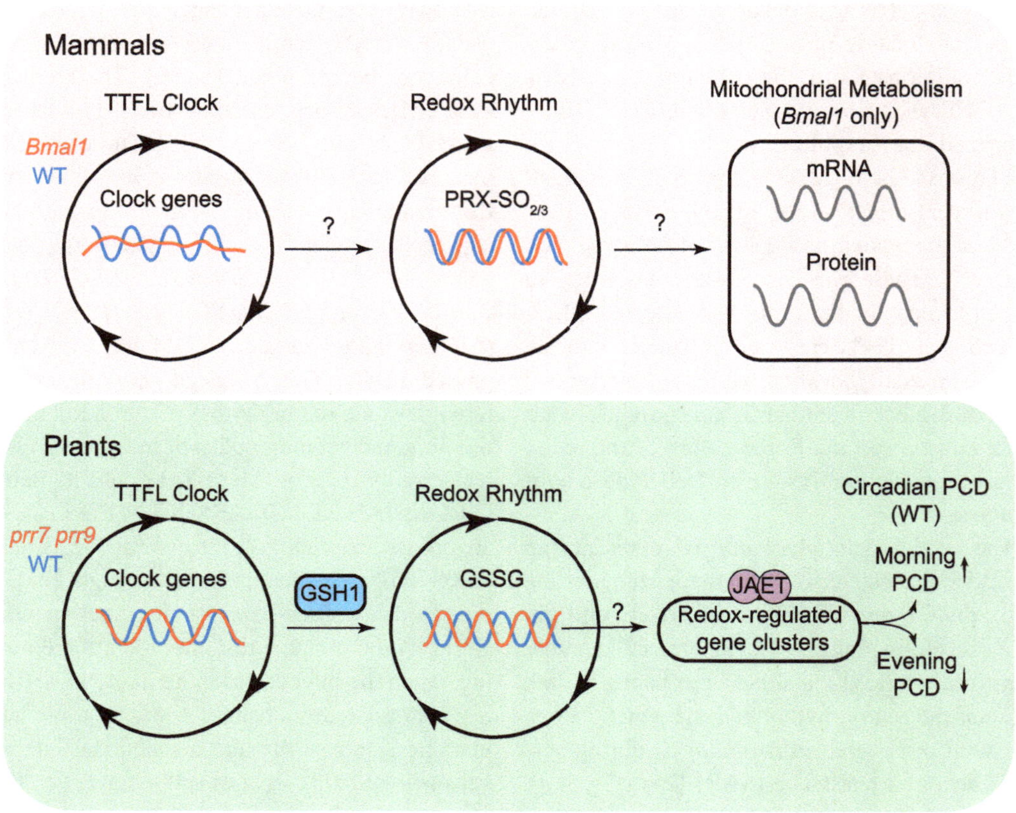

Fig. 4.4 Biological processes regulated by the redox rhythm. In the TTFL clock-defective *Bmal1* mutant mouse cells, the oscillations of PRX-$SO_{2/3}$ persist along with rhythmic mitochondrial metabolism reflected in oscillatory transcripts, proteins, and metabolites. However, there is little to no overlap between oscillating features in *Bmal1* and WT. In the *Arabidopsis prr7 prr9* double mutant, the TTFL clock oscillates with a longer period driving the oscillations of total glutathione through regulation of *GSH1*, but not the oxidized GSSG, which represents the redox rhythm and oscillates with a phase-shift and ~24 h period. Gene clusters with ~24 h period were identified, indicating that the circadian regulation of programmed cell death (PCD) in the WT is through the redox rhythm via the jasmonic acid/ethylene (JA/ET) defense pathway

oscillations. An untargeted metabolomics approach was then used which found that levels of key electron carriers, glutathione and NAD $(P)^+$, as well as tricarboxylic acid (TCA) cycle intermediates oscillate.

To investigate the possible targets of the redox rhythm in mammals, *Bmal1* mutant mice were used [37]. *Bmal1* knockout results in the disruptions of core TTFL oscillation [38] and the behavioral rhythms [39]. Cells from both liver and skin fibroblast tissues of *Bmal1* knockout mice were synchronized and then released to constant conditions for RNA-seq. A significant portion of the transcriptome in *Bmal1* mutant cells was shown to exhibit ~24 h oscillations and the temperature compensation was confirmed in fibroblasts. To further eliminate the possibility of these oscillations being driven by the environment, two sets of cells were synchronized to opposite phases and then released to free running (i.e., constant) conditions. Supporting the autonomy of the oscillations, the cells kept their antiphasic relationship. When proteomic analysis was performed, rhythmic proteins were detected, enriched in metabolism with the notable localization to mitochondria. Rhythmic PRX-$SO_{2/3}$ oscillations were also detected. Overall, the results from both fly and mouse suggest that there might be a metabolism-driven oscillation in parallel to the genetic clock. However, the transcripts rhythmic in either S2 or *Bmal1* cells generally did not oscillate in wildtype (WT), indicating that while circadian metabolism might persist without the periodic input from the TTFL clock, its impact on the physiology in animals has yet to be elucidated.

The difficulty of understanding the physiological impact of the redox oscillations in either metabolically perturbed (lack of consistency) or TTFL-defective (lack of overlap with WT) systems indicated that an alternative approach might be needed, preferably with both the redox rhythm and the genetic clock present. However, under normal conditions, both rhythms oscillate with a ~24 h period, resulting in synchronization. Nevertheless, it has been shown before that coupled oscillators, which are normally phase locked to have the same period, could be distinguished if the period separation was large enough [40]. The *prr7 prr9* double mutant of *Arabidopsis* has significantly longer free running genetic clock period of ~32 h at 22 °C (typical growth temperature) compared to WT. Furthermore, this mutant is temperature overcompensated, i.e., the period lengthens in *prr7 prr9* with increased temperature, reaching ~36 h at 30 °C [41]. This made the *prr7 prr9* double mutant a promising genetic background to investigate the outputs of the redox rhythm. Thus, parallel glutathione and targeted transcriptomic (~700 genes) time course analyses were performed for WT and *prr7 prr9* at 22 °C and 30 °C [42]. In agreement with the redox rhythm having an intrinsic timekeeping machinery, temperature compensated ~24 h-period GSSG oscillations were observed in *prr7 prr9*, albeit with a phase-shift compared to WT (Fig. 4.4). Interestingly, the total glutathione levels (i.e., GSSG+GSH) oscillated with a significantly longer period, likely due to de novo synthesis of GSH that is driven by the genetic clock through transcription of the rate limiting glutathione biosynthesis enzyme, glutamate-cysteine ligase (GSH1). Therefore, while oxidation of GSH to GSSG reflects the cellular redox rhythm, the total glutathione is controlled by the genetic clock (Fig. 4.4).

Co-expression analysis then revealed gene clusters with the temporal signature of GSSG (phase-shifted and ~24 h period in the long period *prr7 prr9* mutant background) identifying them as potential targets of the redox rhythm. The period separation of redox-related candidate genes selected from the clusters, namely *S-NITROSOGLUTATHIONE REDUCTASE 1* (*GSNOR1*) and *G6PD6*, was further confirmed *in planta* by luciferase reporter analysis in comparison to known genetic clock output genes, *CAT2* and *CAT3* encoding catalases [25].

While these redox rhythm-regulated clusters did include metabolic genes, the top gene ontology (GO) term was the programmed cell death (PCD) induced by effector-triggered immunity (ETI). During ETI, a pathogen effector is recognized by its corresponding intracellular receptor, leading to PCD of infected cells. While stronger PCD in the subjective morning was previously reported [43], the mechanism of this gating effect was not explored. In agreement with the redox rhythm-mediated gating, circadian PCD persisted in the TTFL-defective [44] *CCA1-OX* line and the triple mutant of the core clock genes *toc1 lhy cca1* [45]. In contrast, PCD lost time-of-day sensitivity in the mutants of enzymes catalyzing either the second step of the oxidative PPP, *6-phosphogluconolactonase 3* (*pgl3*) (Fig. 4.2) or glutathione biosynthesis, *phytoalexin deficient 2 (pad2)*. Two primary plant defense pathways mediated by the phytohormones, SA and jasmonic acid (JA), were then examined for a mechanistic link between the redox rhythm and PCD. While the *npr1* mutant (the SA pathway) retained the circadian-gated PCD, the *ethylene-insensitive3* (*ein3*) and *ethylene-insensitive3-like 1* (*eil1*) double mutant of the shared JA/ethylene pathway did not. This indicated the gating of PCD by the redox rhythm through JA and/or ethylene, although the precise mechanism remains elusive (Fig. 4.4).

PCD is a logical target for circadian gating by the redox metabolism. Not only is PCD associated with ROS burst [46], but also requires glutathione [47] and significant increase in ATP levels [48, 49]. Therefore, stronger morning PCD would coincide with more favorable cellular environment for execution, when both glutathione [42] and ATP [50] are at the peak. Supporting this hypothesis, treatment of the plants with a cell-permeable glutathione analog, glutathione monoethyl ester (GSHmee), to provide more reducing power [25], abolished the difference between morning and evening responses by increasing evening PCD. The advantage of the regulation by the redox rhythm over the genetic clock lies in its flexible nature. As the sum of the overall cellular metabolism, the redox rhythm is highly sensitive to perturbations and might serve as a central signaling hub to efficiently distribute the metabolic flux between energy-intensive stress responses such as ETI-triggered PCD. It would be interesting to test whether similar energy-intensive stress responses, such as pyroptosis in mammals [51], are similarly gated by the redox rhythm.

4.6 Redox Oscillation or Redox Oscillator?

Despite major advances in elucidating the nature of the redox rhythm, a fundamental question remains: is it a redox *oscillation* or a redox *oscillator*? The lack of mechanistic understanding of the redox rhythm inevitably led to the debates on its significance as a circadian timekeeper or its existence itself [24, 52]. There is substantial evidence [11, 14–16, 18, 19] to argue for the existence of a non-transcriptional autonomous oscillator that drives the redox rhythm. However, whether the rhythm of this non-TTFL oscillator is generated via PRX-$SO_{2/3}$-dependent mechanism and/or the sum of cellular metabolic activities is unknown.

In general, biological oscillators are dissipative systems producing relaxation oscillations. That is, in contrast to the well-known linear harmonic oscillator, such as the simple pendulum at small angles, they are highly nonlinear with a stable limit cycle (i.e., the oscillation pattern that the system will always try to achieve). This feature confers certain robustness to these clocks whereby transient perturbations might alter the phase but will have no lasting effect on the period and amplitude of the oscillator after the relaxation (i.e., return) to the limit cycle. In order to exhibit such oscillations, a network needs three components: (1) negative feedback to reset the oscillation cycle, (2) delays in the system so that the oscillation trajectory "overshoots" and "undershoots" the unstable equilibrium point typically associated with the limit cycle, and (3) sufficient sources of nonlinearities to sustain the oscillations and prevent their decay to a steady state by making the excursions from the equilibrium more pronounced [53]. The nonlinearities in biological oscillators often arise from protein-protein interactions. For example, the key activators for the mammalian circadian clock, BMAL1 and CLOCK, function as a heterodimer [54]. In a more extreme example, the three core proteins for the cyanobacteria circadian clock, KaiA, KaiB, and KaiC are assembled into a large multimeric complex with the KaiC hexamer at the core [55]. This highly nonlinear architecture might stem from the simplicity of bacterial clock, as it will lack the sources of delays associated with more complex clocks in higher organisms which contain numerous regulatory steps, each with a corresponding delay. Finally, the negative feedback in the genetic clocks can also have many sources, from differential binding affinity of KaiC to KaiA and KaiB based on its phosphorylation state [55], to the mixed feedback loop core architecture of fungal and animal clocks (TFs activating the production of their inhibitors) [56, 57], to a complex web of repression steps in higher plants [31].

The redox rhythm is ancient with the origin hypothesized to date close to that of the KaiABC oscillator [16]. Therefore, its core oscillatory mechanism would be expected to resemble that of the bacterial clock, having fewer regulatory steps but with stronger delays/nonlinearities. It is tempting to assume that a purely phosphorylation-driven circadian oscillation, like in cyanobacteria, or the KaiABC oscillator itself (in cyanobacteria) could be the timekeeping mechanism behind the redox rhythm. However, the redox rhythm still persists in the Δ*kaiA* mutant [16] lacking the KaiABC oscillator. Furthermore, the KaiABC oscillator is inextricably tied to the cellular redox/energy metabolism [12, 13, 58] and more recent estimates put the evolution of the antioxidant enzymes to occur as early as 1 billion years before the appearance of KaiABC oscillator [16]. These findings suggest the possibility that a metabolic redox oscillator could have evolved first and then facilitated the evolution of the genetic KaiABC oscillator in cyanobacteria [59].

What, then, could be the rhythm-generating mechanism of this putative redox oscillator? An attempt has been made to create a simple model for the redox rhythm in mammalian adrenal gland, heart, and brown adipose tissues, based on a peroxiredoxin PRX3 and sulfiredoxin (SRX) feedback loop [60]. First, mitochondrial activity leads to H_2O_2 production which is scavenged by PRX3. After certain H_2O_2 threshold, PRX3 gets overoxidized (PRX3-SO_2H), leading to its inactivation and leakage of H_2O_2 to the cytoplasm. Cytosolic H_2O_2 promotes translocation of SRX into mitochondria, so that the latter can reduce PRX3-SO_2H to reactivate PRX3 and remove excessive H_2O_2, restarting the cycle. In this model, the nonlinearities arise from the switch-like fast activation/deactivation of PRX3, and the delay results from the slow accumulation of H_2O_2. While able to generate relaxation oscillations, the PRX-SRX model cannot explain PRX oscillations in organisms lacking corresponding SRX homologs, such as Neurospora, nor the oscillations in red blood cells derived from SRX-knockout mice [18]. Therefore, it is reasonable to consider extending the definition of the redox rhythm beyond PRX-$SO_{2/3}$ oscillations (Fig. 4.5).

It has been already shown that the rhythmic glucose flux through glycolysis and the PPP lies at the core of the redox rhythm of RBCs, which appear to switch between higher metabolism and ROS production during the day and lower metabolism at night [19]. Thus, the circadian energy/redox metabolism mightbe a better candidate for being the redox oscillator than a purely PRX-based system. It is possible to conceptually

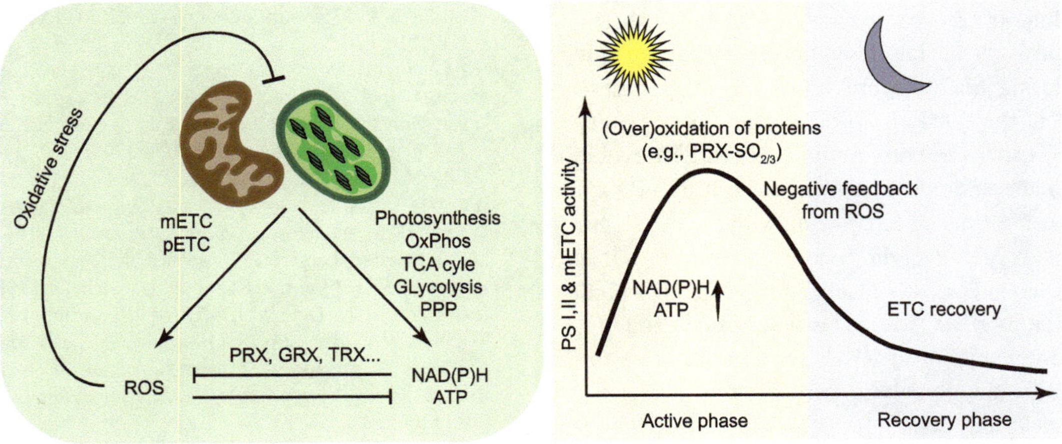

Fig. 4.5 A proposed model for a metabolic redox oscillator. During the active daytime (diurnal animals and plants), the high-metabolic activity will generate energy (ATP) and reducing power via photosynthesis, oxidative phosphorylation (OxPhos), TCA cycle, glycolysis, and the PPP. However, due to high activity of photosystems I and II in the photosynthetic electron transport chain (pETC), as well as mitochondrial ETC (mETC), reactive oxygen species (ROS) will be generated as a byproduct. ROS will be partially neutralized by the antioxidant enzymes such as peroxiredoxins (PRX), glutaredoxins (GRX), and thioredoxins (TRX). As ROS accumulate, they cause overoxidation of antioxidant systems, such as hyper- and overoxidized PRX (PRX-$SO_{2/3}$) as well as ETC components. This initiates negative feedback, which reduces metabolic activity and allows recovery of the ETC machinery

divide the cellular metabolism into two circadian phases: active and recovery. In the active phase, such as during the day for photosynthetic and diurnal organisms, there is a high activity of mitochondrial electron transport chain (mETC) generating ATP as well photosynthetic ETC (pETC) producing both ATP and reducing power in plants [61, 62] (Fig. 4.5). However, these activities would produce ROS as a byproduct. Then, the slow buildup of ROS could lead to inhibition of pETC and mETC to prevent further oxidative damage. This negative feedback does not necessarily require transcription via retrograde signaling to the nucleus and could be through direct modulation of the proteins through oxidation [61, 63, 64] such as via hyperoxidation of PRXs, that the existing model is based on [60] (Fig. 4.5). This mechanism could provide both nonlinearities and the negative feedback necessary for resetting the oscillation. During the recovery phase, the lower oxidative stress would allow the repair of damaged organelles [64], which coincides with the circadian growth and cell division at the organismal level [65, 66]. Such "seesaw" dynamics also aligns naturally with the diurnal energy production, an obvious example of which is photosynthesis occurring only in the presence of light during the day, providing the delay necessary for the ~24 h oscillations. In support of ETC-driven redox rhythm, the earliest evolution of the antioxidant enzymes, such as PRXs, has been estimated to coincide with the development of oxygen-evolving photosynthesis, presumably to counteract the localized oxygen/ROS production by pETC [59]. This would also explain why redox rhythm in RBCs is more sensitive to the PPP and glycolysis perturbations compared to nucleated cells [19, 21, 24], as RBCs lack mitochondria. In the absence of mETC, the RBC redox rhythm would rely on direct regulation of metabolic enzymes by ROS for feedback, such as switching from glycolysis to PPP due to oxidation of GAPDH [67]. It would be interesting to dissect the relative contributions of ETC, glycolysis, and the PPP to the redox rhythm generation in the nucleated cells in the future. The centrality of the active and recovery phases for the mitochondrial function and maintenance has already been proposed [64]. Experiments with direct manipulation of organelle activity, such as using ETC inhibitors, can shed new light on the nature of the redox rhythm and finally establish it as a bona fide redox oscillator.

References

1. de Mairan J-JdO (1729) Observation botanique. Histoire de l'Académie royale des sciences avec les mémoires de mathématique et de physique:35–36
2. de Candolle AP (1832) Physiologie végétale, ou Exposition des forces et des fonctions vitales des végétaux
3. Pittendrigh CS (1954) On temperature independence in the clock system controlling emergence time in DROSOPHILA*†. Proc Natl Acad Sci 40:1018–1029. https://doi.org/10.1073/pnas.40.10.1018
4. Konopka RJ, Benzer S (1971) Clock mutants of Drosophila melanogaster. Proc Natl Acad Sci 68:2112–2116. https://doi.org/10.1073/pnas.68.9.2112
5. Bargiello TA, Young MW (1984) Molecular genetics of a biological clock in Drosophila. Proc Natl Acad Sci USA 81:2142–2146. https://doi.org/10.1073/pnas.81.7.2142
6. Zehring WA, Wheeler DA, Reddy P, Konopka RJ, Kyriacou CP, Rosbash M, Hall JC (1984) P-element transformation with period locus DNA restores rhythmicity to mutant, arrhythmic Drosophila melanogaster. Cell 39:369–376. https://doi.org/10.1016/0092-8674(84)90015-1
7. McClung CR, Fox BA, Dunlap JC (1989) The Neurospora clock gene frequency shares a sequence element with the Drosophila clock gene period. Nature 339:558–562. https://doi.org/10.1038/339558a0
8. Millar AJ, Carré IA, Strayer CA, Chua NH, Kay SA (1995) Circadian clock mutants in Arabidopsis identified by luciferase imaging. Science 267:1161–1163. https://doi.org/10.1126/science.7855595
9. Vitaterna MH, King DP, Chang AM, Kornhauser JM, Lowrey PL, McDonald JD, Dove WF, Pinto LH, Turek FW, Takahashi JS (1994) Mutagenesis and mapping of a mouse gene, clock, essential for circadian behavior. Science 264:719–725. https://doi.org/10.1126/science.8171325
10. Ishiura M, Kutsuna S, Aoki S, Iwasaki H, Andersson CR, Tanabe A, Golden SS, Johnson CH, Kondo T (1998) Expression of a gene cluster kaiABC as a circadian feedback process in cyanobacteria. Science 281:1519–1523. https://doi.org/10.1126/science.281.5382.1519
11. Nakajima M, Imai K, Ito H, Nishiwaki T, Murayama Y, Iwasaki H, Oyama T, Kondo T (2005) Reconstitution of circadian oscillation of cyanobacterial KaiC phosphorylation in vitro. Science

308:414–415. https://doi.org/10.1126/science.1108451
12. Rust MJ, Golden SS, O'Shea EK (2011) Light-driven changes in energy metabolism directly entrain the cyanobacterial circadian oscillator. Science 331:220–223. https://doi.org/10.1126/science.1197243
13. Wood TL, Bridwell-Rabb J, Kim YI, Gao T, Chang YG, LiWang A, Barondeau DP, Golden SS (2010) The KaiA protein of the cyanobacterial circadian oscillator is modulated by a redox-active cofactor. Proc Natl Acad Sci USA 107:5804–5809. https://doi.org/10.1073/pnas.0910141107
14. O'Neill JS, Reddy AB (2011) Circadian clocks in human red blood cells. Nature 469:498–503. https://doi.org/10.1038/nature09702
15. O'Neill JS, van Ooijen G, Dixon LE, Troein C, Corellou F, Bouget FY, Reddy AB, Millar AJ (2011) Circadian rhythms persist without transcription in a eukaryote. Nature 469:554–558. https://doi.org/10.1038/nature09654
16. Edgar RS, Green EW, Zhao Y, van Ooijen G, Olmedo M, Qin X, Xu Y, Pan M, Valekunja UK, Feeney KA et al (2012) Peroxiredoxins are conserved markers of circadian rhythms. Nature 485:459–464. https://doi.org/10.1038/nature11088
17. Olmedo M, O'Neill JS, Edgar RS, Valekunja UK, Reddy AB, Merrow M (2012) Circadian regulation of olfaction and an evolutionarily conserved, nontranscriptional marker in Caenorhabditis elegans. Proc Natl Acad Sci 109:20479–20484. https://doi.org/10.1073/pnas.1211705109
18. Cho C-S, Yoon HJ, Kim JY, Woo HA, Rhee SG (2014) Circadian rhythm of hyperoxidized peroxiredoxin II is determined by hemoglobin autoxidation and the 20S proteasome in red blood cells. Proc Natl Acad Sci 111:12043–12048. https://doi.org/10.1073/pnas.1401100111
19. Ch R, Rey G, Ray S, Jha PK, Driscoll PC, Dos Santos MS, Malik DM, Lach R, Weljie AM, MacRae JI et al (2021) Rhythmic glucose metabolism regulates the redox circadian clockwork in human red blood cells. Nat Commun 12:377. https://doi.org/10.1038/s41467-020-20479-4
20. Cho ES, Cha YH, Kim HS, Kim NH, Yook JI (2018) The pentose phosphate pathway as a potential target for cancer therapy. Biomol Ther (Seoul) 26:29–38. https://doi.org/10.4062/biomolther.2017.179
21. Rey G, Valekunja UK, Feeney KA, Wulund L, Milev NB, Stangherlin A, Ansel-Bollepalli L, Velagapudi V, O'Neill JS, Reddy AB (2016) The pentose phosphate pathway regulates the circadian clock. Cell Metab 24:462–473. https://doi.org/10.1016/j.cmet.2016.07.024
22. Dansen TB, Smits LMM, van Triest MH, de Keizer PLJ, van Leenen D, Koerkamp MG, Szypowska A, Meppelink A, Brenkman AB, Yodoi J et al (2009) Redox-sensitive cysteines bridge p300/CBP-mediated acetylation and FoxO4 activity. Nat Chem Biol 5:664–672. https://doi.org/10.1038/nchembio.194
23. Chorley BN, Campbell MR, Wang X, Karaca M, Sambandan D, Bangura F, Xue P, Pi J, Kleeberger SR, Bell DA (2012) Identification of novel NRF2-regulated genes by ChIP-Seq: influence on retinoid X receptor alpha. Nucleic Acids Res 40:7416–7429. https://doi.org/10.1093/nar/gks409
24. Putker M, Crosby P, Feeney KA, Hoyle NP, Costa ASH, Gaude E, Frezza C, O'Neill JS (2017) Mammalian circadian period, but not phase and amplitude, is robust against redox and metabolic perturbations. Antioxid Redox Signal. https://doi.org/10.1089/ars.2016.6911
25. Zhou M, Wang W, Karapetyan S, Mwimba M, Marques J, Buchler NE, Dong X (2015) Redox rhythm reinforces the circadian clock to gate immune response. Nature 523:472–476. https://doi.org/10.1038/nature14449
26. Fu ZQ, Dong X (2013) Systemic acquired resistance: turning local infection into global defense. Annu Rev Plant Biol 64:839–863. https://doi.org/10.1146/annurev-arplant-042811-105606
27. Tada Y, Spoel SH, Pajerowska-Mukhtar K, Mou Z, Song J, Wang C, Zuo J, Dong X (2008) Plant immunity requires conformational charges of NPR1 via S-Nitrosylation and Thioredoxins. Science 321:952–956. https://doi.org/10.1126/science.1156970
28. Mou Z, Fan W, Dong X (2003) Inducers of plant systemic acquired resistance regulate NPR1 function through redox changes. Cell 113:935–944. https://doi.org/10.1016/S0092-8674(03)00429-X
29. Kumar S, Zavaliev R, Wu Q, Zhou Y, Cheng J, Dillard L, Powers J, Withers J, Zhao J, Guan Z et al (2022) Structural basis of NPR1 in activating plant immunity. Nature 605:561–566. https://doi.org/10.1038/s41586-022-04699-w
30. Más P, Alabadí D, Yanovsky MJ, Oyama T, Kay SA (2003) Dual role of TOC1 in the control of circadian and photomorphogenic responses in Arabidopsis. Plant Cell 15:223–236. https://doi.org/10.1105/tpc.006734
31. Pokhilko A, Mas P, Millar AJ (2013) Modelling the widespread effects of TOC1 signalling on the plant circadian clock and its outputs. BMC Syst Biol 7:23–23. https://doi.org/10.1186/1752-0509-7-23
32. Zhang C, Xie QG, Anderson RG, Ng GN, Seitz NC, Peterson T, McClung CR, McDowell JM, Kong DD, Kwak JM, Lu H (2013) Crosstalk between the circadian clock and innate immunity in Arabidopsis. PLoS Pathog 9
33. Wang W, Barnaby JY, Tada Y, Li H, Tor M, Caldelari D, Lee DU, Fu XD, Dong XN (2011) Timing of plant immune responses by a central circadian regulator. Nature 470:110–U126. https://doi.org/10.1038/nature09766
34. Darlington TK, Wager-Smith K, Ceriani MF, Staknis D, Gekakis N, Thomas DLS, Weitz CJ, Takahashi JS, Kay SA (1998) Closing the

circadian loop: CLOCK-induced transcription of its own inhibitors per and tim. Science 280:1599–1603
35. Saez L, Young MW (1996) Regulation of nuclear entry of the Drosophila clock proteins period and timeless. Neuron 17:911–920. https://doi.org/10.1016/S0896-6273(00)80222-6
36. Rey G, Milev NB, Valekunja UK, Ch R, Ray S, Silva Dos Santos M, Nagy AD, Antrobus R, MacRae JI, Reddy AB (2018) Metabolic oscillations on the circadian time scale in Drosophila cells lacking clock genes. Mol Syst Biol 14:e8376. https://doi.org/10.15252/msb.20188376
37. Ray S, Valekunja UK, Stangherlin A, Howell SA, Snijders AP, Damodaran G, Reddy AB (2020) Circadian rhythms in the absence of the clock gene Bmal1. Science 367:800–806. https://doi.org/10.1126/science.aaw7365
38. Baggs JE, Price TS, DiTacchio L, Panda S, FitzGerald GA, Hogenesch JB (2009) Network features of the mammalian circadian clock. PLoS Biol 7:e1000052. https://doi.org/10.1371/journal.pbio.1000052
39. Bunger MK, Wilsbacher LD, Moran SM, Clendenin C, Radcliffe LA, Hogenesch JB, Simon MC, Takahashi JS, Bradfield CA (2000) *Mop3* is an essential component of the master circadian pacemaker in mammals. Cell 103:1009–1017. https://doi.org/10.1016/S0092-8674(00)00205-1
40. Feillet C, Krusche P, Tamanini F, Janssens RC, Downey MJ, Martin P, Teboul M, Saito S, Lévi FA, Bretschneider T et al (2014) Phase locking and multiple oscillating attractors for the coupled mammalian clock and cell cycle. Proc Natl Acad Sci 111:9828–9833. https://doi.org/10.1073/pnas.1320474111
41. Salomé PA, Weigel D, McClung CR (2010) The role of the Arabidopsis morning loop components CCA1, LHY, PRR7, and PRR9 in temperature compensation. Plant Cell 22:3650–3661. https://doi.org/10.1105/tpc.110.079087
42. Karapetyan S, Mwimba M, Chen T, Yao Z, Dong X (2025) The redox rhythm gates immune-induced cell death distinctly from the genetic clock. Proc Natl Acad Sci 122:e2519251122. https://doi.org/10.1073/pnas.2519251122
43. Korneli C, Danisman S, Staiger D (2014) Differential control of pre-invasive and post-invasive antibacterial defense by the Arabidopsis circadian clock. Plant Cell Physiol 55:1613–1622. https://doi.org/10.1093/pcp/pcu092
44. Wang Z-Y, Tobin EM (1998) Constitutive expression of the *CIRCADIAN CLOCK ASSOCIATED 1* (*CCA1*) gene disrupts circadian rhythms and suppresses its own expression. Cell 93:1207–1217. https://doi.org/10.1016/S0092-8674(00)81464-6
45. Ito S, Nakamichi N, Nakamura Y, Niwa Y, Kato T, Murakami M, Kita M, Mizoguchi T, Niinuma K, Yamashino T, Mizuno T (2007) Genetic linkages between circadian clock-associated components and phytochrome-dependent red light signal transduction in Arabidopsis thaliana. Plant Cell Physiol 48:971–983. https://doi.org/10.1093/pcp/pcm063
46. Torres MA, Dangl JL, Jones JDG (2002) Arabidopsis gp91phox homologues AtrbohD and AtrbohF are required for accumulation of reactive oxygen intermediates in the plant defense response. Proc Natl Acad Sci 99:517–522. https://doi.org/10.1073/pnas.012452499
47. Dubreuil-Maurizi C, Vitecek J, Marty L, Branciard L, Frettinger P, Wendehenne D, Meyer AJ, Mauch F, Poinssot B (2011) Glutathione deficiency of the Arabidopsis mutant pad2-1 affects oxidative stress-related events, defense gene expression, and the hypersensitive response. Plant Physiol 157:2000–2012. https://doi.org/10.1104/pp.111.182667
48. Hatsugai N, Perez Koldenkova V, Imamura H, Noji H, Nagai T (2012) Changes in cytosolic ATP levels and intracellular morphology during bacteria-induced hypersensitive cell death as revealed by real-time fluorescence microscopy imaging. Plant Cell Physiol 53:1768–1775. https://doi.org/10.1093/pcp/pcs119
49. Chen T, Xu G, Mou R, Greene GH, Liu L, Motley J, Dong X (2023) Global translational induction during NLR-mediated immunity in plants is dynamically regulated by CDC123, an ATP-sensitive protein. Cell Host Microbe. https://doi.org/10.1016/j.chom.2023.01.014
50. Cervela-Cardona L, Yoshida T, Zhang Y, Okada M, Fernie A, Mas P (2021) Circadian control of metabolism by the clock component TOC1. Front Plant Sci 12:683516. https://doi.org/10.3389/fpls.2021.683516
51. Billingham LK, Stoolman JS, Vasan K, Rodriguez AE, Poor TA, Szibor M, Jacobs HT, Reczek CR, Rashidi A, Zhang P et al (2022) Mitochondrial electron transport chain is necessary for NLRP3 inflammasome activation. Nat Immunol 23:692–704. https://doi.org/10.1038/s41590-022-01185-3
52. Abruzzi KC, Gobet C, Naef F, Rosbash M (2021) Comment on "circadian rhythms in the absence of the clock gene Bmal1". Science 372:eabf0922. https://doi.org/10.1126/science.abf0922
53. Novák B, Tyson JJ (2008) Design principles of biochemical oscillators. Nat Rev Mol Cell Biol 9:981–991. https://doi.org/10.1038/nrm2530
54. Gekakis N, Staknis D, Nguyen HB, Davis FC, Wilsbacher LD, King DP, Takahashi JS, Weitz CJ (1998) Role of the CLOCK protein in the mammalian circadian mechanism. Science 280:1564–1569. https://doi.org/10.1126/science.280.5369.1564
55. Kageyama H, Nishiwaki T, Nakajima M, Iwasaki H, Oyama T, Kondo T (2006) Cyanobacterial circadian pacemaker: Kai protein complex dynamics in the KaiC phosphorylation cycle in vitro. Mol Cell 23:161–171. https://doi.org/10.1016/j.molcel.2006.05.039
56. Menet JS, Abruzzi KC, Desrochers J, Rodriguez J, Rosbash M (2010) Dynamic PER repression mechanisms in the Drosophila circadian clock: from on-DNA to off-DNA. Genes Dev 24:358–367. https://doi.org/10.1101/gad.1883910

57. Froehlich AC, Loros JJ, Dunlap JC (2003) Rhythmic binding of a WHITE COLLAR-containing complex to the frequency promoter is inhibited by FREQUENCY. Proc Natl Acad Sci 100:5914–5919. https://doi.org/10.1073/pnas.1030057100
58. Pattanayak GK, Lambert G, Bernat K, Rust MJ (2015) Controlling the cyanobacterial clock by synthetically rewiring metabolism. Cell Rep 13:2362–2367. https://doi.org/10.1016/j.celrep.2015.11.031
59. Inupakutika MA, Sengupta S, Devireddy AR, Azad RK, Mittler R (2016) The evolution of reactive oxygen species metabolism. J Exp Bot 67:5933–5943. https://doi.org/10.1093/jxb/erw382
60. del Olmo M, Kramer A, Herzel H (2019) A robust model for circadian redox oscillations. Int J Mol Sci 20. https://doi.org/10.3390/ijms20092368
61. Foyer CH, Hanke G (2022) ROS production and signalling in chloroplasts: cornerstones and evolving concepts. Plant J 111:642–661. https://doi.org/10.1111/tpj.15856
62. Zhao RZ, Jiang S, Zhang L, Yu ZB (2019) Mitochondrial electron transport chain, ROS generation and uncoupling (review). Int J Mol Med 44:3–15. https://doi.org/10.3892/ijmm.2019.4188
63. Sies H, Jones DP (2020) Reactive oxygen species (ROS) as pleiotropic physiological signalling agents. Nat Rev Mol Cell Biol 21:363–383. https://doi.org/10.1038/s41580-020-0230-3
64. Richardson RB, Mailloux RJ (2023) Mitochondria need their sleep: redox, bioenergetics, and temperature regulation of circadian rhythms and the role of cysteine-mediated redox signaling, uncoupling proteins, and substrate cycles. Antioxidants (Basel) 12. https://doi.org/10.3390/antiox12030674
65. Davis W, Endo M, Locke JCW (2022) Spatially specific mechanisms and functions of the plant circadian clock. Plant Physiol 190:938–951. https://doi.org/10.1093/plphys/kiac236
66. Farshadi E, van der Horst GTJ, Chaves I (2020) Molecular links between the circadian clock and the cell cycle. J Mol Biol 432:3515–3524. https://doi.org/10.1016/j.jmb.2020.04.003
67. Talwar D, Miller CG, Grossmann J, Szyrwiel L, Schwecke T, Demichev V, Mikecin Drazic A-M, Mayakonda A, Lutsik P, Veith C et al (2023) The GAPDH redox switch safeguards reductive capacity and enables survival of stressed tumour cells. Nat Metab 5:660–676. https://doi.org/10.1038/s42255-023-00781-3

An Autonomous Clock that Controls Genetic Information Flow and Cellular Proteostasis

5

Bokai Zhu

Abstract

The central dogma of molecular biology, which delineates the flow of genetic information from DNA to RNA to proteins, is fundamental to cellular identity and functionality. This process involves intricate steps of transcription, mRNA processing, translation, and protein processing, each of which has been studied extensively over the past decades. Despite significant advancements in our understanding of these individual steps, the mechanisms by which these processes are spatiotemporally coordinated within each cell remain elusive. Efficient and precise transfer of genetic information is critical; any disruptions or "traffic jams" along this pathway including cryptic transcription, aberrant mRNA splicing, ribosome stalling, and protein misfolding and aggregation can lead to severe consequences such as aging and various diseases. Recently, our group has unveiled a cell-autonomous 12-h clock in mammalian cells that modulates the rhythmic activity of the entire genetic information flow. This discovery challenges the existing paradigm that genetic information flow is functioning at steady state but rather suggests that it oscillates with a robust 12-h cycle under normal physiological conditions, peaking at times of transition at dawn and dusk when metabolic stress is heightened. In this chapter, I will delve into the regulation, function, and evolutionary origins of the 12-h clock, and further elaborate how the 12-h clock can be utilized as a discovery tool to uncover hidden principles governing the spatiotemporal coordination of genetic information flow. Special attention will be given to the role of phase separation in facilitating this coordination, highlighting its significance in ensuring the seamless flow of genetic information across various cellular compartments and organelles.

Keywords

Autonomous clocks · 12-h ultradian rhythms · Genetic information flow · Cellular proteostasis · XBP1 · Unfolded protein response · IRE1α · Nuclear speckles · Liquid–liquid phase separation (LLPS) · SON · mRNA processing · ER/Golgi membrane homeostasis · Lipid metabolism · Circadian rhythms · Vehicle-cargo hypothesis

B. Zhu (✉)
Aging Institute of UPMC, University of Pittsburgh School of Medicine, Pittsburgh, PA, USA

Pittsburgh Liver Research Center, University of Pittsburgh, Pittsburgh, PA, USA

Division of Endocrinology and Metabolism, Department of Medicine, School of Medicine, University of Pittsburgh, Pittsburgh, PA, USA
e-mail: bzhu@pitt.edu

M. G. Aydogan (ed.), *Biological Time Control*, Advances in Experimental Medicine and Biology 1521,
https://doi.org/10.1007/978-3-032-36189-9_5

5.1 12-h Ultradian Rhythms Are Prevalent in Mammals, Including Humans

Unbeknownst to many, in addition to the ~24-h circadian rhythms, gene expression and metabolic rhythms cycling with an approximately 12-h period also exist in mammals, including humans. The presence of ~12-h ultradian rhythms of gene expression in mice was first comprehensively revealed in a seminal study by Hughes and colleagues in 2009 [1]. Utilizing high-temporal-resolution microarray with 1-h interval over a span of 48 h, this study unveiled ~200 genes cycling robustly with a 12-h period in mouse liver under constant darkness condition [1]. This initial number, however, was later recognized as an underestimation due to the limitations of the statistical methods and microarray technology available at the time. Nonetheless, this study made several notable discoveries regarding the ~12-h rhythms [1]. First, it showed that the peaks of these gene expression rhythms are predominantly aligned with the day's "rush hours"—shortly after sunrise (CT0; CT stands for constant time, and CT0 is around 7 am) and sunset (CT12), suggesting a synchronization of 12-h rhythms with the natural transitions between the day and night. Second, it was revealed that these genes are significantly enriched in unfolded protein response (UPR), suggesting a potential role of these rhythms in regulating proteostasis and stress response. Lastly, it demonstrated that restricting feeding to an 8-h window during the day disrupts many of these 12-h rhythms of gene expression, pointing to a link between feeding patterns and the regulation of these ultradian rhythms.

While the number of hepatic ~12 h genes in mice was originally thought to be relatively small (~200), with the advent of high-temporal-resolution RNA-seq and powerful new mathematical tools (e.g. RAIN [2]) and parameter-free signal processing tools (e.g., eigenvalue/pencil, autocorrelation, wavelet analysis that are capable of unmasking ultradian from superimposed/co-expressed circadian oscillations [3–7]), we have shown the real number of bona fide ~12 h genes in mouse liver to be around two to three thousand [8–11]. Besides liver, abundant 12 h transcriptomes were also identified in brown and white adipose tissues, muscle, adrenal gland, aorta, lung, liver, heart, and kidney in mice [12]. Expanding beyond rodents, El-Athman [13] and colleagues conducted a comprehensive analysis of diurnal transcriptome data across 64 tissues from olive baboons [14], employing the harmonic regression technique. This study unveiled a widespread prevalence of 12 h rhythms in the transcriptomes of peripheral tissues, including bone marrow, mesenteric white adipose tissue, mesenteric lymph nodes, smooth muscle, esophagus, kidney, prostate, pancreas, and skin [13]. Remarkably, certain brain regions—notably the hippocampus and cerebellum—also displayed extensive 12 h gene expression rhythms. In contrast, the suprachiasmatic nucleus (SCN), which harbors the central circadian clock pacemaker [15], is largely devoid of 12 h transcriptome [13]. Although the implications of these discoveries are yet to be fully understood, they imply a potential link between the 12 h gene expression rhythms in the primate hippocampus and the dual peaks of neural activity that facilitate memory acquisition during the day and memory consolidation at night [16, 17]. Beyond the transcriptome, robust ~12 h rhythms have also been documented in the hepatic proteome and metabolome of mice, encompassing approximately 20–30% of detected proteins and metabolites [18]. Importantly, a recent study combining integrative multi-omics profiling with kinetic mathematical modeling demonstrated that a substantial fraction of liver metabolic pathways, particularly those involved in lipid and carbohydrate metabolism, are governed predominantly by ~12 h rather than canonical ~24 h circadian rhythms [19]. These findings highlight a close functional integration between the 12 h UPR program and fundamental metabolic processes, suggesting that the 12 h oscillator serves as a key temporal regulator of cellular metabolism. Beyond the molecular rhythms, 12 h rhythms of energy expenditure [4, 20] and locomotor activity [5, 6] have also been observed in rodents. A very recent study further reported 12 h rhythms of the temporal dynamics and composition of ocular surface microbiota in mice [21].

Maybe the most important question is whether ~12-h rhythms of gene expression and activity also exist in humans. In two recent studies, we identified prevalent ~12-h rhythms of UPR gene expression in peripheral white blood cells [22] and dorsolateral prefrontal cortex region [23] in humans. Intriguingly, human subjects with schizophrenia lose ~12-h rhythms of UPR gene expression in the cortex [23]. In an independent study, analyzing longitudinal actigraphy recording of human motor activity also revealed ~12-h oscillations [24]. More importantly, in all three studies, human participants lived under free-living or free-living mimics conditions, indicating that ~12-h rhythms in humans are likely robust and refractory to daily environmental fluctuations [22].

5.2 The Vehicle-Cargo Hypothesis Delineates the Distinct Functions of 12-h Versus the Circadian Rhythms

In the initial microarray study by Hughes et al., UPR was identified as the sole significantly enriched pathway for the ~200 12-h hepatic genes in mice liver [1]. However, subsequent Gene Ontology (GO) analyses of the more comprehensive mouse hepatic 12-h transcriptome assayed by RNA-seq revealed enriched pathways spanning the entire spectrum of the genetic information flow. This includes transcription initiation and elongation, mRNA processing and export, ribosome biogenesis, translation initiation, protein folding, processing, and sorting in the endoplasmic reticulum (ER) and Golgi, encompassing both anabolic and catabolic processes (Fig. 5.1) [9–11]. Strikingly, a closer examination of the acrophases of ~12-h genes revealed a 1 ~ 2-h delay in the peaking time of genes involved in transcription and mRNA processing, at CT1 (~8 am) and CT13, compared to genes participating in the protein secretory pathway at CT2–3 and CT14–15. This timing difference aligns perfectly with the unidirectional sequential flow of genetic information [10].

The flow of genetic information is tightly coupled to the metabolic state of the cells, thus it is not surprising that 12-h hepatic genes involved in various metabolic pathways essential to mRNA and protein homeostasis were also prevalent [25]. These include genes involved in purine and pyrimidine de novo synthesis, which provides building blocks for nucleic acid synthesis; hexosamine and UDP-GlcNAc biosynthesis that supplies substrates for protein N-linked glycosylation in ER and Golgi, and glycophospholipids and sphingolipids synthesis, which regulates ER/Golgi homeostasis by regulating membrane lipid composition and fluidity (Fig. 5.1) [26]. Importantly, the genetic information flow and metabolic gene signatures are also enriched in 12-h transcriptomes of additional tissues in mice and peripheral white blood cells in humans [27, 28].

To describe the distinct functions of ~12-h-versus circadian rhythms, we introduced the vehicle-cargo hypothesis [10, 29]. This conjecture argues that the 12-h clock functions to boost the overall capacity for gene expression and processing during the two ‘rush hours’ of the day at dawn and dusk. This ultradian rhythm adjusts the pace of mRNA and protein processing to align with the 12-h cycles of metabolic stress, thus serving as the ‘vehicle’ for the genetic information flow. Conversely, the circadian rhythm is responsible for specifying which gene products are processed at each rush hour, functioning as the “cargo.” To draw a parallel to everyday life, the 12-h biological rhythm is akin to the management of a highway’s capacity to handle daily traffic surges, such as by opening the high occupancy vehicle lanes during the busiest times twice a day, while the circadian clock is likened to deciding which cars go on the HOV lane at each time (Fig. 5.2).

Beyond simple gene expression, further support for the existence of a 12-h cycle of genetic information flow comes from the discovery of a ~12-h cycle of mRNA intron retention (i.e.,

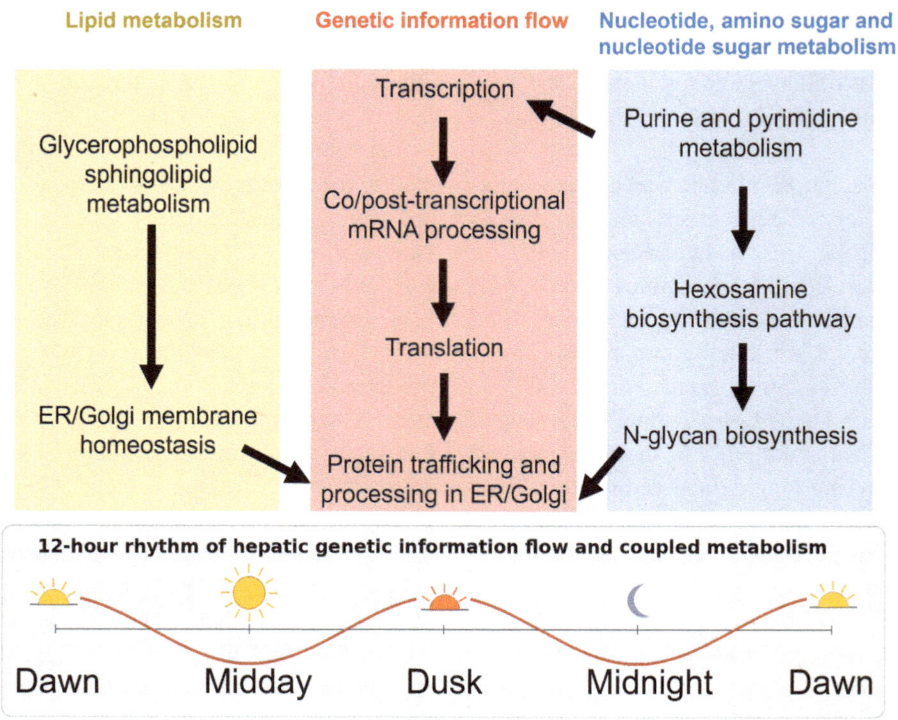

Fig. 5.1 12-h rhythms coordinate genetic information flow and supporting metabolic pathways. Schematic illustrating the coordinated ~12-h oscillations of hepatic gene expression involved in genetic information flow, including transcription, co–/post-transcriptional mRNA processing, translation, and protein trafficking and processing within the Endoplasmic Reticulum (ER) and Golgi apparatus. These processes are temporally coupled to rhythmic metabolic pathways that provide the structural and energetic resources required for efficient gene expression, including glycerophospholipid and sphingolipid metabolism, which support ER/Golgi membrane homeostasis, as well as Purine/Pyrimidine, Hexosamine, and N-glycan biosynthetic pathways, which supply precursors for nucleic acid synthesis, mRNA processing, protein synthesis, and protein maturation. Together, these coordinated oscillations establish a ~12-h rhythm of hepatic genetic information flow and its associated metabolic infrastructure

introns are temporarily retained rather than spliced out) events in both mouse liver and human peripheral white blood cells, peaking at dawn and dusk and synchronized to the expression of mRNA splicing genes [22]. These oscillations reinforce that the 12-h clock regulates not only transcription but also the efficiency and timing of mRNA maturation, thereby coordinating multiple successive steps in gene expression. Intriguingly, analysis of mRNAs with significant intron retention in human white blood cells reveals a pronounced enrichment for immune-related genes, particularly those involved in antigen presentation via MHC class I molecules [22]. These findings not only reinforce the vehicle-cargo hypothesis but also highlight that the primary 'cargo' transported by this 12-h "vehicle" in human white blood cells is largely comprised of immune genes critical for antigen presentation. Given the dependence of antigen presentation on the ER and Golgi apparatus [30], these results suggest a crucial role for the human 12-h rhythms in optimizing the efficiency of antigen presentation by immune cells, a process of significant interest in cancer immunology.

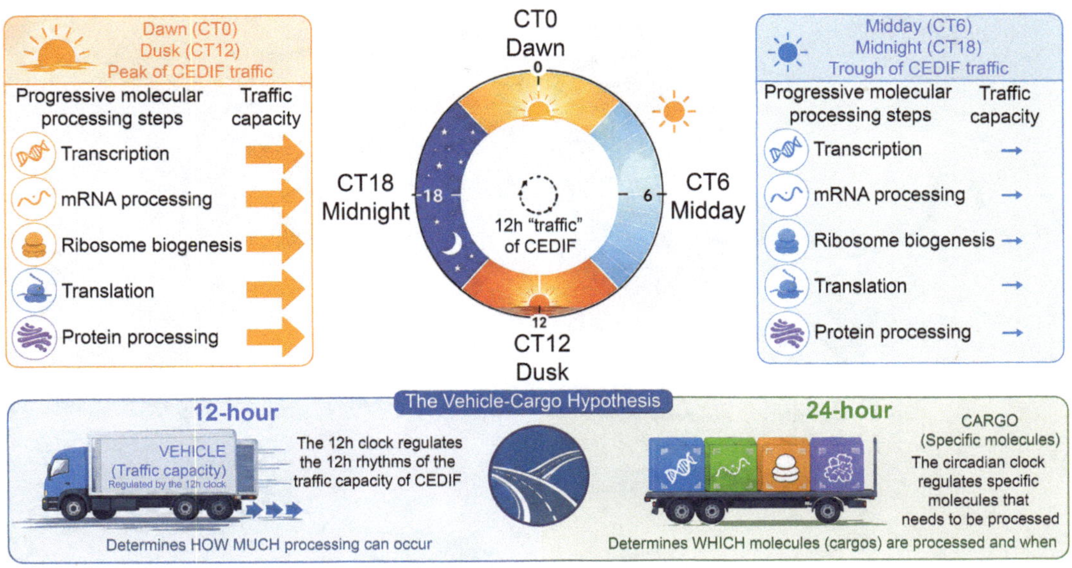

Fig. 5.2 Vehicle-cargo model of temporal regulation by the mammalian 12-h and circadian clocks. The mammalian 12-h clock generates two daily peaks in the capacity for cellular genetic information flow (CEDIF), occurring around dawn (CT0) and dusk (CT12), and two troughs around midday (CT6) and midnight (CT18). During peak phases, transcription, mRNA processing, ribosome biogenesis, translation, and protein processing pathways are coordinately upregulated, increasing the overall throughput of molecular processing. In contrast, these processes are reduced during trough phases, resulting in lower processing capacity. The central diagram illustrates the recurring ~12-h oscillation in CEDIF traffic capacity across the day. The lower panel depicts the vehicle-cargo hypothesis, in which the 12-h clock and the circadian clock regulate complementary aspects of gene expression. The 12-h clock functions as a vehicle, determining the overall processing capacity of the cellular gene expression machinery—analogous to controlling the number of available lanes on a highway or the carrying capacity of a truck. In contrast, the circadian clock functions as a cargo selector, determining which specific transcripts, proteins, and biological pathways are processed at a given time of day. Together, these two oscillatory systems coordinate both the quantity and identity of molecular cargo moving through successive steps of genetic information flow, thereby optimizing cellular homeostasis and adaptation to predictable daily demands

5.3 Evidence Supporting the Existence of an *Xbp1*-Dependent 12-h Clock, Separate from the Canonical Circadian Clock

The origin and mechanisms of regulation of 12-h ultradian rhythms are poorly understood and spark various hypotheses over the years [31]. Initially, it was proposed that these rhythms are not established by a cell-autonomous oscillator, but rather by the combined effects of circadian clock and feeding cues [1, 32]. This idea was supported by findings showing that a number of ~12-h proteostasis gene expression were lost in CRY1/CRY2 double knockout mice (with abrogated circadian clock), as well as in wild-type mice under altered feeding schedules [32]. Another hypothesis posited that 12-h rhythms could be driven by two circadian transcription factors with anti-phasic transcriptional activity and binding cooperativity in a cell-autonomous manner [33].

However, recent studies have challenged these views, suggesting instead that mammals possess a cell-intrinsic 12-h clock that independently regulates a significant portion of 12-h rhythms of gene expression (Fig. 5.3a, b) [4, 9–11]. At the core of the mammalian 12-h clock is the UPR transcription factor spliced form of X-box binding protein 1 (XBP1s), which sustains cell-autonomous ~12-h oscillations in its own

A

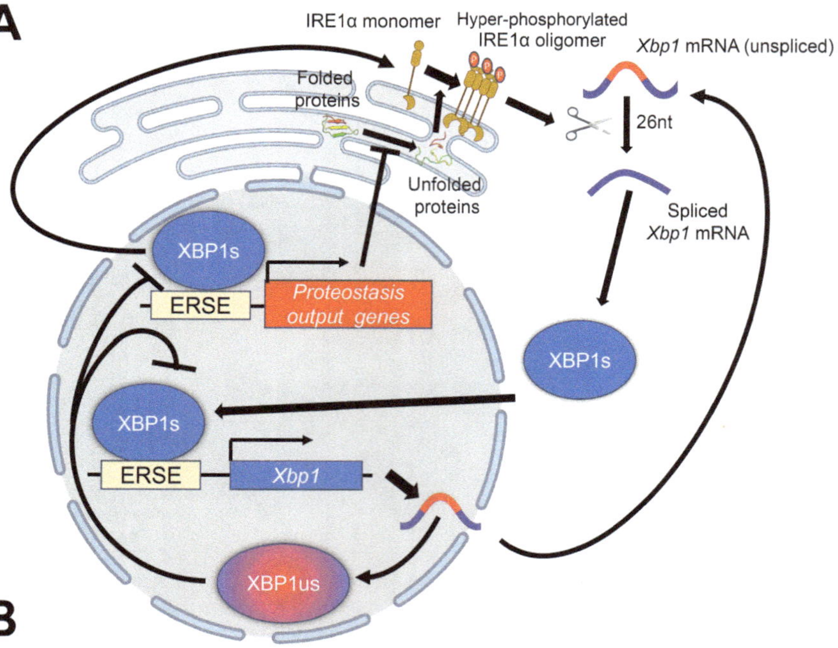

B

Molecules	Phase (ZT/CT)	Reference
p-IRE1α (Ser724) protein	~0.2	29, 52
Xbp1s mRNA	~0.4	10, 12
XBP1s protein	~1	29, 52
Proteostasis output mRNA	1~3	10
Ire1α mRNA	~2.1	10, 52
Xbp1us mRNA	~4	10, 12
XBP1us protein	N/A	
Proteostasis output protein (HSP40, HSP60, etc)	6~8	4
Total IRE1α protein	~7.4	29, 52
Folded protein	N/A	
Unfolded/misfolded protein	N/A	

Fig. 5.3 Transcriptional regulatory network of the mammalian 12-h clock. (**a**) Schematic illustrating the core transcriptional network underlying the mammalian 12-h oscillator. Accumulation of unfolded proteins in the endoplasmic reticulum (ER) promotes oligomerization and autophosphorylation of IRE1α, activating its endoribonuclease activity and inducing unconventional splicing of *Xbp1* mRNA to generate the active transcription factor XBP1s. XBP1s transactivates a broad set of proteostasis genes through ER stress response elements (ERSEs), including molecular chaperones and protein-folding factors, while simultaneously inducing the expression of *Xbp1* and *Ire1α*, thereby establishing a positive feedforward circuit. The oscillator is constrained by two negative feedback mechanisms. First, unspliced *Xbp1* (*Xbp1us*) gives rise to XBP1us protein, which antagonizes XBP1s activity and promotes its degradation. Second, XBP1s-induced chaperones, such as HSPA5 (BiP/GRP78), facilitate protein folding and suppress IRE1α oligomerization and activation, thereby reducing *Xbp1* mRNA splicing. Together, these coupled feedforward and feedback loops generate and sustain cell-autonomous ~12-h oscillations in proteostasis gene expression and ER homeostasis. (**b**) Peak phases of key molecular components within the hepatic 12-h clock network in mouse liver. Listed are the approximate circadian/circatidal times (ZT/CT) at which individual transcripts and proteins reach maximal abundance or activity. Phase information is currently unavailable (N/A) for XBP1us protein and for the bulk cellular pools of folded and unfolded/misfolded proteins. Abbreviations: *ZT* zeitgeber time, *CT* circadian time

expression as well as that of hundreds of direct target genes [9, 10]. This oscillator is maintained through interconnected positive feedforward and negative feedback loops. In the feedforward arm, XBP1s transcriptionally activates both *Xbp1* and *Ire1α* [9, 10], while IRE1α catalyzes the unconventional splicing of unspliced *Xbp1* (*Xbp1us*) mRNA to generate XBP1s [10, 34–36]. Conversely, in the negative feedback arm, XBP1us functions as a repressor of XBP1s activity, thereby limiting pathway activation and contributing to rhythmic oscillations [37] (Fig. 5.3a). Additional negative feedback mechanisms might involve XBP1s-mediated upregulation of proteostasis output genes, such as heat shock proteins assisting in protein folding. These proteins help alleviate the proteotoxic stress in the ER and thus prevent the signaling that leads to the oligomerization of IRE1α and *Xbp1* splicing (Fig. 5.3a). Consistent with this model, the phases of *Xbp1s* and *Xbp1us* mRNA are separated by ~four hours, with the former peaking at CT0 and the latter around CT4 (Fig. 5.3b) [10, 27]. In addition, the peak of heat shock protein expression (and other proteins involved in UPR) are largely between CT6 and CT8, a time coinciding with the trough of p-IRE1α and XBP1s level (Fig. 5.3b) [8]. XBP1s-dependent transcription is further reinforced by the coactivator Steroid Receptor Coactivator-3 (SRC-3) [20] and the Complex of Proteins Associated with Set1 (COMPASS) histone H3K4 methyltransferase complex [38], which cooperatively promote rhythmic chromatin activation at XBP1s target genes. Disruption of either pathway markedly impairs hepatic ~12-h-transcriptomic and metabolomic oscillations, underscoring their critical roles in maintaining the mammalian 12-h clock [20, 38].

This model of 12-h clock regulation is supported by numerous experimental evidence including (1) the enriched pathways of ~12-h-transcriptome are distinct from those of circadian rhythms in both human and mice [9–11, 28]; (2) the mathematical and statistical independence of the detection of ~12-h and circadian rhythms in both human and mice [3, 6, 8, 28, 39]; (3) the majority of hepatic 12-h rhythms of gene expression are impaired in XBP1 liver-specific knockout ($XBP1^{LKO}$) mice [9, 10], but are maintained in BMAL1 (the master regulator of circadian clock in mammals) knockout mice [11]; (4) circadian rhythms of core circadian clock and output genes are largely maintained in $XBP1^{LKO}$ mice [9, 10]; (5) 12-h chromatin occupancy of XBP1s were identified on 12-h, but not circadian genes promoters [8–10], (6) cell-autonomous 12-h XBP1s and proteostasis expression is maintained in *Bmal1* knockdown mouse embryonic fibroblasts (MEFs) after synchronization with very low amount of ER stress inducer tunicamycin [4, 10], and (7) the loss of 12-h rhythms of mRNA intron retention rhythms in the liver of $XBP1^{LKO}$ mice [28]. The discrepancies observed between BMAL1 and CRY1/CRY2 knockout mice on the presence or absence of 12-h rhythms may be attributed to the fact that CRY proteins can modulate overall proteostasis independently from its canonical function in circadian control, via protecting against proteome imbalance and proteotoxic stress [40].

Since XBP1s only physically associates with the promoter regions of roughly six hundred 12-h output genes [10], it indicates the existence of additional transcription regulators participating in 12-h clock control. Using in silico analyses, a tripartite regulatory network involving three main groups of transcription factors was proposed to be central to the mammalian 12-h clock [27]. These include the E26 transformation specific (ETS) family (notably GABPA), the Basic Leucine Zipper Domain (bZIP) family (including XBP1s, ATF6, and CRELD2), and the nuclear transcription factor Y (NFY) family (comprising NFYA and NFYB) [27]. Interestingly, knocking down GABPA in MEFs lengthened the period of *Manf* promoter-driven luciferase oscillation (*Manf* is a robust 12-h clock output gene) to more than 14 h [10]. Future research endeavors employing both novel genetic cellular and mouse models and mathematical modeling will be crucial in uncovering the full gene regulatory network governing the mammalian 12-h clock.

5.4 Subcellular Mechanisms of Genetic Information Flow Control Via the 12-h Clock

Genetic transfer of information within a cell involves relaying information across spatially separated compartments, facilitated by both lipid membranes and cytoskeletal structures for organization and stability. Beyond these structured organelles, cells also contain membraneless organelles that maintain their integrity and dynamic architecture without conventional organizational features. These structures, known collectively as biomolecular condensates, are formed through phase separation—a process mainly driven by weak, multivalent interactions among proteins' intrinsically disordered regions (IDRs) [41–45]. This mechanism, akin to the liquid–liquid phase separation (LLPS) between oil and water, allows these condensates to exist as droplet-like condensates within the cell and play key roles in organizing the dynamic genetic information flow from DNA to RNA to protein [46, 47]. Examples of such condensates include P-granules [48] and stress granules [49] in the cytoplasm, and the nucleolus [50], transcriptional condensates [51], Cajal bodies [52], and nuclear speckles [53–56] in the nucleus. Together, they are crucial in organizing key cellular processes, from chromatin regulation and transcription to alternative splicing, by acting as hubs for the dynamic regulation of genetic information [46, 47]. This IDR-mediated phase separation highlights a fundamental principle in cellular organization, emphasizing the importance of these liquid-like structures in facilitating a coherent and efficient flow of genetic information.

In addition to biomolecular condensates composed of membrane-less assemblies comprising proteins or nucleic acids, lipid bilayer membranes can also experience phase transitions between ordered and disordered states [57]. These transitions are influenced by the lipid composition, specifically the ratio of saturated to unsaturated acyl chains, which directly affects membrane fluidity [58]. Changes in this fluidity can sense metabolic stress, transmitting signals that adjust the rate of genetic information transfer, particularly through the protein secretion pathways in the ER and Golgi. For instance, a higher ratio of saturated lipids in the membrane can induce lipid bilayer stress, triggering the activation of the UPR [59]. In this section, I will discuss in detail two mechanisms by which the 12-h clock modulates genetic information flow: controlling the LLPS dynamics of nuclear speckle and the phase behavior of lipid bilayers.

5.4.1 12-h Clock Control of Genetic Information Flow Via Fine-Tuning Nuclear Speckles LLPS Dynamics

The critical function of XBP1s in regulating proteostasis has been extensively studied within the context of the UPR and ER stress [60, 61]. Yet, the mechanisms by which XBP1s impacts mRNA metabolism and links it to protein homeostasis remain unclear. Our recent research uncovered an XBP1-SON axis that couples the 12-h rhythms of mRNA and protein metabolism (Fig. 5.4, *top right*) [62]. *Son*, encoding a key structural component of nuclear speckles—membrane-less organelles within the nucleus crucial for various mRNA metabolic processes [53, 63–70]—emerges as a direct transcriptional output of the 12-h clock, displaying pronounced XBP1s-dependent 12-h oscillations at both mRNA and protein level [62]. As the scaffold protein of nuclear speckle condensates, the 12-h rhythm of SON expression subsequently affects the LLPS dynamics of nuclear speckles, causing them to switch between a spherical/istagnant and a dispersed/supfluid state with a 12-h interval [62]. The 12-h nuclear speckle LLPS dynamics further drives a global 12-h rhythm of nuclear speckle–chromatin interactions, uncoupled from the expression of individual genes [62]. In the hepatocytes of XBP1 liver-specific knockout mice, the oscillations of SON expression, nuclear speckle LLPS, and interactions with chromatin were not only weakened but also exhibited a shortened period of approximately 10 h

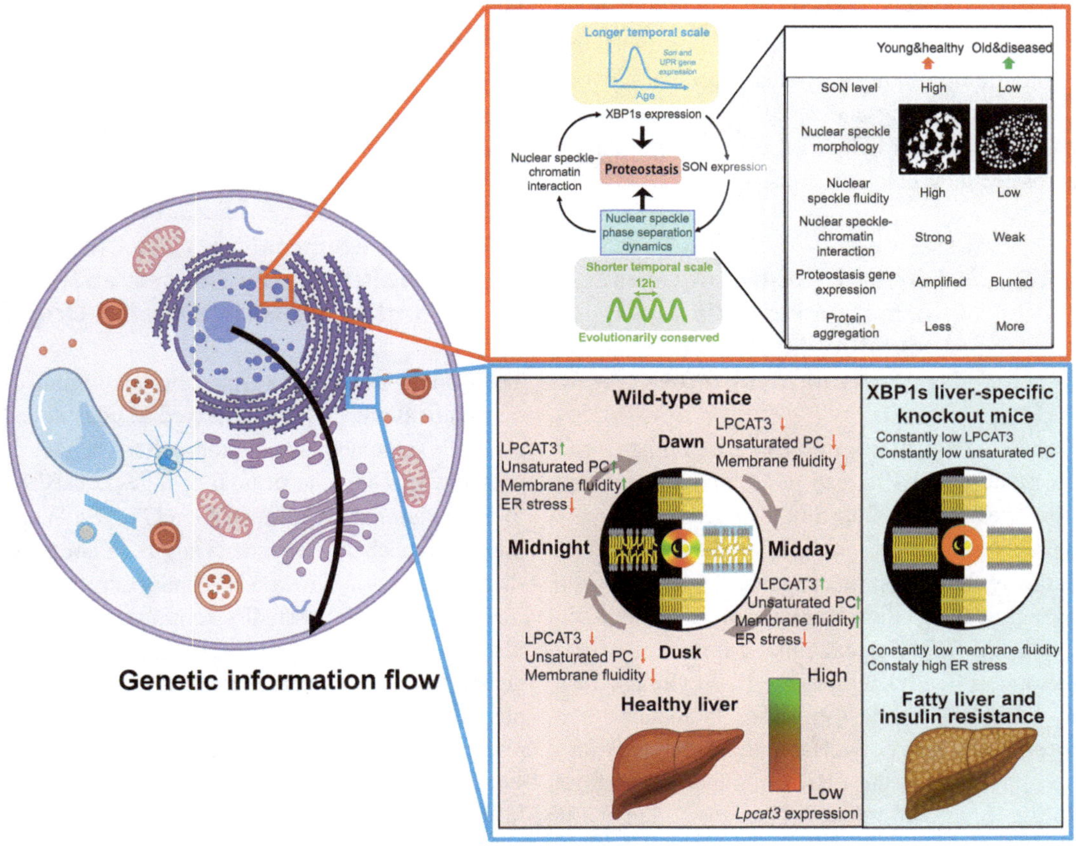

Fig. 5.4 Mammalian 12-h clock coordinates genetic information flow through regulation of nuclear speckle dynamics and ER/Golgi membrane homeostasis. The 12-h clock promotes efficient genetic information flow by synchronizing two complementary cellular systems: nuclear speckle phase behavior in the nucleus and membrane lipid remodeling in the secretory pathway. *Top right*, a model illustrating how the XBP1s–SON axis regulates nuclear speckle function. XBP1s drives rhythmic *Son* expression, resulting in ~12-h oscillations in nuclear speckle liquid–liquid phase separation (LLPS) dynamics and speckle–chromatin interactions. These oscillations enhance the expression of proteostasis genes and maintain protein homeostasis. During aging and disease, reduced SON expression is associated with decreased speckle fluidity, weakened speckle–chromatin interactions, blunted proteostasis gene expression, and increased protein aggregation. *Bottom right*, a model illustrating how the XBP1s-dependent 12-h clock regulates rhythmic expression of *Lpcat3*, a key enzyme controlling phosphatidylcholine (PC) remodeling. Oscillations in *Lpcat3* expression drive corresponding rhythms in polyunsaturated PC abundance, ER membrane fluidity, and ER stress across the day, thereby maintaining hepatic metabolic homeostasis. Liver-specific ablation of *Xbp1* disrupts these rhythms, resulting in constitutively low LPCAT3 expression, reduced membrane fluidity, chronic ER stress, and the development of fatty liver and insulin resistance. Together, these findings suggest that the mammalian 12-h clock coordinates both the nuclear and membrane environments required for efficient genetic information flow and cellular proteostasis

[62]. Moderately increasing SON levels was found to enhance the fluidity and diffuseness of nuclear speckles, boost their association with chromatin, elevate expression of genes involved in proteostasis, and reduce protein aggregation. Conversely, decreasing SON expression yields opposite effects (Fig. 5.4, *top right*). The observation that SON overexpression also increases XBP1s level points to a positive feedforward loop that ties nuclear speckle dynamics to the regulation of mRNA and protein homeostasis [62]. These findings indicate that proteostasis-

related gene expression is highly sensitive to changes in nuclear speckle phase behavior, highlighting a tight functional coupling between mRNA processing and protein homeostasis that promotes efficient flow of genetic information across cellular compartments.

5.4.2 12-h Clock Control of Genetic Information Flow Via Modulating Lipid Phase Behaviors of Membranes

Besides regulating mRNA metabolism and proteostasis, XBP1s also play important roles in lipid metabolism [71]. Studies over recent years have shown that activating XBP1s and other UPR pathways helps to protect against hepatic steatosis by influencing the composition of membrane lipids [72–75]. For example, expression of the Lysophosphatidylcholine Acyltransferase 3 (*Lpcat3*) gene, which facilitates the incorporation of polyunsaturated fatty acids into phospholipids in the ER membrane, is observed during high-diet-induced ER stress and UPR [4, 76]. Such lipid incorporation enhances ER membrane fluidity, reducing hepatic inflammation and mitigating ER stress [76]. Notably, *Lpcat3* mRNA and levels of various 2-Lyso-phosphatidylcholines (PC) species (LPCAT3 conjugates 2-Lyso-PC and unsaturated Acyl-CoA to form PC) exhibit strong 12-h rhythms in the mouse liver (Fig. 5.4, *bottom right*), alongside rhythms in the expression of fatty acid modifying enzymes like *Scd1* and *Elovl6* [4, 9, 18]. These rhythmic changes in lipid composition affect the fluidity of cellular membranes, impacting information transfer across lipid bilayers and indicating a broader influence on systemic metabolism that remains to be fully explored.

In mice with liver-specific XBP1 deletion, the 12-h rhythm of *Lpcat3* expression is disrupted, leading to decreased levels of polyunsaturated phospholipids, reduced membrane fluidity, and impaired lipid metabolism (Fig. 5.4, *bottom right*) [9]. This results in accelerated development of non-alcoholic fatty liver disease (NAFLD) and liver aging, along with glucose intolerance and hyperinsulinemia [9]. Since the impairment of 12-h transcriptome precedes the manifestation of the NAFLD phenotype in XBP1^{LKO} mice, it suggests that the loss of 12-h clock is likely a driver of NAFLD.

5.5 The Mammalian 12-h Clock Is Evolutionarily Conserved and May Be Circatidal in Origin

The 12-h rhythms found in mammals reminisce of the circatidal rhythms observed in coastal and estuarine animals, which align their behaviors with the 12.4-h cycle of tidal changes caused by the gravitational pull of the Moon [77, 78]. While the exact mechanisms underlying these circatidal rhythms in marine life are still being explored, an increasing number of studies support the idea of a distinct circatidal clock that operates independently from the circadian clock [79]. For example, in the crustacean sea louse *Eurydice pulchra*, research has shown that the 12-h circatidal timing system is separate from its circadian counterpart [80]. In circatidal mangrove crickets, 12-h rhythms persist in constant darkness, even after the removal of the optic lobe. Furthermore, these 12-h rhythms are intact even when the expression of circadian clock components is reduced through siRNA-mediated knockdown [81–83]. The circadian and circatidal clocks in marine animals are also entrained by distinct environmental cues: while the former is entrained by light-dark cycles, the latter is unresponsive to light cues but sensitive to vibration of the water in artificial laboratory settings [84].

Comparative transcriptome analyses between mammals and marine species possessing a circatidal clock have shed light on the evolutionary origins of mammalian 12-h rhythms. Striking similarities were observed when comparing the ~12-h transcriptomes of circatidal animals with that of mammals. For instance, in *Eurydice pulchra,* distinct circatidal rhythms were previously identified for 10 mitochondrial DNA (mtDNA)-encoded protein-coding genes (*Mt-Nd*1 ~ 6, *Mt-Cox*1 ~ 3 and *Cytb*) [85]. Intriguingly, these 12-h rhythmic transcription of mtDNA-encoded genes is also preserved in the liver of mice, suggesting a conserved evolutionary mechanism that underpins these rhythms across diverse species [4]. Further comparing the 12-h transcriptomes of mouse liver [10] and human peripheral white blood cells [28] with the circatidal transcriptome of the sea anemone *Aiptasia diaphana* [86] *revealed statistically significant overlap of genes implicated in mRNA and protein homeostasis* [28]. 12-h rhythms of gene expression involved in genetic information flow were also identified in the circatidal mangrove cricket [87] *and more recently in* a deep-sea hot vent shrimp [88]. These findings collectively indicate that the 12-h clock observed in mammals may have its evolutionary roots in the circatidal clocks of coastal and estuarine animals [4, 10, 18]. This theory gains additional support from recent findings, which highlight a greater degree of conservation in the chromosomal locations of 12-h genes [89] and that genes with 12-h rhythms tend to have an older evolutionary origin compared to those exhibiting circadian rhythms [90].

What are the evolutionary benefits of the 12-h clock for land-dwelling mammals, which no longer inhabit coastal regions, is an important question. A possible explanation is that this clock has been co-opted to navigate the metabolic shifts that occur at dawn and dusk in land-dwelling animals [18]. Specifically, at subjective dawn (ZT12 in mice), a period marked by an energy deficit from overnight fasting and sleep-related energy expenditure, there is a heightened need for catabolic processes to generate energy; conversely, at subjective dusk (ZT0 in mice), the surplus energy from daytime feeding, coupled with a reduction in activity as diurnal animals begin to rest, prompts anabolic activities for energy storage. Both conditions—energy surplus and deficit—trigger the IRE1α-XBP1s signaling pathway to activate the UPR to manage these shifts [27, 91–93]. Thus, the 12-h clock enhances metabolic efficiency, preparing organisms for the energetic demands at the start and end of the day. So, in a nutshell, we reason that the 12-h clock is responding to and synchronized to the absolute change in energy ($|\Delta E|$), while the circadian clock is aligned to the energy level itself (E).

5.6 Conclusion and Future Perspectives

Genes with similar functions often share similar temporal profiles, a phenomenon known as temporal-functional relationship [94]. Thus, we conjecture that by studying the 12-h clock, we can uncover hidden principles of genetic information flow and proteostasis control. The demonstrated role of the XBP1s-SON axis exemplifies the potential of leveraging the 12-h clock as a tool for biological discovery, offering a glimpse into novel mechanisms of proteostasis control. Future research directions of interest include decoding the epigenetic codes that regulate genetic information flow, dissecting the mutual influences of the 12-h clock and time-restricted feeding on the dynamics of energy expenditure oscillation, elucidating the relationship between nucleotide metabolism and nuclear speckle LLPS in proteostasis, investigating how tRNA homeostasis affects specific genetic information flow pathways, and identifying paracrine factors capable of synchronizing genetic information flow across cellular populations. Pursuing these research paths promises to not only deepen our understanding of the fundamental mechanisms governing genetic information flow, but also to identify novel therapeutic targets for disease management.

Acknowledgments I apologize for any potential omission of relevant works and citations due to space constraints. B.Z. is supported by the National Institute of General Medical Sciences of the National Institutes of Health under Award Number DP2GM140924 and R35 GM157978. B.Z has no conflict of interest to declare.

References

1. Hughes ME et al (2009) Harmonics of circadian gene transcription in mammals. PLoS Genet 5:e1000442. https://doi.org/10.1371/journal.pgen.1000442
2. Thaben PF, Westermark PO (2014) Detecting rhythms in time series with RAIN. J Biol Rhythm 29:391–400. https://doi.org/10.1177/0748730414553029
3. Antoulas AC et al (2018) A novel mathematical method for disclosing oscillations in gene transcription: a comparative study. PLoS One 13:e0198503. https://doi.org/10.1371/journal.pone.0198503

4. Zhu B et al (2017) A cell-autonomous mammalian 12 hr clock coordinates metabolic and stress rhythms. Cell Metab 25:1305–1319 e1309. https://doi.org/10.1016/j.cmet.2017.05.004
5. Hasanpour M, Mitricheva E, Logothetis N, Noori HR (2021) Intensive longitudinal characterization of multidimensional biobehavioral dynamics in laboratory rats. Cell Rep 35:108987. https://doi.org/10.1016/j.celrep.2021.108987
6. Kembro JM et al (2023) A dynamically coherent pattern of rhythms that matches between distant species across the evolutionary scale. Sci Rep 13:5326. https://doi.org/10.1038/s41598-023-32286-0
7. Psomas A et al (2023) Co-expression of diurnal and ultradian rhythms in the plasma metabolome of common voles (Microtus arvalis). FASEB J 37:e22827. https://doi.org/10.1096/fj.202201585R
8. Zhu B et al (2017) A cell-autonomous mammalian 12 hr clock coordinates metabolic and stress rhythms. Cell Metab 25:1305–1319. e1309
9. Meng H et al (2020) XBP1 links the 12-hour clock to NAFLD and regulation of membrane fluidity and lipid homeostasis. Nat Commun 11:6215. https://doi.org/10.1038/s41467-020-20028-z
10. Pan Y et al (2020) 12-h clock regulation of genetic information flow by XBP1s. PLoS Biol 18:e3000580. https://doi.org/10.1371/journal.pbio.3000580
11. Zhu B, Liu S (2023) Preservation of ~12-h ultradian rhythms of gene expression of mRNA and protein metabolism in the absence of canonical circadian clock. Front Physiol 14. https://doi.org/10.3389/fphys.2023.1195001
12. Zhu B (2020) Decoding the function and regulation of the mammalian 12-h clock. J Mol Cell Biol 12:752–758
13. El-Athman R, Knezevic D, Fuhr L, Relogio A (2019) A computational analysis of alternative splicing across mammalian tissues reveals circadian and ultradian rhythms in splicing events. Int J Mol Sci 20. https://doi.org/10.3390/ijms20163977
14. Mure LS et al (2018) Diurnal transcriptome atlas of a primate across major neural and peripheral tissues. Science 359:1232. https://doi.org/10.1126/science.aao0318
15. Hoon Kim Y, Lazar MA (2020) Transcriptional control of circadian rhythms and metabolism: a matter of time and space. Endocr Rev. https://doi.org/10.1210/endrev/bnaa014
16. Koyanagi I et al (2019) Memory consolidation during sleep and adult hippocampal neurogenesis. Neural Regen Res 14:20–23. https://doi.org/10.4103/1673-5374.243695
17. Nadel L, Hupbach A, Gomez R, Newman-Smith K (2012) Memory formation, consolidation and transformation. Neurosci Biobehav Rev 36:1640–1645. https://doi.org/10.1016/j.neubiorev.2012.03.001
18. Zhu B, Dacso CC, O'Malley BW (2018) Unveiling "Musica universalis" of the cell: a brief history of biological 12-hour rhythms. J Endocr Soc 2:727–752. https://doi.org/10.1210/js.2018-00113
19. Matz-Soja M et al (2026) Modeling the dynamics of hepatic metabolism: the predominance of 12-hour rhythmicity in metabolic adaptation. Cell Mol Life Sci. https://doi.org/10.1007/s00018-025-06046-4
20. Meng H et al (2022) Defining the mammalian coactivation of hepatic 12-h clock and lipid metabolism. Cell Rep 38:110491. https://doi.org/10.1016/j.celrep.2022.110491
21. Jiao X, Li Z (2023) Temporal dynamics and composition of ocular surface microbiota in C57BL/6J mice: uncovering a 12h ultradian rhythm. Front Cell Infect Microbiol 13:1244454. https://doi.org/10.3389/fcimb.2023.1244454
22. Zhu B et al (2024) Evidence for ~12-h ultradian gene programs in humans. npj Biol Timing Sleep 1:4. https://doi.org/10.1038/s44323-024-00005-1
23. Scott MR et al (2023) Twelve-hour rhythms in transcript expression within the human dorsolateral prefrontal cortex are altered in schizophrenia. PLoS Biol 21:e3001688. https://doi.org/10.1371/journal.pbio.3001688
24. Maczák B, Gingl Z, Vadai G (2024) General spectral characteristics of human activity and its inherent scale-free fluctuations. Sci Rep 14:2604. https://doi.org/10.1038/s41598-024-52905-8
25. Pan Y et al (2019) 12h-clock control of central dogma information flow by XBP1s. bioRxiv:559039. https://doi.org/10.1101/559039
26. McMaster CR (2001) Lipid metabolism and vesicle trafficking: more than just greasing the transport machinery. Biochem Cell Biol 79:681–692. https://doi.org/10.1139/o01-139
27. Zhu B (2020) Decoding the function and regulation of the mammalian 12h-clock. J Mol Cell Biol. https://doi.org/10.1093/jmcb/mjaa021
28. Zhu B et al (2023) Evidence for conservation of primordial ~12-hour ultradian gene programs in humans under free-living conditions. bioRxiv:2023.2005.2002.539021. https://doi.org/10.1101/2023.05.02.539021
29. Ballance H, Zhu B (2021) Revealing the hidden reality of the mammalian 12-h ultradian rhythms. Cell Mol Life Sci. https://doi.org/10.1007/s00018-020-03730-5
30. Hewitt EW (2003) The MHC class I antigen presentation pathway: strategies for viral immune evasion. Immunology 110:163–169. https://doi.org/10.1046/j.1365-2567.2003.01738.x
31. Emery P, Gachon F (2025) Biological rhythms: living your life, one half-day at a time. npj Biol Timing Sleep 2:21. https://doi.org/10.1038/s44323-025-00037-1
32. Cretenet G, Le Clech M, Gachon F (2010) Circadian clock-coordinated 12 Hr period rhythmic activation of the IRE1alpha pathway controls lipid metabolism in mouse liver. Cell Metab 11:47–57. https://doi.org/10.1016/j.cmet.2009.11.002
33. Westermark PO, Herzel H (2013) Mechanism for 12 hr rhythm generation by the circadian clock. Cell Rep 3:1228–1238. https://doi.org/10.1016/j.celrep.2013.03.013
34. Yoshida H, Matsui T, Yamamoto A, Okada T, Mori K (2001) XBP1 mRNA is induced by ATF6 and spliced

by IRE1 in response to ER stress to produce a highly active transcription factor. Cell 107:881–891. https://doi.org/10.1016/s0092-8674(01)00611-0
35. Acosta-Alvear D et al (2007) XBP1 controls diverse cell type- and condition-specific transcriptional regulatory networks. Mol Cell 27:53–66. https://doi.org/10.1016/j.molcel.2007.06.011
36. Uemura A, Oku M, Mori K, Yoshida H (2009) Unconventional splicing of XBP1 mRNA occurs in the cytoplasm during the mammalian unfolded protein response. J Cell Sci 122:2877–2886. https://doi.org/10.1242/jcs.040584
37. Yoshida H, Oku M, Suzuki M, Mori K (2006) pXBP1(U) encoded in XBP1 pre-mRNA negatively regulates unfolded protein response activator pXBP1(S) in mammalian ER stress response. J Cell Biol 172:565–575. https://doi.org/10.1083/jcb.200508145
38. Kubra S et al (2025) The SET1/COMPASS subunit RBBP5 orchestrates epigenetic control of global proteostasis and the 12h oscillator to safeguard metabolic and cellular homeostasis. bioRxiv:2024.2009.2013.612812. https://doi.org/10.1101/2024.09.13.612812
39. Zhu B, Dacso CC, O'Malley BW (2018) Unveiling "Musica universalis" of the cell: a brief history of biological 12-hour rhythms. J Endocr Soc 2:727–752
40. Wong DCS et al (2022) CRYPTOCHROMES promote daily protein homeostasis. EMBO J 41:e108883. https://doi.org/10.15252/embj.2021108883
41. Lyon AS, Peeples WB, Rosen MK (2021) A framework for understanding the functions of biomolecular condensates across scales. Nat Rev Mol Cell Biol 22:215–235. https://doi.org/10.1038/s41580-020-00303-z
42. Folkmann AW, Putnam A, Lee CF, Seydoux G (2021) Regulation of biomolecular condensates by interfacial protein clusters. Science 373:1218–1224. https://doi.org/10.1126/science.abg7071
43. Sabari BR, Dall'Agnese A, Young RA (2020) Biomolecular condensates in the nucleus. Trends Biochem Sci 45:961–977. https://doi.org/10.1016/j.tibs.2020.06.007
44. Gouveia B et al (2022) Capillary forces generated by biomolecular condensates. Nature 609:255–264. https://doi.org/10.1038/s41586-022-05138-6
45. Conti BA, Oppikofer M (2022) Biomolecular condensates: new opportunities for drug discovery and RNA therapeutics. Trends Pharmacol Sci 43:820–837. https://doi.org/10.1016/j.tips.2022.07.001
46. Tsang B, Pritišanac I, Scherer SW, Moses AM, Forman-Kay JD (2020) Phase separation as a missing mechanism for interpretation of disease mutations. Cell 183:1742–1756. https://doi.org/10.1016/j.cell.2020.11.050
47. Kato M, McKnight SL (2018) A solid-state conceptualization of information transfer from gene to message to protein. Annu Rev Biochem 87:351–390. https://doi.org/10.1146/annurev-biochem-061516-044700
48. Wang JT, Seydoux G (2014) P granules. Curr Biol 24:R637–r638. https://doi.org/10.1016/j.cub.2014.06.018
49. Protter DSW, Parker R (2016) Principles and properties of stress granules. Trends Cell Biol 26:668–679. https://doi.org/10.1016/j.tcb.2016.05.004
50. Maccaroni K, La Torre M, Burla R, Saggio I (2022) Phase separation in the nucleus and at the nuclear periphery during post-mitotic nuclear envelope reformation. Cells 11. https://doi.org/10.3390/cells11111749
51. Demmerle J, Hao S, Cai D (2023) Transcriptional condensates and phase separation: condensing information across scales and mechanisms. Nucleus 14:2213551. https://doi.org/10.1080/19491034.2023.2213551
52. Nizami Z, Deryusheva S, Gall JG (2010) The Cajal body and histone locus body. Cold Spring Harb Perspect Biol 2:a000653. https://doi.org/10.1101/cshperspect.a000653
53. Liao SE, Regev O (2020) Splicing at the phase-separated nuclear speckle interface: a model. Nucleic Acids Res 49:636–645. https://doi.org/10.1093/nar/gkaa1209
54. Chen X, Zhu B (2026) Decoding the functions of nuclear speckles in neurodegeneration. Trends Neurosci 49:292
55. Namba H, Mir S, Groce EL, Alexander KA (2026) Nuclear speckles: a fundamental layer of gene regulation. Trends Cell Biol
56. Masanta S, Wakimian K, Cieśla M (2026) Phosphorylation-controlled nuclear speckle dynamics regulate splicing. Trends Biochem Sci. https://doi.org/10.1016/j.tibs.2026.05.004
57. Elson EL, Fried E, Dolbow JE, Genin GM (2010) Phase separation in biological membranes: integration of theory and experiment. Annu Rev Biophys 39:207–226. https://doi.org/10.1146/annurev.biophys.093008.131238
58. Ballweg S, Ernst R (2017) Control of membrane fluidity: the OLE pathway in focus. Biol Chem 398:215–228. https://doi.org/10.1515/hsz-2016-0277
59. Volmer R, van der Ploeg K, Ron D (2013) Membrane lipid saturation activates endoplasmic reticulum unfolded protein response transducers through their transmembrane domains. Proc Natl Acad Sci 110:4628–4633. https://doi.org/10.1073/pnas.1217611110
60. Sriburi R, Jackowski S, Mori K, Brewer JW (2004) XBP1: a link between the unfolded protein response, lipid biosynthesis, and biogenesis of the endoplasmic reticulum. J Cell Biol 167:35–41. https://doi.org/10.1083/jcb.200406136
61. Hetz C (2012) The unfolded protein response: controlling cell fate decisions under ER stress and beyond. Nat Rev Mol Cell Biol 13:89–102. https://doi.org/10.1038/nrm3270
62. Dion W et al (2022) Four-dimensional nuclear speckle phase separation dynamics regulate proteostasis. Sci Adv 8:eabl4150
63. Ilik İA et al (2020) SON and SRRM2 are essential for nuclear speckle formation. elife 9:e60579. https://doi.org/10.7554/eLife.60579
64. Sharma A, Takata H, Shibahara K, Bubulya A, Bubulya PA (2010) Son is essential for nuclear speckle

organization and cell cycle progression. Mol Biol Cell 21:650–663. https://doi.org/10.1091/mbc.E09-02-0126
65. Spector DL, Lamond AI (2011) Nuclear speckles. Cold Spring Harb Perspect Biol 3. https://doi.org/10.1101/cshperspect.a000646
66. Kim J, Khanna N, Belmont AS (2019) Transcription amplification by nuclear speckle association. bioRxiv:604298. https://doi.org/10.1101/604298
67. Hasenson SE, Shav-Tal Y (2020) Speculating on the roles of nuclear speckles: how RNA-protein nuclear assemblies affect gene expression. BioEssays 42: e2000104. https://doi.org/10.1002/bies.202000104
68. Dion W et al (2025) SON-dependent nuclear speckle rehabilitation alleviates proteinopathies. Nat Commun 16:7065. https://doi.org/10.1038/s41467-025-62242-7
69. Małszycki M et al (2026) Nuclear speckles enable processing of RNA from GC-rich isochores. Cell 189:2024–2039.e2025. https://doi.org/10.1016/j.cell.2026.01.011
70. Bhat P et al (2024) Genome organization around nuclear speckles drives mRNA splicing efficiency. Nature 629:1165–1173. https://doi.org/10.1038/s41586-024-07429-6
71. Moncan M et al (2021) Regulation of lipid metabolism by the unfolded protein response. J Cell Mol Med 25: 1359–1370. https://doi.org/10.1111/jcmm.16255
72. Yamamoto K et al (2010) Induction of liver steatosis and lipid droplet formation in ATF6alpha-knockout mice burdened with pharmacological endoplasmic reticulum stress. Mol Biol Cell 21:2975–2986. https://doi.org/10.1091/mbc.E09-02-0133
73. Rutkowski DT et al (2008) UPR pathways combine to prevent hepatic steatosis caused by ER stress-mediated suppression of transcriptional master regulators. Dev Cell 15:829–840. https://doi.org/10.1016/j.devcel.2008.10.015
74. Zhang K et al (2011) The unfolded protein response transducer IRE1alpha prevents ER stress-induced hepatic steatosis. EMBO J 30:1357–1375. https://doi.org/10.1038/emboj.2011.52
75. Herrema H et al (2016) XBP1s is an anti-lipogenic protein. J Biol Chem 291:17394–17404. https://doi.org/10.1074/jbc.M116.728949
76. Rong X et al (2013) LXRs regulate ER stress and inflammation through dynamic modulation of membrane phospholipid composition. Cell Metab 18:685–697. https://doi.org/10.1016/j.cmet.2013.10.002
77. Wilcockson D, Zhang L (2008) Circatidal clocks. Curr Biol 18:R753–R755. https://doi.org/10.1016/j.cub.2008.06.041
78. Andreatta G, Tessmar-Raible K (2020) The still dark side of the moon: molecular mechanisms of lunar-controlled rhythms and clocks. J Mol Biol 432:3525–3546. https://doi.org/10.1016/j.jmb.2020.03.009
79. Barford E (2013) Biological clocks defy circadian rhythms. Nature. https://doi.org/10.1038/nature.2013.13833
80. Zhang L et al (2013) Dissociation of circadian and circatidal timekeeping in the marine crustacean Eurydice pulchra. Curr Biol 23:1863–1873. https://doi.org/10.1016/j.cub.2013.08.038
81. Takekata H, Matsuura Y, Goto SG, Satoh A, Numata H (2012) RNAi of the circadian clock gene period disrupts the circadian rhythm but not the circatidal rhythm in the mangrove cricket. Biol Lett 8:488–491. https://doi.org/10.1098/rsbl.2012.0079
82. Takekata H, Numata H, Shiga S, Goto SG (2014) Silencing the circadian clock gene clock using RNAi reveals dissociation of the circatidal clock from the circadian clock in the mangrove cricket. J Insect Physiol 68:16–22. https://doi.org/10.1016/j.jinsphys.2014.06.012
83. Takekata H, Numata H, Shiga S (2014) The circatidal rhythm persists without the optic lobe in the mangrove cricket Apteronemobius asahinai. J Biol Rhythm 29: 28–37. https://doi.org/10.1177/0748730413516309
84. Kwiatkowski ER, Schnytzer Y, Rosenthal JJC, Emery P (2023) Behavioral circatidal rhythms require Bmal1 in Parhyale hawaiensis. Curr Biol 33:1867–1882.e1865. https://doi.org/10.1016/j.cub.2023.03.015
85. O'Neill JS et al (2015) Metabolic molecular markers of the tidal clock in the marine crustacean Eurydice pulchra. Curr Biol 25:R326–R327. https://doi.org/10.1016/j.cub.2015.02.052
86. Sorek M et al (2018) Setting the pace: host rhythmic behaviour and gene expression patterns in the facultatively symbiotic cnidarian Aiptasia are determined largely by Symbiodinium. Microbiome 6:83. https://doi.org/10.1186/s40168-018-0465-9
87. Satoh A, Terai Y (2019) Circatidal gene expression in the mangrove cricket Apteronemobius asahinai. Sci Rep 9:3719. https://doi.org/10.1038/s41598-019-40197-2
88. Zhang H et al (2024) Circatidal control of gene expression in the deep-sea hot vent shrimp *Shinkaicaris leurokolos*. bioRxiv:2024.2001.2012.575359. https://doi.org/10.1101/2024.01.12.575359
89. Genov N et al (2019) A multi-layered study on harmonic oscillations in mammalian genomics and proteomics. Int J Mol Sci 20. https://doi.org/10.3390/ijms20184585
90. Castellana S et al (2018) Systematic analysis of mouse genome reveals distinct evolutionary and functional properties among circadian and ultradian genes. Front Physiol 9:1178. https://doi.org/10.3389/fphys.2018.01178
91. Deng Y et al (2013) The Xbp1s/GalE axis links ER stress to postprandial hepatic metabolism. J Clin Invest 123:455–468. https://doi.org/10.1172/JCI62819
92. Shao M et al (2014) Hepatic IRE1alpha regulates fasting-induced metabolic adaptive programs through the XBP1s-PPARalpha axis signalling. Nat Commun 5:3528. https://doi.org/10.1038/ncomms4528
93. Lipson KL et al (2006) Regulation of insulin biosynthesis in pancreatic beta cells by an endoplasmic reticulum-resident protein kinase IRE1. Cell Metab 4: 245–254. https://doi.org/10.1016/j.cmet.2006.07.007
94. Zhu B (2022) Logic of the temporal compartmentalization of the hepatic metabolic cycle. Physiology 37. https://doi.org/10.1152/physiol.00003.2022

Autonomous Oscillations of NF-κB

6

Shah Md Toufiqur Rahman and Myong-Hee Sung

Abstract

The nuclear factor κB (NF-κB) signaling pathway plays a crucial role in regulating immune responses, inflammation, and cell survival. After the discovery of NF-κB by Ranjan Sen and David Baltimore as a nuclear factor that binds near the κ light-chain gene in activated B-cells [1], extensive research has been conducted to uncover key biochemical mechanisms underlying the pathway and the downstream genes regulated by NF-κB. The core regulatory module of NF-κB/IκB (inhibitor of NF-κB)/IKK (IκB kinase) has also been a leading example of cell signaling pathways that can produce oscillatory activities, particularly in response to the pro-inflammatory cytokine tumor necrosis factor α (TNF-α). Despite this status, the existence and function of NF-κB oscillations was a contentious topic in the early years. In this chapter, we touch on the history, progression, and outlook of the quest to decipher the properties and biological function of NF-κB oscillations.

Keywords

Signaling dynamics · NF-kappaB · Single-cell biology · Live-cell microscopy · Quantitative systems biology · Feedback control · Oscillations · Asynchronous cell behavior · Inflammatory signaling · Immune regulation

S. Md Toufiqur Rahman · M.-H. Sung (✉)
Laboratory of Molecular Biology and Immunology, National Institute on Aging, National Institutes of Health, Baltimore, MD, USA
e-mail: sungm@nih.gov

6.1 How It All Began: Path to the Discovery of Autonomous NF-κB Oscillations

The potential for oscillation of NF-κB stems from the delayed feedback structure of the core NF-κB/IκB/IKK module, where the IKK-initiated degradation of IκBs releases NF-κB from the IκB-bound cytoplasmic complex such that it accumulates in the nucleus and transcriptionally activates negative feedback genes, most notably IκBα (Fig. 6.1a). Newly synthesized IκB proteins inhibit NF-κB by binding and restoring the steady state localization to the cytoplasm, but persistent IKK activity may re-initiate another cycle of nuclear translocation of NF-κB.

Ironically, the initial study by Hoffmann et al. that formally considered the potential for oscillations suggested that the multiple inhibitory proteins (IκBα, IκBβ, IκBε) collectively antagonize the propensity for NF-κB to oscillate, as if uncontrolled NF-κB oscillations might be functionally detrimental [2]. Nevertheless, this work by Alexander Hoffmann and colleagues was the first interdisciplinary exposition combining mathematical modeling, mouse genetics, and bulk biochemical methods, which motivated theoreticians

M. G. Aydogan (ed.), *Biological Time Control*, Advances in Experimental Medicine and Biology 1521,
https://doi.org/10.1007/978-3-032-36189-9_6

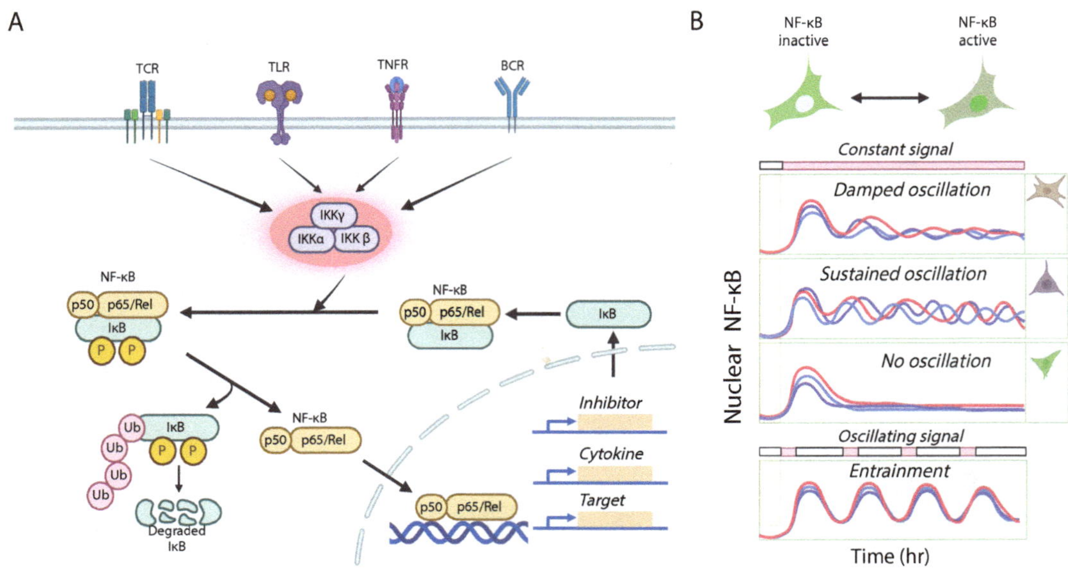

Fig. 6.1 NF-κB signaling: the canonical signaling network and oscillatory dynamics of nuclear localization. (**a**) Canonical NF-κB signaling pathway in most cell types has the core module consisting of NF-κB, IκBs, and the IKK complex. External signals, such as Toll-like receptor (TLR) agonists, T- or B-cell receptor (TCR/BCR) agonists, or TNF-α, activate the kinase activity of IKK complex, which initiates the phosphorylation/ubiquitination-mediated degradation of IκBs by the proteasome, and nuclear accumulation of (the constantly shuttling) NF-κB dimers. NF-κB binds to cognate DNA motifs in the epigenome and activates the transcription of numerous target genes. One of the immediate early genes, IκBα, is a rapid but delayed negative feedback loop that binds to NF-κB and restores the cytoplasmic localization, but sustained IKK activity can repeat the next cycle of nuclear translocation of NF-κB. (**b**) In the presence of a constant stimulating signal, NF-κB nuclear translocation can exhibit different oscillatory dynamics, including damped, sustained, or non-oscillatory behavior, depending on the stimulus and cell type. Oscillatory signaling dynamics of NF-κB are asynchronous among individual cells and can become synchronized with the stimulating signal if the frequency of the external signal allows time intervals that exceed the refractory period of the NF-κB signaling system. The three-color lines represent different single-cell responses under the same stimulating condition

and microscopists to examine the NF-κB system at the single-cell level.

In a pioneering microscopy study of single live cells expressing fluorescently tagged RelA/p65 and IκBα, Michael White and colleagues showed that individual cells can have oscillations of NF-κB with various dampness depending on the cell type (neuroblastoma, HeLa) [3]. Later came convincing evidence that TNF-α does induce sustained NF-κB oscillations and that this is not an artifact of overexpressing a fluorescent fusion protein of RelA (a canonical NF-κB subunit, also known as p65), from imaging the endogenous RelA in knock-in mouse embryonic fibroblasts (MEFs) [4]. This helped settle the debate and established that NF-κB oscillation is a physiological property of the molecular network. With this clarification about NF-κB signaling as a naturally occurring oscillator, numerous studies have ensued to explore the dynamic properties of NF-κB signaling and the cell contexts where oscillations are observed (Fig. 6.1b). Much attention has also been given to the biological function of oscillatory signaling, but this remains an enigmatic topic and will be discussed further in the last section. While it is now known that other stimuli or cell contexts do not lead to oscillations in NF-κB signaling [5–7], this chapter will not discuss these features but will focus on properties of NF-κB as an autonomous biochemical oscillator.

6.2 Cell Type Specificity of NF-κB Oscillations

Both experimental results and simulations of computational models showed that IκBα is key in driving the autonomous oscillations of NF-κB by the activation-induced inhibition of NF-κB, which provides a strong negative feedback loop with delay. The other inhibitory proteins (IκBβ and IκBε) were found to reduce the system's oscillatory potential in genetic knockout experiments [8]. Mechanistically, IκBε has a delay in the transcriptional induction relative to IκBα, which causes two negative feedback loops to be out-of-phase, thereby promoting the dampening of the oscillatory dynamics of NF-κB at the population level [9, 10]. Mathematical models have demonstrated that different NF-κB dynamic behaviors can emerge by adjusting key parameters [4, 7, 9, 11–13]. For example, different cell types may have distinct composition of basal upstream factors or wiring of regulators, thus exhibiting different patterns of response dynamics to the same stimulus, as discussed below.

Both the neuroblastoma cells (SK-N-AS) used in the first live-cell imaging experiment and the GFP-p65 knock-in MEFs tracking the endogenous RelA produced several cycles of nuclear NF-κB occupancy in sustained oscillations in response to TNF-α stimulation, with a period of 100–130 min [3, 4]. However, in HeLa cells, under the same conditions, the oscillations were much less pronounced [3, 14, 15]. Another study compared different cancer cell lines and found the dynamics of NF-κB signaling to vary across the cell lines stimulated under the same conditions [16]. Interestingly, the oscillatory dynamics of NF-κB induced by TNF-α are distinct between primary cells and immortalized cells derived from the same tissue [17]. Primary cells exhibited slower and weaker NF-κB responses compared to immortalized cells after TNF-α stimulation.

6.3 Stimulus Specificity of NF-κB Oscillations

Interestingly, NF-κB oscillations have so far been found only for a limited number of stimuli, although numerous ligands are known to activate NF-κB. While most data on NF-κB oscillations are from TNF-α as the stimulus, a few studies have reported that oscillations may also arise from IL-1β. On the other hand, stimulation with LPS, a TLR4 ligand that elicits a downstream signaling cascade partly shared with IL-1, has been shown to induce non-oscillatory NF-κB dynamics in mouse fibroblasts [18, 19], mouse macrophage-like RAW264.7 cells [20], and primary bone marrow-derived macrophages (BMDMs) [17, 21], with various response durations depending on the cell type. Moreover, the signaling dynamics are remarkably ligand-specific, to the extent that distinct patterns of NF-κB response dynamics can be used to distinguish between different TLR ligands in BMDMs [7]. Among TLR ligands, R848, a ligand of the endosomal TLR7, produces oscillatory dynamics of NF-κB in macrophages with some cell-to-cell variability [6, 7].

NF-κB signaling also exhibits distinct dynamic patterns at different ligand concentrations. For example, independent studies have shown that as the TNF-α dose increases, a larger proportion of cells respond [22, 23]. Mathematical models incorporating stochasticity in receptor activation and negative feedback mechanisms have successfully recapitulated the heterogeneous behavior [22]. Using live imaging with microfluidic controls, Savas Tay's group demonstrated that a prolonged, weak stimulation activates fewer cells compared to a short, strong pulse [24], supporting the concept of an activation threshold influenced by cellular noise. Robin Lee and colleagues showed that this activation threshold varies among individual cells within the population [25].

6.4 Cell-to-Cell Variability of NF-κB Oscillations

Even in homogeneous cell populations, single-cell NF-κB signaling dynamics can show significant cell-to-cell heterogeneity in response to stimuli, much like other signaling processes. This variability can result from non-genetic differences, such as stochastic fluctuations in signaling reactions (intrinsic noise) or subtle differences in cell states [26–28]. NF-κB single-cell responses are regulated by various signaling modules, such as the dual-delayed negative feedback motif [2, 9], temporal patterns of IKK activation [29, 30], and a multipoint negative feedback mechanism mediated by A20 and IκBα [31]. Variations in any of these modules can lead to cell-to-cell variability. Heterogeneity in TNFR1 receptor expression also contributes to the variability in TNF-α-induced NF-κB activation among individual cells [32]. Single-cell variability was also observed regarding the refractory state of the cells when stimulated with TNF-α pulses at intervals shorter than the natural oscillation period of 100 min [33].

Despite this heterogeneity, certain dynamic features of NF-κB signaling remain robust. For instance, a study using live-cell imaging and mathematical modeling found that the period of NF-κB oscillations remains consistent across a population, even under different ligands and doses, although it may vary between individual cells [34]. Overall, while NF-κB responses within a cell population are heterogeneous, this variability follows predictable patterns, providing valuable insights into how each population collectively responds to inflammatory signals in a stereotyped manner.

6.5 Entrainment of NF-κB Oscillations and Impact on Gene Regulation

Advances in microfluidic devices have allowed the investigation of NF-κB oscillations in more complex stimulation regimes. For example, periodic cytokine stimulations under microfluidic control are hypothesized to mimic some secretion patterns of individual cytokine-producing cells within tissues in vivo [24, 35]. If these stimulus pulses are given at certain frequencies, NF-κB oscillations can be entrained to show synchronized cycles of nuclear localization. This phenomenon has been demonstrated with 3T3 mouse fibroblast cells, where oscillations were synchronized in response to TNF-α pulses [24]. Synchronized oscillations, entrained by such an external periodic forcing, produced a higher transcriptional output of immediate early genes than naturally occurring asynchronous damped oscillations [24]. Since transcription is typically measured by bulk assays, an increased total output from synchronized NF-κB activity can be generally explained by the reduced variability in the timing of transcriptional bursts among single cells.

Extensive data have suggested that oscillatory NF-κB signaling dynamics are essential for determining the timing and specificity of gene expression, allowing cells to mount appropriate transcriptional responses to external signals [2–4, 6, 36–39]. Through single-cell live imaging and endpoint RNA-seq analysis of the same imaged cells, Markus Covert and colleagues [40] explored the impact of natural cell-to-cell heterogeneity in NF-κB dynamics on gene expression using RAW 264.7 cells stimulated with LPS. They identified multiple distinct cytokine expression patterns highly correlated with specific NF-κB signaling dynamics, establishing a functional role for NF-κB dynamics in shaping cellular phenotypes at the single-cell resolution.

6.6 Other Environmental Factors

Little data are available regarding whether NF-κB signaling dynamics may vary under different physiological tissue conditions such as a febrile temperature, mechanical stress, or hypoxia. NF-κB dynamics are influenced by temperature within the mammalian physiological range (34–41 °C), although most live-cell imaging data are from 37 °C. Harper and colleagues [41] showed that a higher temperature leads to a higher frequency of NF-κB oscillations in TNF-α stimulated cells. In live-cell imaging of both cell lines and primary fibroblasts, the period of oscillation was shortened down to 60 min at 40 °C. The simulations of their mathematical model suggested that this temperature sensitivity cannot be explained by the accelerated biochemical reactions expected from higher temperatures and is mediated, at least in part, by the A20 negative feedback gene. The effect of temperature on oscillation frequency was confirmed by Heltberg and co-workers, where a controlled microfluidic setup and a series of temperature manipulations showed that an oscillation in temperature itself can entrain NF-κB oscillations [42]. Heat shock experiments using non-physiological temperatures such as 43 °C, relevant for heat therapy, revealed a crosstalk with heat shock factor (HSF1) pathway, where the recovery from exposure to heat shock rendered cells differentially sensitive to subsequent inflammatory challenges to TNF-α versus IL-1β [43].

6.7 Technical Caveats in the Study of NF-κB Oscillations

While our understanding of NF-κB oscillatory dynamics has greatly benefited from live monitoring of single cells expressing fluorescent fusion proteins, concerns over artifacts arise when the fluorescent proteins are overexpressed [3]. Hence, it has become desirable to adopt more refined cell systems for imaging, such as lentiviral clonal cell lines, stable RelA fluorescent fusion expression in RelA knockout 3T3 cells (to remove endogenous expression), and primary cells from transgenic mouse models. Knock-in mouse lines have also been developed to monitor endogenous NF-κB dynamics [4, 17, 44, 45], which avoid potential distortions seen in cell lines or transgenic animals. Still, it is unclear how much of the existing knowledge from transformed cell lines would be verified with primary cells of the endogenous knock-in reporter mice, since a comprehensive comparison across these experimental systems remains lacking.

Another technical caveat from a common practice during live-cell imaging involves the method to label nuclei, either with DNA dyes such as Hoechst or expressing fluorescently tagged nuclear proteins like histones, for nuclear segmentation. Both methods may disrupt signaling dynamics—imaging of Hoechst requires cell illumination at a phototoxic wavelength, which can compromise cell health and affect native signaling, and the chemical dye is also known to intercalate DNA and alter the chromatin dynamics [46]. Fluorescently labeled histones may also modify the chromatin structure, potentially distorting enhancer activity and gene expression. A promising yet challenging solution to this issue is the use of brightfield imaging combined with deep learning techniques to segment nuclear areas without the drawbacks of dyes or fluorescent proteins.

6.8 Cross-Talks with Master Temporal Programs: Interactions with the Circadian Clock, the Cell Cycle and Beyond

Various biological oscillators across different timescales act as master temporal programs to mediate essential cell physiological processes. This raises the possibility that the otherwise autonomous NF-κB oscillations may have co-evolved to interface and interact functionally with other biological oscillators for concordant timekeeping. Here, we discuss such inter-oscillator crosstalk mechanisms for a glimpse of the complex regulatory networks that maintain homeostasis.

The circadian clock, which produces a 24-h cycle regulating processes such as sleep-wake, hormone release, and metabolism, consists of a network of genes and proteins that generate oscillations in their expression [47–50]. Genetic and pharmacological approaches have demonstrated that the canonical NF-κB subunit, RelA, alters core clock gene expression in human U2OS cells [51]. Activation of RelA shortened the circadian period and reduced the amplitude. RelA competes with the clock repressor CRY1 protein for binding to the E-boxes of clock genes, thereby modifying the circadian rhythm. NF-κB oscillations can also synchronize with circadian cycles, leading to variations in NF-κB target gene expression at different times of the day [52, 53]. In addition, the CLOCK protein has been reported to enhance the transcriptional activity of NF-κB [54]. These data suggest an intricate cross-regulation between circadian oscillations and NF-κB (Fig. 6.2a). Hence, disruptions in the circadian rhythm may impact NF-κB signaling,

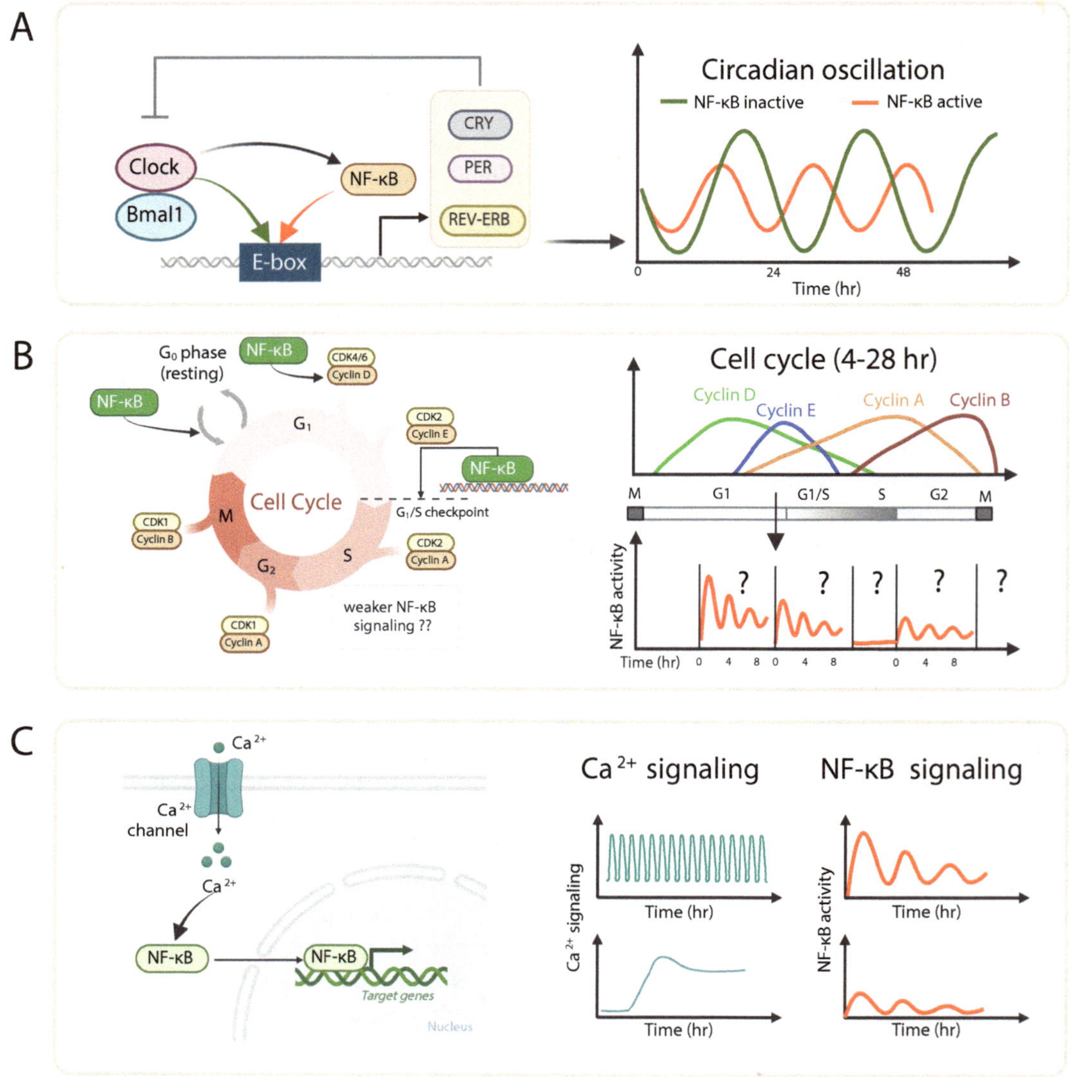

Fig. 6.2 Interactions with other biological oscillators. (**a**) Schematic of the core circadian network and its interactions with NF-κB. The circadian clock is regulated by Clock and Bmal1. The activated Clock heterodimer binds to E-boxes in the promoters of target genes, controlling the expression of hundreds of genes, including PER, CRY, and REV-ERB, which, in turn, negatively regulate the formation of the Clock heterodimer complex. The RelA subunit of NF-κB competes with the clock repressor protein CRY1 for binding to the E-boxes of clock genes, thereby influencing the circadian oscillation. The green and red lines indicate circadian oscillations associated with inactive and active NF-κB states, respectively. (**b**) Schematic of the cell cycle and crosstalk with NF-κB. Most non-dividing mammalian cells enter a quiescent state (G0) after exiting the G1 phase of the cell cycle. The diagram illustrates the cyclin-CDK complexes involved at various stages of the cell cycle and their interactions with NF-κB. NF-κB plays a key role in cell cycle regulation, particularly at the G0/G1 and G1/S phase transitions. Less is known about the role of cell cycle on NF-κB signaling. In HeLa reporter cells, NF-κB signaling was weaker during the S phase compared to G1, G1/S, and G2 phases. However, more data from physiological reporter systems are needed to clarify how NF-κB signaling might differ between cell cycle stages. (**c**) Simplified schematic of Ca^{2+} ion signaling and its interaction with NF-κB signaling. Oscillatory dynamics of Ca^{2+} signaling have been reported to activate NF-κB signaling more effectively than sustained Ca^{2+} signaling.

and vice versa, contributing to the development of diseases such as inflammation, autoimmune disorders, obesity, and cancer [55, 56]. Aligning immune responses with the body's circadian clock may optimize the defense against infections while minimizing tissue damage [57].

The cell cycle is another major temporal program characterized by distinct phases (G0, G1, S, G2, and M) that ensure error-free cycles of DNA replication and segregation. These phases are regulated by the oscillatory and phase-specific expression and activity of cyclins and cyclin-dependent kinases (CDKs) [58–61]. NF-κB plays a role in cell cycle regulation, particularly in the G0/G1 and G1/S phase transitions (Fig. 6.2b). The first link between NF-κB and the cell cycle was the discovery that NF-κB DNA-binding activity increases during the G1/S transition in mouse fibroblasts [62]. NF-κB activates the transcription of cyclin D1, a key regulator of G1 checkpoint [63]. In other contexts such as DNA damage responses, NF-κB may induce cell cycle arrest, where activated NF-κB regulates the expression of genes like p21, a cyclin-dependent kinase inhibitor that halts cell cycle progression [64–66]. Overexpression of RelA or c-Rel was found to halt the G1/S transition in pro-B cells and HeLa cells [67, 68]. Furthermore, experiments and computer simulations by Michael White's lab [69] demonstrated that the canonical NF-κB signaling could be reduced if the timing of inflammatory stimuli falls close to the S phase of the cell cycle (Fig. 6.2b). The interaction between cell cycle regulators E2F-1/E2F-4 and NF-κB at different cell cycle checkpoints enables cells to either commit to divide or respond to inflammatory signals in various ways. The interplay between NF-κB signaling and the cell cycle seems crucial for maintaining cellular homeostasis and preventing uncontrolled cell growth, which can lead to cancer.

Calcium ions (Ca^{2+}) serve as vital second messengers in various cellular processes, such as muscle contraction, neurotransmitter release, and gene expression [70]. NF-κB activity is dynamically regulated by the frequency and cumulative duration of Ca^{2+} oscillations, through Ca^{2+}-induced calcineurin signaling [71] (Fig. 6.2c). In Jurkat cells, NF-κB activity was correlated with Ca^{2+} oscillation frequencies ranging from 0.56 to 10 mHz [72]. Similar correlations were observed in various cell types, including human aortic endothelial cells and porcine bronchial epithelial cells [73]. Oscillatory Ca^{2+} signals were more effective at activating NF-κB than sustained elevations of Ca^{2+}, with NF-κB capable of distinguishing between different amplitudes and durations of Ca^{2+} transients [74] (Fig. 6.2c). In lymphocytes, Ca^{2+} entry is required for efficient NF-κB activation following TCR and BCR engagement [75], with c-Rel expression strongly regulated by Ca^{2+}-dependent mechanisms [76, 77]. Thus, dysregulation of these Ca^{2+} and NF-κB interactions may contribute to immune disorders and inflammation.

The interplay between NF-κB oscillations and other biological oscillators highlights the complexity of cellular signaling networks. Better understanding of these interactions may pave the way for novel therapeutic strategies aimed at restoring the balance in the interconnected oscillators that ultimately improve health outcomes.

6.9 Challenges and Remaining Questions

Although two decades of intense quantitative investigations about NF-κB dynamics have uncovered fascinating insights, many important questions are still left unanswered. Given that there is seemingly a limited number of biological situations (certain stimuli and cell types, etc.) that can produce sustained oscillations in NF-κB signaling, what are the relevant features of oscillations that have provided a selection advantage during evolution? Is it the temporal pattern of amplitudes that are reached in different cycles? Is it the temporally distinct time windows of NF-κB interactions with the cognate chromatin? Is it the need for an "off" period between consecutive encounters with chromatin between cycles? Does the asynchrony of oscillations between individual cells matter for functionality?

One robust feature of NF-κB oscillations, which has been remarkably reproducible across different cellular contexts and laboratories, is their period. With appropriate stimuli and cell types that produce these oscillations, the period has consistently been 90–100 min at 37 °C. This contrasts with other features such as how the oscillations dampen, which vary widely among different contexts. It is possible that experimental conditions tested so far may not have exhausted all endogenous factors that could modify the period. If so, the identification of such a perturbation might reveal an important insight into the function of the period of oscillations. It is conceptually plausible that, if the interval between two consecutive interactions of NF-κB with target genomic loci is lengthened, as would be if the period of oscillations is increased, then the chromatin resetting or remodeling may proceed to varying degrees at target sites and ultimately affect nearby gene regulation.

Although numerous studies have examined the impact of NF-κB signaling dynamics on gene regulation, we still do not have a solid understanding about this important topic, largely because of technical limitations. Combined analysis of NF-κB signaling and gene regulatory consequences is typically measured by bulk assays (such as RT-qPCR) and/or, more seriously, in endpoint assays (even in the highest resolution single-cell, single-molecule RNA FISH) [4, 15, 24, 36], which makes it impossible to link specific patterns of single-cell signaling dynamics with concomitant kinetics of target gene expression in the same cells. A recently developed technique, termed Live-seq, allowed live-cell imaging and simultaneous probing of near-transcriptomic profiling at a few selected time points during the time course of LPS-induced responses in a macrophage reporter cell line [78]. Although the methodology involves specialized instrumentation and cell viability issues from the extraction of cytoplasmic content, it is an important step toward a non-invasive simultaneous monitoring of both NF-κB signaling and transcriptional kinetics in individual cells.

In conclusion, the autonomous oscillations of NF-κB, driven by the regulation of negative feedback proteins (IκBs and A20), is a critical feature that shapes gene expression and cellular responses to stimuli. The dynamics of these oscillations—whether in their amplitude, periodicity, damping, or synchronization—are finely tuned by factors specific to cell types and environmental conditions. However, the available data so far are mostly derived from a few cell types and a limited set of conditions. New regimes of oscillatory dynamics may arise in unexplored cell types such as primary immune cell populations or in untested conditions such as hypoxia, hypothermia, mechanical cues, or cellular senescence. Understanding the determinants of oscillatory patterns in NF-κB signaling in a variety of pathophysiological conditions will open new avenues for therapeutic interventions in diseases affected by dysregulated immune responses, including cancer, chronic inflammation, and autoimmune disorders.

References

1. Sen R, Baltimore D (1986) Multiple nuclear factors interact with the immunoglobulin enhancer sequences. Cell 46(5):705–716
2. Hoffmann A, Levchenko A, Scott ML, Baltimore D (2002) The IkappaB-NF-kappaB signaling module: temporal control and selective gene activation. Science 298(5596):1241–1245
3. Nelson DE, Ihekwaba AE, Elliott M, Johnson JR, Gibney CA, Foreman BE et al (2004) Oscillations in NF-kappaB signaling control the dynamics of gene expression. Science 306(5696):704–708
4. Sung MH, Salvatore L, De Lorenzi R, Indrawan A, Pasparakis M, Hager GL et al (2009) Sustained oscillations of NF-κB produce distinct genome scanning and gene expression profiles. PLoS One 4(9): e7163
5. Adelaja A, Taylor B, Sheu KM, Liu Y, Luecke S, Hoffmann A (2021) Six distinct NFkappaB signaling codons convey discrete information to distinguish stimuli and enable appropriate macrophage responses. Immunity 54(5):916–30.e7
6. Martin EW, Pacholewska A, Patel H, Dashora H, Sung MH (2020) Integrative analysis suggests cell type-specific decoding of NF-kappaB dynamics. Sci Signal 13(620):eaax7195
7. Rahman SMT, Singh A, Lowe S, Aqdas M, Jiang K, Vaidehi Narayanan H et al (2024) Co-imaging of RelA and c-Rel reveals features of NF-κB signaling for ligand discrimination. Cell Rep 43(3):113940

8. Hoffmann A, Levchenko A, Scott ML, Baltimore D (2002) The IκB-NF-κB signaling module: temporal control and selective gene activation. Science 298(5596):1241–1245
9. Kearns JD, Basak S, Werner SL, Huang CS, Hoffmann A (2006) IκBε provides negative feedback to control NF-κB oscillations, signaling dynamics, and inflammatory gene expression. J Cell Biol 173(5):659–664
10. Paszek P, Ryan S, Ashall L, Sillitoe K, Harper CV, Spiller DG et al (2010) Population robustness arising from cellular heterogeneity. Proc Natl Acad Sci USA 107(25):11644–11649
11. Mothes J, Busse D, Kofahl B, Wolf J (2015) Sources of dynamic variability in NF-κB signal transduction: a mechanistic model. BioEssays 37(4):452–462
12. Gaudet S, Zambrano S, De Toma I, Piffer A, Bianchi ME, Agresti A (2015) NF-kB oscillations translate into functionally related patterns of gene expression. elife 5:e09100
13. Krishna S, Jensen MH, Sneppen K (2006) Minimal model of spiky oscillations in NF-κB signaling. Proc Natl Acad Sci USA 103(29):10840–10845
14. Gaudet S, Ankers JM, Awais R, Jones NA, Boyd J, Ryan S et al (2016) Dynamic NF-kB and E2F interactions control the priority and timing of inflammatory signalling and cell proliferation. elife 5:e10473
15. Lee REC, Walker SR, Savery K, Frank DA, Gaudet S (2014) Fold change of nuclear NF-κB determines TNF-induced transcription in single cells. Mol Cell 53(6):867–879
16. Bartfeld S, Hess S, Bauer B, Machuy N, Ogilvie LA, Schuchhardt J et al (2010) High-throughput and single-cell imaging of NF-κB oscillations using monoclonal cell lines. BMC Cell Biol 11:21
17. Rahman SMT, Aqdas M, Martin EW, Ardori FT, Songkiatisak P, Oh K-S et al (2022) Double knockin mice show NF-κB trajectories in immune signaling and aging. Cell Rep 41(8):111682
18. Gutschow MV, Mason JC, Lane KM, Maayan I, Hughey JJ, Bajar BT et al (2019) Combinatorial processing of bacterial and host-derived innate immune stimuli at the single-cell level. Mol Biol Cell 30(2):282–292
19. Lee TK, Denny EM, Sanghvi JC, Gaston JE, Maynard ND, Hughey JJ et al (2009) A noisy paracrine signal determines the cellular NF-kappaB response to lipopolysaccharide. Sci Signal 2(93):ra65
20. Sung M-H, Li N, Lao Q, Gottschalk RA, Hager GL, Fraser ID (2014) Switching of the relative dominance between feedback mechanisms in lipopolysaccharide-induced NF-κB signaling. Sci Signal 7(308):ra6
21. Cheng QJ, Ohta S, Sheu KM, Spreafico R, Adelaja A, Taylor B et al (2021) NF-κB dynamics determine the stimulus specificity of epigenomic reprogramming in macrophages. Science 372(6548):1349–1353
22. Tay S, Hughey JJ, Lee TK, Lipniacki T, Quake SR, Covert MW (2010) Single-cell NF-κB dynamics reveal digital activation and analogue information processing. Nature 466(7303):267–271
23. Turner DA, Paszek P, Woodcock DJ, Nelson DE, Horton CA, Wang Y et al (2010) Physiological levels of TNFα stimulation induce stochastic dynamics of NF-κB responses in single living cells. J Cell Sci 123(16):2834–2843
24. Kellogg RA, Tay S (2015) Noise facilitates transcriptional control under dynamic inputs. Cell 160(3): 381–392
25. Zhang Q, Gupta S, Schipper DL, Kowalczyk GJ, Mancini AE, Faeder JR et al (2017) NF-κB dynamics discriminate between TNF doses in single cells. Cell Syst 5(6):638–45.e5
26. Satija R, Shalek AK (2014) Heterogeneity in immune responses: from populations to single cells. Trends Immunol 35(5):219–229
27. Snijder B, Pelkmans L (2011) Origins of regulated cell-to-cell variability. Nat Rev Mol Cell Biol 12(2): 119–125
28. Spencer SL, Gaudet S, Albeck JG, Burke JM, Sorger PK (2009) Non-genetic origins of cell-to-cell variability in TRAIL-induced apoptosis. Nature 459(7245):428–432
29. Cruz JA, Mokashi CS, Kowalczyk GJ, Guo Y, Zhang Q, Gupta S et al (2021) A variable-gain stochastic pooling motif mediates information transfer from receptor assemblies into NF-κB. Sci Adv 7(30): eabi9410
30. Werner SL, Barken D, Hoffmann A (2005) Stimulus specificity of gene expression programs determined by temporal control of IKK activity. Science 309(5742): 1857–1861
31. Werner SL, Kearns JD, Zadorozhnaya V, Lynch C, O'Dea E, Boldin MP et al (2008) Encoding NF-κB temporal control in response to TNF: distinct roles for the negative regulators IκBα and A20. Genes Dev 22(15):2093–2101
32. Cheng CT, Hsiao JC, Hoffmann A, Tu HL (2024) TNFR1 mediates heterogeneity in single-cell NF-κB activation. iScience 27(4):109486
33. Adamson A, Boddington C, Downton P, Rowe W, Bagnall J, Lam C et al (2016) Signal transduction controls heterogeneous NF-κB dynamics and target gene expression through cytokine-specific refractory states. Nat Commun 7:12057
34. Hughey JJ, Gutschow MV, Bajar BT, Covert MW (2015) Single-cell variation leads to population invariance in NF-kappaB signaling dynamics. Mol Biol Cell 26(3):583–590
35. Yde P, Mengel B, Jensen MH, Krishna S, Trusina A (2011) Modeling the NF-κB mediated inflammatory response predicts cytokine waves in tissue. BMC Syst Biol 5:1–9
36. Lane K, Van Valen D, DeFelice MM, Macklin DN, Kudo T, Jaimovich A et al (2017) Measuring signaling and RNA-seq in the same cell links gene expression to dynamic patterns of NF-κB activation. Cell Syst 4(4): 458–69.e5
37. Son M, Frank T, Holst-Hansen T, Wang AG, Junkin M, Kashaf SS et al (2022) Spatiotemporal

NF-κB dynamics encodes the position, amplitude, and duration of local immune inputs. Sci Adv 8:eabn6240
38. Wong VC, Mathew S, Ramji R, Gaudet S, Miller-Jensen K (2019) Fold-change detection of NF-κB at target genes with different transcript outputs. Biophys J 116(4):709–724
39. Zambrano S, De Toma I, Piffer A, Bianchi ME, Agresti A (2016) NF-kappaB oscillations translate into functionally related patterns of gene expression. elife 5:e09100
40. Lane K, Van Valen D, DeFelice M, Macklin D, Kudo T, Jaimovich A et al (2017) Measuring signaling and RNA-seq in the same cell links gene expression to dynamic patterns of NF-κB activation. Cell Syst 4(4): 458–469.e5
41. Harper CV, Woodcock DJ, Lam C, Garcia-Albornoz-M, Adamson A, Ashall L et al (2018) Temperature regulates NF-κB dynamics and function through timing of A20 transcription. Proc Natl Acad Sci USA 115(22):E5243–E52E9
42. Heltberg M, von Borries M, Bendix PM, Oddershede LB, Jensen MH (2022) Temperature controls onset and period of NF-κB oscillations and can lead to chaotic dynamics. Front Cell Dev Biol 10:910738
43. Paszek A, Kardyńska M, Bagnall J, Śmieja J, Spiller DG, Widłak P et al (2020) Heat shock response regulates stimulus-specificity and sensitivity of the pro-inflammatory NF-κB signalling. Cell Commun Signal 18(1):77
44. De Lorenzi R, Gareus R, Fengler S, Pasparakis M (2009) GFP-p65 knock-in mice as a tool to study NF-κB dynamics in vivo. Genesis 47(5):323–329
45. Adelaja A, Taylor B, Sheu KM, Liu Y, Luecke S, Hoffmann A (2021) Six distinct NFκB signaling codons convey discrete information to distinguish stimuli and enable appropriate macrophage responses. Immunity 54(5):916–30.e7
46. Ku H, Park G, Goo J, Lee J, Park TL, Shim H et al (2022) Effects of transcription-dependent physical perturbations on the chromosome dynamics in living cells. Front Cell Dev Biol 10:822026
47. Patke A, Young MW, Axelrod S (2020) Molecular mechanisms and physiological importance of circadian rhythms. Nat Rev Mol Cell Biol 21(2):67–84
48. Ko CH, Takahashi JS (2006) Molecular components of the mammalian circadian clock. Hum Mol Genet 15 (suppl_2):R271–R2R7
49. Buhr ED, Takahashi JS (2013) Molecular components of the mammalian circadian clock. In: Circadian Clocks. Springer, Heidelberg, pp 3–27
50. Fagiani F, Di Marino D, Romagnoli A, Travelli C, Voltan D, Di Cesare Mannelli L et al (2022) Molecular regulations of circadian rhythm and implications for physiology and diseases. Signal Transduct Target Ther 7(1):41
51. Shen Y, Endale M, Wang W, Morris AR, Francey LJ, Harold RL et al (2021) NF-κB modifies the mammalian circadian clock through interaction with the core clock protein BMAL1. PLoS Genet 17(11):e1009933
52. Cecon E, Fernandes PA, Pinato L, Ferreira ZS, Markus RP (2010) Daily variation of constitutively activated nuclear factor kappa B (NFKB) in rat pineal gland. Chronobiol Int 27(1):52–67
53. Tanenhaus AK, Zhang J, Yin JC (2012) In vivo circadian oscillation of dCREB2 and NF-κB activity in the Drosophila nervous system. PLoS One 7(10):e45130
54. Spengler ML, Kuropatwinski KK, Comas M, Gasparian AV, Fedtsova N, Gleiberman AS et al (2012) Core circadian protein CLOCK is a positive regulator of NF-kappaB-mediated transcription. Proc Natl Acad Sci USA 109(37):E2457–E2465
55. Comas M, Gordon CJ, Oliver BG, Stow NW, King G, Sharma P et al (2017) A circadian based inflammatory response–implications for respiratory disease and treatment. Sleep Sci Pract 1:1–19
56. Pekovic-Vaughan V, Gibbs J, Yoshitane H, Yang N, Pathiranage D, Guo B et al (2014) The circadian clock regulates rhythmic activation of the NRF2/glutathione-mediated antioxidant defense pathway to modulate pulmonary fibrosis. Genes Dev 28(6):548–560
57. Haspel JA, Anafi R, Brown MK, Cermakian N, Depner C, Desplats P et al (2020) Perfect timing: circadian rhythms, sleep, and immunity—an NIH workshop summary. JCI Insight 5(1):e131487
58. Coutts AS, Weston L (2016) Cell cycle oscillators. Springer
59. Haase SB, Reed SI (1999) Evidence that a free-running oscillator drives G1 events in the budding yeast cell cycle. Nature 401(6751):394–397
60. Pomerening JR, Kim SY, Ferrell JE (2005) Systems-level dissection of the cell-cycle oscillator: bypassing positive feedback produces damped oscillations. Cell 122(4):565–578
61. Orlando DA, Lin CY, Bernard A, Wang JY, Socolar JE, Iversen ES et al (2008) Global control of cell-cycle transcription by coupled CDK and network oscillators. Nature 453(7197):944–947
62. Baldwin AS Jr, Azizkhan JC, Jensen DE, Beg AA, Coodly LR (1991) Induction of NF-κB DNA-binding activity during the G0-to-G1 transition in mouse fibroblasts. Mol Cell Biol 11(10):4943–4951
63. Hinz M, Krappmann D, Eichten A, Heder A, Scheidereit C, Strauss M (1999) NF-κB function in growth control: regulation of cyclin D1 expression and G0/G1-to-S-phase transition. Mol Cell Biol 19:2690
64. Joyce D, Albanese C, Steer J, Fu M, Bouzahzah B, Pestell RG (2001) NF-κB and cell-cycle regulation: the cyclin connection. Cytokine Growth Factor Rev 12(1): 73–90
65. Basile JR, Eichten A, Zacny V, Munger K (2003) NF-kappaB-mediated induction of p21(Cip1/Waf1) by tumor necrosis factor alpha induces growth arrest and cytoprotection in normal human keratinocytes. Mol Cancer Res 1(4):262–270
66. Nicolae CM, O'Connor MJ, Constantin D, Moldovan GL (2018) NFkappaB regulates p21 expression and controls DNA damage-induced leukemic differentiation. Oncogene 37(27):3647–3656

67. Bash J, Zong W-X, Gélinas C (1997) c-Rel arrests the proliferation of HeLa cells and affects critical regulators of the G1/S-phase transition. Mol Cell Biol 17(11):6526–6536
68. Sheehy AM, Schlissel MS (1999) Overexpression of RelA causes G1 arrest and apoptosis in a pro-B cell line. J Biol Chem 274(13):8708–8716
69. Ankers JM, Awais R, Jones NA, Boyd J, Ryan S, Adamson AD et al (2016) Dynamic NF-κB and E2F interactions control the priority and timing of inflammatory signalling and cell proliferation. elife 5:e10473
70. Bootman MD (2012) Calcium signaling. Cold Spring Harb Perspect Biol 4(7):a011171
71. Frantz B, Nordby EC, Bren G, Steffan N, Paya CV, Kincaid RL et al (1994) Calcineurin acts in synergy with PMA to inactivate I kappa B/MAD3, an inhibitor of NF-kappa B. EMBO J 13(4):861–870
72. Dolmetsch RE, Xu K, Lewis RS (1998) Calcium oscillations increase the efficiency and specificity of gene expression. Nature 392(6679):933–936
73. Hu Q, Deshpande S, Irani K, Ziegelstein RC (1999) [Ca(2+)](i) oscillation frequency regulates agonist-stimulated NF-κB transcriptional activity. J Biol Chem 274(48):33995–33998
74. Dolmetsch RE, Lewis RS, Goodnow CC, Healy JI (1997) Differential activation of transcription factors induced by Ca2+ response amplitude and duration. Nature 386(6627):855–858
75. Liu X, Berry CT, Ruthel G, Madara JJ, MacGillivray K, Gray CM et al (2016) T cell receptor-induced nuclear factor κB (NF-κB) signaling and transcriptional activation are regulated by STIM1- and Orai1-mediated calcium entry. J Biol Chem 291(16):8440–8452
76. Nolz JC, Fernandez-Zapico ME, Billadeau DD (2007) TCR/CD28-stimulated actin dynamics are required for NFAT1-mediated transcription of c-rel leading to CD28 response element activation. J Immunol 179(2):1104–1112
77. Grumont R, Lock P, Mollinari M, Shannon FM, Moore A, Gerondakis S (2004) The mitogen-induced increase in T cell size involves PKC and NFAT activation of Rel/NF-κB-dependent c-myc expression. Immunity 21(1):19–30
78. Chen W, Guillaume-Gentil O, Rainer PY, Gabelein CG, Saelens W, Gardeux V et al (2022) Live-seq enables temporal transcriptomic recording of single cells. Nature 608(7924):733–740

Synthetic Cell-Based Approaches to Dissect Autonomous Cellular Oscillators: Recent Applications on the Cell Cycle

7

Yeonghoon Kim, Franco Tavella, and Qiong Yang

Abstract

Cell-free extracts have proven to be a tremendous tool for studying cell cycle regulation. This chapter reviews recent advances that expand upon this platform by incorporating synthetic cell-like compartments to encapsulate functional cell cycle regulators. These synthetic approaches allow for a programmable reconstitution of self-sustained autonomous oscillators, including that of mitosis, exhibiting a diverse range of biochemical and physical properties. The high throughput and manipulability of this system provide a comprehensive tool to systematically analyze the single-cell dynamics of this autonomous cell-cycle oscillator across broader yet finer-grained parametric, temporal, and spatial scales. We highlight key applications from the past few years and demonstrate how extending the well-established extract-based research with synthetic systems opens new avenues for tackling unexplored questions regarding cell cycle oscillations and their emergent phenomena.

Keywords

Cell-free extract systems · Water-in-oil microemulsions · Positive and negative feedback loops · Relaxation oscillator · Cyclin B-Cdk1 network · PP2A-Greatwall pathway · Bistable switch · High-throughput single-cell perturbation and analysis · Parameter space mapping · Saddle-node bifurcation · Mitotic waves · Spatiotemporal synchronization · Robustness and tunability · Temperature scaling · Energy flow

7.1 Introduction

The eukaryotic cell cycle is a fundamental cell-autonomous oscillatory process that governs cell proliferation [1]. To ensure accurate transmission of genetic information to daughter cells, cell cycle regulators temporally compartmentalize cellular functions into sequential, functionally distinct phases.

The periodicity of mitotic spindle assembly and chromosome segregation is primarily driven by two central regulators that form the core negative feedback loop of the cell cycle circuit (Fig. 7.1a): the cyclin B–cyclin-dependent kinase

Y. Kim · F. Tavella
Department of Biophysics, University of Michigan, Ann Arbor, MI, USA

Q. Yang (✉)
Department of Biophysics, University of Michigan, Ann Arbor, MI, USA

Department of Physics, University of Michigan, Ann Arbor, MI, USA
e-mail: qiongy@umich.edu

M. G. Aydogan (ed.), *Biological Time Control*, Advances in Experimental Medicine and Biology 1521,
https://doi.org/10.1007/978-3-032-36189-9_7

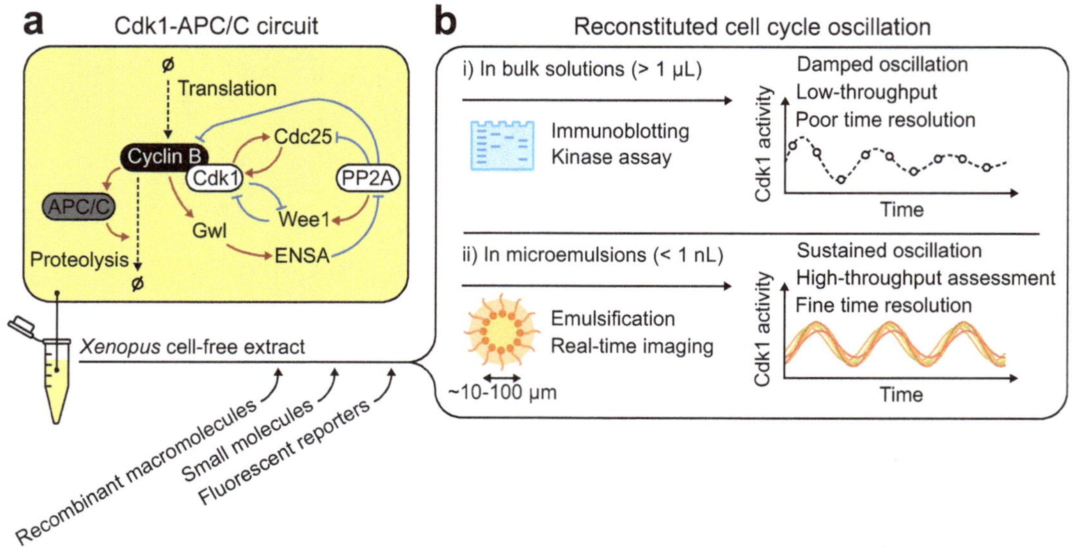

Fig. 7.1 (**a**) The Cdk1–APC/C cell cycle circuit and its dynamical behaviors can be reconstituted in cell-free environments using *Xenopus laevis* egg extracts. By modulating the functions of individual circuit constituents, cell cycle oscillations can be adapted to a diverse range of dynamical scenarios. The system's response to circuit perturbations can be monitored using reporters that are easily integrated into extracts. (**b**) Conventional biochemical assays using extracts prepared in bulk solutions often suffer from low throughput, poor time resolution, and damping. In contrast, encapsulating extracts in cell-sized synthetic compartments enables the production of large populations of independent, sustained oscillators, amenable to high-throughput and fine time-resolved measurements

1 (Cdk1) complex and the anaphase-promoting complex/cyclosome (APC/C). Cdk1, activated by binding to its regulatory subunit cyclin B, drives the transition from G2 to M phase by phosphorylating a myriad of mitotic substrates, including at least 200 direct substrates [2]. APC/C, a multi-subunit E3 ubiquitin ligase, activated through phosphorylation by active Cdk1 and binding of its coactivator Cdc20, mediates the ubiquitination of cyclins for their subsequent proteasomal degradation. Thus, accumulated Cdk1 activity eventually results in cyclin degradation followed by inactivation of Cdk1 itself. With a time delay of 15 min between Cdk1 activation and cyclin degradation [3], the negative feedback drives mitotic exit and a return to interphase, preparing the cell for the next round of cyclin synthesis.

Abrupt and hysteretic activation of Cdk1 at mitotic entry ensures that cellular functions during the G2 and M phases are qualitatively distinct and that progression through these phases is irreversible [4]. This switch-like behavior is partially regulated by the Cdk1-inhibitory kinase Wee1 and its antagonistic phosphatase Cdc25, both of which are phosphorylated by Cdk1. Their functions constitute double-negative feedback (Cdk1/Wee1) and double-positive feedback (Cdk1/Cdc25) that drive the bistable activation of cyclin B–Cdk1.

Opposing Cdk1's kinase activity is the protein phosphatase 2A (PP2A), which operates as a complementary off switch of mitosis by dephosphorylating many of Cdk1's targets [5]. Moreover, PP2A inhibits Cdk1 indirectly by suppressing its self-activation and, presumably, also directly by dephosphorylating Cdk1 at the Thr-161 site [6]. On the other hand, the Greatwall-Endosulfine (Gwl-ENSA) pathway is responsible for the regulation of PP2A by Cdk1 [7, 8]. In this pathway, cyclin B–Cdk1 activates Gwl kinase, leading to the phosphorylation of ENSA, a stoichiometric inhibitor of PP2A. The direct and indirect mutual inhibition between

Cdk1 and PP2A creates multiple feedback motifs that collectively contribute to the diversity of clock functions beyond periodicity, such as robustness and tunability [9, 10].

7.2 Cell-Free Extracts for Mechanistic Study of the Autonomous Mitotic Oscillator

The current understanding of the biochemistry behind cell cycle oscillations has been largely shaped by studies using cell-free extracts of activated eggs or early embryos. The early embryonic cell cycle lacks typical features found in mature cells, including gap phases, checkpoint controls, cell growth, and zygotic gene transcription [11]. Despite the functional minimalism, cell-free extracts retain the periodic transition between interphase and mitosis, with the Cdk1 regulatory network and its downstream regulations, such as chromosomal and nuclear dynamics, remaining qualitatively intact. The extract system thus has served as a tractable top-down framework for quantitatively interrogating the functional constituents of cell cycle regulation, providing previous genetic studies of the yeast cell cycle [12, 13] with richer biochemical contexts. Furthermore, experiments in egg and embryonic extracts laid the groundwork for predictive mathematical models of cell cycle control [14–16], contributing to a principled explanation for general autonomous oscillations in biology that have underlying mechanistic commonalities [17, 18].

One of the earliest uses of cell-free egg extracts in the study of cell cycle regulation was in 1983 by Lohka and Masui [19]. They used cytoplasmic extracts from *Rana pipiens* eggs with demembranated *Xenopus laevis* sperm chromatin for cell-free reconstitution of nuclear envelope formation and chromatin decondensation.

Over the following years, several variants of *Xenopus* cell-free extracts, including cycling, metaphase-arrested, and interphase-arrested extracts, were developed to examine the roles and functions of the key mitotic regulators. Extracts derived from artificially activated eggs (cycling extracts) demonstrated the ability to produce multiple autonomous cycles of chromatin condensation and decondensation [20]. Even after RNase treatments that stopped the cell cycle, adding exogenous cyclin B mRNA could restore the oscillatory behavior of the Cdk1 kinase activity. This indicates that cyclin B synthesis constitutes the essential anabolic process required for the cell cycle, underscoring its role as a mitotic trigger. The use of extracts in cytostatic factor (CSF)-mediated metaphase arrest (CSF-arrested extracts) revealed that cyclin B degradation, on the other hand, is an essential prerequisite for mitotic exit [21]. This was shown by providing CSF-arrested extracts with proteolysis-resistant cyclin B, which prevents the cell cycle from being released from mitotic arrest. Several other studies used extracts released from metaphase and subsequently arrested in interphase (interphase extracts), where the endogenous cyclin B level is kept low and constant. Separate experiments using interphase extracts demonstrated that the introduction of either exogenously loaded cyclin B–Cdk1 [22] or recombinant nondegradable cyclin B that activates Cdk1 [3] triggers cyclin degradation after a time delay of approximately 15 minutes. With the knowledge that the ubiquitin-dependent selective degradation of cyclin is attributed to the E3 ligase activity of APC/C [23, 24], these findings confirm the existence of a time-delayed negative feedback loop that autonomously drives oscillations and demonstrate that its function is critically coupled to the synthesis and degradation of cyclin B. This biochemical mechanism of periodicity centered on cyclin B turnover is consistent with findings from cell-free system studies of other species, such as starfish [25] and sea urchins [26].

Research on Cdc25 and Wee1, whose regulatory functions were first established through genetic studies of fission yeasts in the 1980s [27, 28] and later examined using cell-free extracts in the 1990s [29, 30], provided experimental validation of the autocatalytic activity of Cdk1 predicted by mathematical models [14]. Using CSF-arrested extracts, Kumagai and Dunphy observed that Cdc25 is rapidly phosphorylated upon mitotic entry, concurrent with an increase in its phosphatase activity [29]. Together with the presence of the Cdk1 consensus phosphorylation sequence in the N-terminal domain of Cdc25, this observation illuminated the existence of a Cdc25-mediated positive feedback mechanism that regulates Cdk1 activation. Another study demonstrated that adding exogenous Wee1 in its active, unphosphorylated form to CSF-arrested extracts leads to inhibitory phosphorylation of Cdk1 on Tyr-15, thereby delaying Cdk1 activation in a dose-dependent manner [30]. Conversely, during mitosis, the hyperphosphorylation of Wee1, which downregulates its kinase activity, correlates with increased Cdk1 activity. This observation identifies a double-negative feedback loop (functionally equivalent to positive feedback) between Cdk1 and Wee1.

The synergy between mathematical modeling and experimentation played an essential role in developing a unified systems-level understanding of cell cycle control. The pioneering mathematical work of Novák and Tyson [14–16] from the early 1990s explained the abrupt Cdk1 activation [31] as a manifestation of bistability mediated by the positive feedback loops involving Wee1 and Cdc25. In addition, it was suggested that the circuit could exhibit hysteresis in Cdk1 activity [32]. These qualitative predictions were confirmed a decade later by model-driven experiments using *Xenopus* extracts [33, 34]. In agreement with theoretical predictions, disrupting the positive feedback loops of Wee1 and Cdc25 causes oscillations to deviate significantly from the behavior typical of relaxation oscillators [35]. These modeling and experimental studies underscore the role of the Cdk1/Wee1/Cdc25 bistable switch in directing the cell cycle through alternating two discrete states (interphase and mitosis), ensuring faithful temporal separation of DNA replication and segregation. More detailed biochemical models have illuminated the less-appreciated roles of other regulatory components of mitotic timing. A notable example is PP2A, whose systems-level roles have been elucidated through combined efforts of modeling and experimental works. Early extract-based research on this phosphatase [5, 7, 8] led to several modeling efforts that generated testable predictions, suggesting that Gwl and PP2A constitute an additional bistable switch [36, 37]. Additional experiments later confirmed these model predictions [38, 39].

7.3 Synthetic Cells for High-Throughput Experimentation of Cell Cycle Dynamics

Despite the well-recognized success of cell cycle studies using extracts, they still present several limitations and ongoing opportunities for improvement. The studies discussed relied on samples prepared in a bulk solution and analyzed with labor-intensive methods, such as immunoblotting and kinase assays. These conventional approaches have drawbacks of poor time resolution and low throughput, which limit the time-resolved and statistically robust analyses of periodic behaviors. In addition, cellular oscillations reconstituted in such large, well-mixed volumes tend to desynchronize rapidly, resulting in a few damped cycles rather than many self-sustained ones [35, 40]. This rapid desynchronization may obscure the full autonomous potential intrinsic to many cellular oscillators. Furthermore, the absence of cell-like compartmentalization necessitates cautious interpretation of the observed dynamics, since finite size and spatial organization can critically influence biochemical kinetics [41–43].

7.3.1 Mitotic Oscillations Reconstituted in Microemulsions

The use of cell mimetics, including vesicles and microemulsions, for isolating reactions provides unique advantages in addressing these problems. The use of artificial cell configurations has a rich history encompassing a wide range of applications, ranging from single-cell RNA sequencing [44] to high-throughput drug screening [45]. Several studies conducted in the past decade have further shown that the Belousov–Zhabotinsky (BZ) reaction [46] and synthetic transcriptional oscillators [47] can be encapsulated within cell-sized emulsions without compromising their essential oscillatory functions, allowing exploration of a larger dynamical space with higher throughput and more physiologically relevant geometric constraints.

Extending this line of work, Guan et al. encapsulated *Xenopus laevis* cell-free cycling extracts within water-in-oil microemulsion droplets to reconstitute mitotic oscillations (Fig. 7.1b) [48, 49]. Emulsification occurs spontaneously when aqueous reaction mixtures, solvents, and surfactants are appropriately proportioned. In this work, simple vortexing was sufficient to accelerate the generation of a large ensemble of independent stochastic oscillators that emulated the dimensions of typical somatic or early embryonic cells (10–300 μm in diameter). The oscillation time courses within individual droplets were simultaneously quantified in real time using multicolor fluorescent reporters for cyclin B degradation, Cdk1 activity, and downstream mitotic events such as nuclear envelope dynamics, which could be readily incorporated into cell-free reaction mixtures. Remarkably, oscillations in partitioned extracts demonstrated greater robustness than those observed in bulk solutions, sustaining over 30–40 cycles and over multiple days.

7.3.2 Use of Microfluidics for Programmed Production of Synthetic Cells

Microfluidic techniques have been shown to further improve the process of preparing these minimal synthetic cells. In Sun et al. [50], droplets were generated using microfluidic devices that implemented pressure-controlled flow focusing (Fig. 7.2a), one of the most commonly employed configurations in droplet microfluidics [51]. The encapsulated cycling extracts maintained their capacity to undergo multiple cell cycles, enabling long-term time-lapse fluorescence imaging. The microfluidic chips employed in this study could be fabricated in-house following well-established soft-lithography protocols [52]. Low-cost, handheld droplet analyzers were also developed to enhance accessibility for educational and laboratory applications [53]. Thus, while still accessible to most laboratories, these methods can provide better automation for the droplet production process and the subsequent imaging, data processing, and quantification of cellular dynamics.

Microfluidic techniques not only improve automation and throughput but also provide better control over encapsulation size, monodispersity, and packing density [54]. Consequently, they can be particularly useful for studying the compartment-size dependence of encapsulated processes. Indeed, microfluidic encapsulation has been utilized to investigate how the sizes of organelles, such as mitotic spindles [55, 56] and nuclei [57], scale with cell size. Observations from synthetic cells across a broad range of cell dimensions provide direct evidence that the availability of assembly components determines the sizes and growth rates of organelles. Furthermore, molecular mechanisms through which cells regulate scaling factors have been investigated using microfluidic techniques. For example, a study by Brownlee and Heald revealed that importin α localizes to membranes only when encapsulants mimic the compositions of the *Xenopus* egg plasma membrane [58]. Physiological spindle size scaling was observed exclusively in artificial cells with realistic lipid boundaries, contrasting with cells with passivated boundaries that lacked importin α localization. This observation underlined the critical role of this adapter protein and its localization mechanism as a cell size sensor, exemplifying the utility of synthetic cell-based setups in investigating the scaling of subcellular structures. Remarkably, nuclear size acts as a critical regulator of the early embryonic cell cycle: The timing of midblastula transition (MBT), which marks the onset of zygotic transcription, is thought to be triggered in response to the exponential increase in the nuclear-to-cytoplasmic ratio during reductive cleavage divisions [59]. Here, the artificial cell setup enables flexibility in manipulating both cell and nuclear sizes

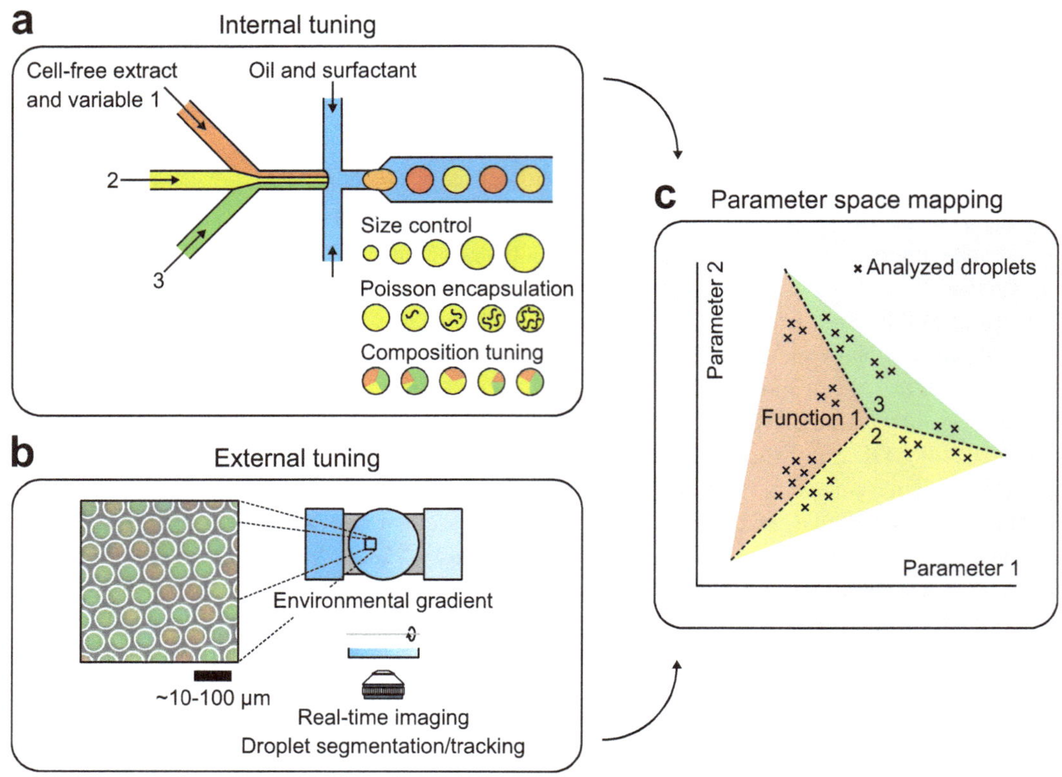

Fig. 7.2 (**a**) A multichannel microfluidic device enables precise control of emulsion size and composition. By adjusting the flow rates at individual inlets, droplets can be generated with multiple components mixed in varying ratios. This setup can also produce emulsions containing varying numbers of discrete objects, such as chromatin strands. (**b**) An array of microemulsion droplets (inset) can be further exposed to external environmental parameter variations, such as an illustrated gradient maintained between two conditions (e.g., a temperature gradient created between cold and hot thermal baths). (**c**) The high degree of manipulability and throughput provided by the synthetic cell platform enables comprehensive parameter space mapping for systematic exploration and characterization of diverse dynamical behaviors in oscillators

and can help delineate how the timing of the biochemical process is regulated in relation to spatial information [60].

Another advantage of microfluidics lies in its capacity to fine-tune the biochemistry of encapsulated reactions in a programmable manner. For example, a microfluidic configuration with multiple inlets for aqueous phases (Fig. 7.2a) allows for systematic tuning of droplet composition by varying the pressure profiles applied to the inlets. Droplets that are heterogeneous in their functions can be indexed based on coinjected inert fluorescent barcode signals or dye concentration levels; thus, their corresponding coordinates in the biochemical parameter space can be effectively located. This multichannel setup can also be integrated with external devices such as concentration [61] or temperature [62] gradient generators, which provide additional dials to modulate dynamics (Fig. 7.2b).

A notable demonstration of this parameter scanning capability is found in a study of synthetic transcriptional oscillators. Genot et al. explored, using multichannel microfluidics, droplets encapsulating DNA switches or predator–prey DNA oscillators with varied kinetic specifications of their DNA replication or catalytic steps [63]. High-resolution dynamical landscapes were constructed in the parameter space, harnessing the large population and functional diversity of the realized oscillators. These landscapes allowed for the identification of optimal parameter regimes, prediction of unforeseen bifurcation structures, and also validation of models, illustrating how the high throughput and manipulability of the artificial cell system can aid in the de novo design and mechanistic dissection of biochemical oscillators (Fig. 7.2c).

7.4 Recent Applications

Taken together, cell-free extracts and synthetic cells offer a platform for in-depth exploration of cell cycle dynamics across extended parametric (e.g., multichannel compositional tuning), temporal (elimination of damped oscillations), and spatial (precise tuning of cellular and subcellular compartment sizes and localization) dimensions beyond the grasp of conventional approaches (Fig. 7.3). The versatility of these methodologies enhances hypothesis testing and modeling, deepening the understanding of autonomous biochemical oscillations and their emergent phenomena. The following sections present two particular examples from recent mitotic oscillator studies that showcase their strengths.

7.4.1 Mapping the Dynamical Landscape That Regulates the Autonomous Mitotic Oscillator

Biological oscillators are continually exposed to environmental and external perturbations. The kinetic and equilibrium properties of biochemical circuits, even under normal physiology, constantly adjust themselves in response to environmental variations. While the architecture of a circuit, as a discrete characterization of a dynamical system, is often a crucial determinant of the circuit's functions [9, 10], such ubiquitous continuous parametric variations introduce additional complexity: A dynamical system, although well-defined at the circuit level, can exhibit multiple qualitatively distinct behaviors across different parameter domains, which may involve bifurcations.

In the cell cycle, such functional diversity readily demonstrates its physiological relevance. During early embryogenesis, cells experience rapid shifts in their division rates, possibly including permanent cell cycle arrests induced by controlled senescence [64]. Furthermore, there is more direct experimental evidence that the mitotic system operates near bifurcation points. For example, the critical slowing down of the cell cycle near the activation threshold of cyclin is deemed indicative of the system's proximity to a saddle-node bifurcation [33]. Near critical points, system behaviors are highly sensitive to even small fluctuations in parameters, making consistent quantitative assessments of dynamics challenging [65]. This calls for a global mapping of the full dynamical potential of the cell cycle circuit across an extended parameter space, instead of studying the system under a finite number of discrete, specific parameter choices with limited predictive power.

Experimental implementation of parameter scanning involves several technical complexities related to manipulability and efficiency. While some noteworthy studies have characterized bifurcation structures in bacterial population oscillations [66] and in vivo cell cycles [67], synthetic cell-based approaches further accelerate the mapping of continuous, full-range dynamical landscapes (Fig. 7.3a).

7.4.1.1 Cytoplasmic Dilution

Many physiologically relevant perturbations are global in nature. Recent studies using cell-free extracts have examined how cell cycle oscillations respond to such perturbations affecting multiple circuit components simultaneously. Cytoplasmic dilution is one such global perturbation that cells can experience. For instance, mammalian cells undergo considerable volume changes throughout the cell cycle—with volume peaking at a 30% increase during metaphase and decreasing before anaphase—while maintaining their dry mass [68]. Notably, this mitotic swelling and corresponding reduction in cell density do not disrupt cell cycle progression.

To study how such robustness is established, Jin et al. inspected cell cycle dynamics in microemulsion-encapsulated *Xenopus* cycling

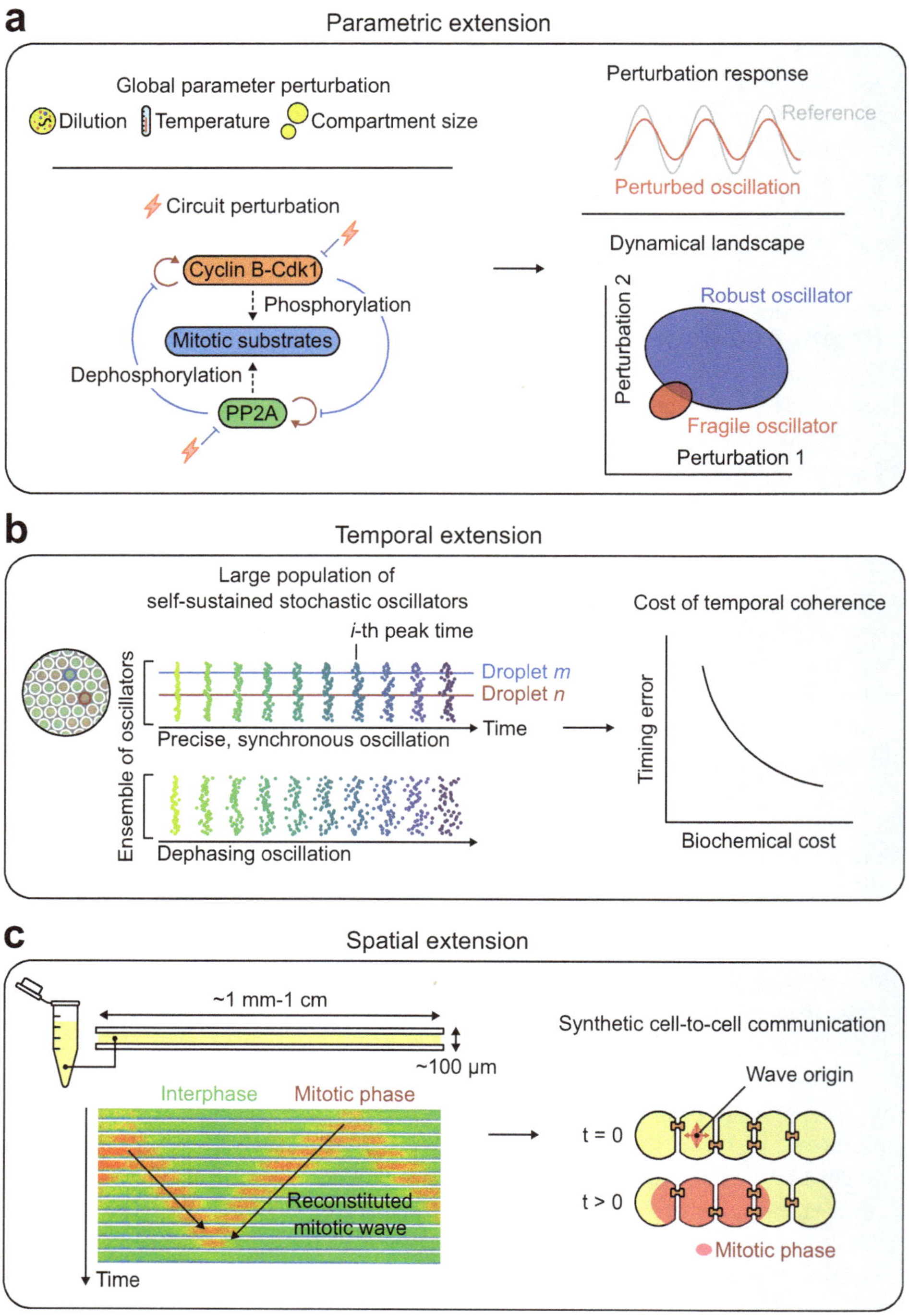

Fig. 7.3 Cell-free extracts and synthetic cells open up opportunities for systematic exploration of (**a**) dynamical landscapes, (**b**) time-dependent and stochastic behaviors, and (**c**) spatial organization and coordination of cell cycle regulation

extracts prepared across a wide, finely resolved range of cytoplasmic dilutions (0–146% of natural density) [69]. The resilience of the cell cycle to cytoplasmic density changes turned out to persist even under more extreme perturbations, with oscillations sustained across 20–122% of the natural cytoplasmic density. This robustness, which could be reconstituted in a cell-free environment, was interpreted as a systems-level property of the cell cycle circuit itself, rather than being reliant on specific biochemical regulatory mechanisms. While it is theoretically anticipated that interlinking positive and negative feedback loops should play an important role in establishing such functional robustness [9], their work further demonstrated that Wee1 and Cdc25 did not critically contribute to the robustness of the cell cycle to dilution. Instead, the study proposed that the PP2A/Gwl feedback loop may serve as a key circuit motif that enhances robustness.

7.4.1.2 Temperature

Temperature variation is another pervasive form of perturbation that modifies the global rate of biochemical reactions. However, it is not fully understood how the kinetic pathways composing the cell cycle collectively respond to temperature variation. The critical role of temperature is recognized in early embryos, where organisms, especially ectotherms, lack fully developed mechanisms to handle temperature fluctuations.

Rombouts et al. addressed this issue, using both *Xenopus* embryos and extracts in a synthetic cell setting [62]. Synthetic cells were placed on an imaging stage in steady-state heat conduction between two thermal reservoirs, which actively generated a temperature gradient between 16 and 32 oC. The period and waveform responses of Cdk1 oscillation were characterized for a continuous and fine-grained coverage of temperature. The experiment revealed that cyclin synthesis, which influences the rising phase of Cdk1 activity, and APC/C activity, which affects the falling phase, respond in opposite directions to temperature changes. The balance between these factors was theorized to ultimately determine the viable temperature range for cellular functions. The work demonstrated the versatility of this synthetic setup, capable of accommodating cell-external modes of altering dynamics.

7.4.1.3 Circuit Perturbations

The synthetic cell methodology also allows for the examination of the effects of localized perturbations, offering valuable insights into how specific topological motifs of circuits correlate with their functionality. Li et al. explored the significance of positive feedback loops involving Wee1/Cdc25 or PP2A in establishing various systems-level properties of oscillators [70]. The work employed a multichannel microfluidics setup to map a detailed landscape of oscillatory features across varying activity levels of mitotic regulators and cyclin B synthesis rates. In particular, the findings of this study revealed that while Wee1 could be nearly completely inhibited without compromising oscillatory behavior, inhibiting PP2A immediately exacerbated the system's oscillatory potential. A systematic characterization of the role of PP2A demonstrated batch-dependent and qualitatively distinct dynamical response patterns of the system to its inhibition. This experimental framework provided sufficient resolution to compare their findings with predictions from analytical models [3, 16], attributing the observed qualitative variations in dynamics to endogenous fluctuations in Gwl levels.

7.4.2 Examining Spatiotemporal Coordination of Mitotic Oscillations

The ability of encapsulated extracts to undergo multiple undamped cycles for a long time presents an interesting research direction regarding the synchronization of oscillators. Rapid synchronous embryonic cycles represent a key feature of cell cleavages in many organisms, particularly anamniotes [59, 71]. While significant attention has been directed toward the mechanisms leading to the loss of synchrony at the onset of MBT, there is relatively less emphasis on how this synchrony is maintained during cleavages. Due to the small physical dimensions of cells and the dilute concentrations of regulatory molecules, cell cycles can be highly susceptible to both intrinsic and extrinsic fluctuations [72]. Furthermore, in the case of *Xenopus* embryos, there appears to be a mechanism to rectify mitotic timing defects that potentially accumulate before gastrulation [73], indicating that such untimely decoherence before critical developmental transitions is a likely scenario that cells can face and have evolved to tolerate.

7.4.2.1 Energy Flux and Synchrony

Such synchrony may be merely an apparent phenomenon, reflecting a high precision of individual autonomous oscillators. The precision of general biochemical circuits incurs a cost in the form of more free energy being dissipated [74, 75]. In a theoretical study, the phase fluctuations of stochastic oscillators, quantified in terms of the phase diffusion constant, were shown to be influenced by the free-energy dissipation associated with their molecular circuits [76]. Specifically, the phase diffusion constant and the energy dissipation per cycle were identified to have an inverse relation, which was universally parameterized regardless of the microscopic details of the circuits.

Therefore, the loss of synchrony is expected to be suppressed at a certain energy expenditure for biological oscillators. Some preliminary evidence supporting this theoretical claim emerged from the study of the in vitro KaiC phosphorylation cycle [76–78]. However, general experimental quantification of the proposed trade-off relationship remains challenging. Systematic modulation of energy levels complicates the examination. In addition, rigorous validation of this statistical assertion should be based on a large population of oscillators, which may be difficult to implement experimentally. Furthermore, oscillations need to be self-sustained for a sufficient number of cycles for the statistical observables, such as the peak time diffusion constant, to be properly defined. These challenges can be uniquely addressed by the approach involving synthetic cells, where initially well-mixed, synchronized biological oscillators are partitioned into numerous independent microemulsions (Fig. 7.3b).

Li et al. observed a quadratic dependence of peak time variance on time [70], in contrast to the theoretically postulated linear growth [76]. Although their research did not primarily focus on the role of energy dissipation in dephasing, this finding prompted further exploration of the relationship by Wang et al. [79]. In this work, using two-channel microfluidic devices, they adjusted ATP levels in droplets containing otherwise intact cycling extracts. They demonstrated that this quadratic dependence could be attributed to the partitioning effect described in the earliest study of encapsulated biochemical oscillators [47]. Following systematic calibration to eliminate this artifact, they found that, as ATP levels approached a certain upper threshold where the system was fully arrested in mitosis, the population of the stochastic oscillator exhibited a slower dephasing rate, consistent with theoretical predictions. This study exemplified how this synthetic cell setting and its throughput can serve as a platform for testing stochastic models, providing new perspectives through which cell cycle regulation can be understood.

7.4.2.2 Mitotic Waves

The second way to physically understand mitotic synchrony in early embryogenesis involves the behavior of cell cycle oscillations in space, which typically results in waves [80]. Waves arising from the diffusive coupling of local bistable dynamics can be responsible for the spatiotemporal coordination of mitotic oscillations, especially in large embryos (e.g., *Xenopus* zygote, 1.2 mm in diameter; *Drosophila*, 0.5 mm along the major axis). Research on starfish oocytes [81] and frog embryos [80] has shown that time-dependent Cdk1 gradients likely signal surface contraction waves (cortical dynamics that occur before cytokinesis), propagating at a speed of 1 μm/s. In fruit fly embryos, there is more direct evidence of mitotic waves, observed as Cdk1 activation that propagates throughout the entire syncytium at a significantly faster speed, for which the phase wave mechanism was suggested to be responsible [82]. While multiple mechanisms seem to account for the diverse wave patterns observed across different organisms and physiological conditions, the connection between these mechanisms is only loosely understood. Additionally, the possible biological implications of this physical mode of signal transfer remain unexplored. The spatial extension of an extract-based setup can be readily implemented, as demonstrated in quasi-1D tubes [80, 83], thin layers [84], and microemulsions [85], offering a valuable platform to explore these questions surrounding mitotic waves in greater detail (Fig. 7.3c).

Puls et al. investigated waves in cell-free extracts prepared in 1D configurations [86]. They directly visualized Cdk1 activity using a recently developed FRET sensor [87], differentiating this work from other studies that relied on downstream events of Cdk1 oscillations, such as nuclear envelope breakdown [80, 83] and microtubule assembly [88], as indicators of cell cycle states. They demonstrated that the autonomy of spatially distributed oscillators can give rise to spontaneous formation of waves, even in the absence of nuclei serving as pacemakers as suggested in earlier studies [83, 84], drawing an analogy to autowaves observed in chemical systems such as the BZ reaction medium. Moreover, the reconstituted cell cycles in tubes persisted for over ten cycles and lasted over 24 hours, enabling a comprehensive characterization of the temporal transitions in wave dynamics. The initial oscillations were highly synchronous across space, resembling fast phase waves observed in *Drosophila*. However, this synchronicity was transient and gradually evolved into typical trigger waves after a finite amount of time. This work outlined a potential link to connect seemingly different synchronization phenomena within a consistent mechanistic framework.

A recent work by Huang et al. suggested one potential biological implication of this physical mechanism [88]. Their research, utilizing *Xenopus* extracts, demonstrated that the speed of mitotic trigger waves remained nearly constant across a wide range of extract dilutions, which opposed the common assumption of an optimal operational range of cytoplasmic concentrations. The robustness was attributed to the opposing and counterbalancing effects of dilution on the reaction rates and cytoplasmic viscosity. This physical mechanism of mitotic synchronization may have broader implications for understanding the functional reliability underlying signal relay occurring in a variety of cellular and tissue-level contexts, including the propagation of regulated cell death [89, 90]. Their work also reaffirms the power of this cell-free system in the mechanistic study of the emergent functions of cell cycle regulation.

7.5 Concluding Remarks

Despite its long history and recognized success, the cell-free extract system continues to present intriguing research opportunities. Extracts in cell-like microcompartments enable oscillations to be reconstituted with prolonged lifetimes and minimal damping, thereby facilitating a mechanistic investigation of the autonomous subcellular periodicity that underlies the cell cycle, as well as the autonomy of the mitotic oscillator itself [91–95]. Microfluidic techniques offer further advantages in automation, throughput, and precision in producing these synthetic cells, providing insights that traditional approaches using extracts in large, homogenized volumes fail to address. Recent studies highlight how synthetic biology approaches can accelerate the exploration of the dynamical landscapes and emergent behaviors of cell cycle regulation, which serve essential functions in cellular physiology.

Such dynamical behaviors of cell cycle regulators are often fundamentally associated with the way that they interact, which is reduced to abstract network motifs dedicated to processing specific logics. The cell cycle circuit incorporates diverse structural elements: Cdk1–APC/C negative feedback, counteracting regulatory inputs from Wee1/Cdc25, and positive feedback loops surrounding their respective regulators Cdk1 and PP2A. This diverse set of motifs, along with their manipulability in the extract system, provides an archetypal framework for examining the general design principles governing biological oscillators, into which knowledge from systems biology can be integrated.

While this chapter focused mainly on cell cycle oscillations, subsequent studies could expand the opportunities to incorporate general autonomous oscillations in biology. Despite the diverse contexts in which they occur, biological oscillators share structural similarities in their underlying molecular regulatory networks. The presence of negative feedback loops is considered essential for establishing periodicity [17]. Motifs such as positive feedback [9] and antagonistic regulation [10], although not intrinsically responsible for periodicity, are found in numerous biological oscillatory networks and are also evolutionarily conserved [96–98]. Theoretical explorations of circuit design space suggest that several systems-level properties common to oscillators, such as robustness and tunability, originate from these shared design patterns [10, 99]. Their systematic experimental validation could be achieved based on the framework presented in this chapter.

The ultimate objective of synthetic biology approaches should be to reconstruct the core features of oscillations using minimal components. Such bottom-up approaches allow for a more controlled and interpretable comparison between experimental observations and principled insights. In recent decades, synthetic biology has made significant strides in implementing and characterizing genetic oscillators [100], yet progress in reconstituting post-translational oscillators reassembled from functional units has remained limited. The multidimensional nature and functional diversity of proteins not only complicate bottom-up approaches but also allow richer dynamical patterns to emerge. For instance, proteins can enable faster kinetics compared to gene regulation with transcriptional and translational delays [101], and their support for rapid responses is a crucial requirement for the development of relaxation-like oscillations and traveling waves. Thus, it can be argued that the existing studies presented here have a clear direction for future expansion. The current lack of widely recognized successes in this pursuit, on the one hand, implies that there remains ample potential to further deepen our understanding of autonomous cellular oscillators.

References

1. Morgan DO (2007) The cell cycle: principles of control. New Science Press, New York
2. Massacci G, Perfetto L, Sacco F (2023) The cyclin-dependent kinase 1: more than a cell cycle regulator. Br J Cancer 129(11):1707–1716
3. Yang Q, Ferrell JE Jr (2013) The Cdk1–APC/C cell cycle oscillator circuit functions as a time-delayed, ultrasensitive switch. Nat Cell Biol 15(5):519–525
4. Novak B, Tyson JJ, Gyorffy B, Csikasz-Nagy A (2007) Irreversible cell-cycle transitions are due to systems-level feedback. Nat Cell Biol 9(7):724–728
5. Mochida S, Ikeo S, Gannon J, Hunt T (2009) Regulated activity of PP2A –B55δ is crucial for controlling entry into and exit from mitosis in Xenopus egg extracts. EMBO J 28(18):2777–2785
6. Millward TA, Zolnierowicz S, Hemmings BA (1999) Regulation of protein kinase cascades by protein phosphatase 2a. Trends Biochem Sci 24(5):186–191
7. Mochida S, Maslen SL, Skehel M, Hunt T (2010) Greatwall phosphorylates an inhibitor of protein phosphatase 2a that is essential for mitosis. Science 330(6011):1670–1673
8. Gharbi-Ayachi A, Labbé J-C, Burgess A, Vigneron S, Strub J-M, Brioudes E, Van-Dorsselaer A, Castro A, Lorca T (2010) The substrate of Greatwall kinase, arpp19, controls mitosis by inhibiting protein phosphatase 2a. Science, 330(6011):1673–1677
9. Yu-Chen T Tsai, Choi YS, Ma W, Pomerening JR, Tang C, and Ferrell JE Jr (2008) Robust, tunable biological oscillations from interlinked positive and negative feedback loops. Science 321(5885): 126–129
10. Li Z, Liu S, Yang Q (2017) Incoherent inputs enhance the robustness of biological oscillators. Cell Syst 5(1):72–81
11. Patrick H O'farrell, Stumpff J, Su TT (2004) Embryonic cleavage cycles: how is a mouse like a fly? Curr Biol 14(1):R35–R45
12. Hartwell LH, Mortimer RK, Culotti J, Culotti M (1973) Genetic control of the cell division cycle in yeast: V. genetic analysis of CDC mutants. Genetics 74(2):267
13. Hartwell LH, Culotti J, Pringle JR, Reid BJ (1974) Genetic control of the cell division cycle in yeast: a model to account for the order of cell cycle events is deduced from the phenotypes of yeast mutants. Science 183(4120):46–51

14. Tyson JJ (1991) Modeling the cell division cycle: cdc2 and cyclin interactions. Proc Natl Acad Sci 88(16):7328–7332
15. Novak B, Tyson JJ (1993a) Modeling the cell division cycle: M-phase trigger, oscillations, and size control. J Theor Biol 165(1):101–134
16. Novak B, Tyson JJ (1993b) Numerical analysis of a comprehensive model of m-phase control in Xenopus oocyte extracts and intact embryos. J Cell Sci 106(4): 1153–1168
17. Novák B, Tyson JJ (2008) Design principles of biochemical oscillators. Nat Rev Mol Cell Biol 9(12): 981–991
18. Ferrell JE, Tsai Y-CT, Yang Q (2011) Modeling the cell cycle: why do certain circuits oscillate? Cell 144(6):874–885
19. Lohka MJ, Masui Y (1983) Formation in vitro of sperm pronuclei and mitotic chromosomes induced by amphibian ooplasmic components. Science 220(4598):719–721
20. Murray AW, Kirschner MW (1989) Cyclin synthesis drives the early embryonic cell cycle. Nature 339(6222):275–280
21. Murray AW, Solomon MJ, Kirschner MW (1989) The role of cyclin synthesis and degradation in the control of maturation promoting factor activity. Nature 339(6222):280–286
22. Félix M-A, Labbé J-C, Dorée M, Hunt T, Karsenti E (1990) Triggering of cyclin degradation in interphase extracts of amphibian eggs by cdc2 kinase. Nature 346(6282):379–382
23. Glotzer M, Murray AW, Kirschner MW (1991) Cyclin is degraded by the ubiquitin pathway. Nature 349(6305):132–138
24. Sudakin V, Ganoth DVORAH, Dahan A, Heller H, Hershko J, Luca FC, Ruderman JV, Hershko A (1995) The cyclosome, a large complex containing cyclin-selective ubiquitin ligase activity, targets cyclins for destruction at the end of mitosis. Mol Biol Cell 6(2):185–197
25. Standart N, Minshull J, Pines J, Hunt T (1987) Cyclin synthesis, modification and destruction during meiotic maturation of the starfish oocyte. Dev Biol 124(1):248–258
26. Minshull J, Pines J, Golsteyn R, Standart N, Mackie S, Colman A, Blow J, Ruderman JV, Wu M, Hunt T (1989) The role of cyclin synthesis, modification and destruction in the control of cell division. J Cell Sci 1989(Supplement_12):77–97
27. Russell P, Nurse P (1986) cdc25+ functions as an inducer in the mitotic control of fission yeast. Cell 45(1):145–153
28. Russell P, Nurse P (1987) Negative regulation of mitosis by Wee1+, a gene encoding a protein kinase homolog. Cell 49(4):559–567
29. Kumagai A, Dunphy WG (1992) Regulation of the cdc25 protein during the cell cycle in Xenopus extracts. Cell 70(1):139–151
30. Mueller PR, Coleman TR, Dunphy WG (1995) Cell cycle regulation of a Xenopus Wee1-like kinase. Mol Biol Cell 6(1):119–134
31. Solomon MJ, Glotzer M, Lee TH, Philippe M, Kirschner MW (1990) Cyclin activation of p34cdc2. Cell 63(5):1013–1024
32. Thron CD (1996) A model for a bistable biochemical trigger of mitosis. Biophys Chem 57(2–3):239–251
33. Sha W, Moore J, Chen K, Lassaletta AD, Yi C-S, Tyson JJ, Sible JC (2003) Hysteresis drives cell-cycle transitions in Xenopus laevis egg extracts. Proc Natl Acad Sci 100(3):975–980
34. Pomerening JR, Sontag ED, Ferrell JE Jr (2003) Building a cell cycle oscillator: hysteresis and bistability in the activation of cdc2. Nat Cell Biol 5(4):346–351
35. Pomerening JR, Kim SY, Ferrell JE (2005) Systems-level dissection of the cell-cycle oscillator: bypassing positive feedback produces damped oscillations. Cell 122(4):565–578
36. Zhang T, Tyson JJ, Novák B (2013) Role for regulated phosphatase activity in generating mitotic oscillations in Xenopus cell-free extracts. Proc Natl Acad Sci 110(51):20539–20544
37. Vinod PK, Novak B (2015) Model scenarios for switch-like mitotic transitions. FEBS Lett 589(6): 667–671
38. Mochida S, Rata S, Hino H, Nagai T, Novák B (2016) Two bistable switches govern m phase entry. Curr Biol 26(24):3361–3367
39. Kamenz J, Gelens L, Ferrell JE (2021) Bistable, biphasic regulation of PP2A -B55 accounts for the dynamics of mitotic substrate phosphorylation. Curr Biol 31(4):794–808
40. Nakajima M, Imai K, Ito H, Nishiwaki T, Murayama Y, Iwasaki H, Oyama T, Kondo T (2005) Reconstitution of circadian oscillation of cyanobacterial KaiC phosphorylation in vitro. Science 308(5720):414–415
41. Meyers J, Craig J, Odde DJ (2006) Potential for control of signaling pathways via cell size and shape. Curr Biol 16(17):1685–1693
42. Lizana L, Bauer B, Orwar O (2008) Controlling the rates of biochemical reactions and signaling networks by shape and volume changes. Proc Natl Acad Sci 105(11):099–4104
43. Matsuura T, Hosoda K, Kazuta Y, Ichihashi N, Suzuki H, Yomo T (2012) Effects of compartment size on the kinetics of intracompartmental multimeric protein synthesis. ACS Synth Biol 1(9):431–437
44. Macosko EZ, Basu A, Satija R, Nemesh J, Shekhar K, Goldman M, Tirosh I, Bialas AR, Kamitaki N, Martersteck EM, et al (2015) Highly parallel genome-wide expression profiling of individual cells using nanoliter droplets. Cell 161(5): 1202–1214
45. Brouzes E, Medkova M, Savenelli N, Marran D, Twardowski M, Hutchison JB, Rothberg JM, Link

DR, Perrimon N, Samuels ML (2009) Droplet microfluidic technology for single-cell high-throughput screening. Proc Natl Acad Sci 106(34): 14195–14200
46. Epstein IR, Vanag VK, Balazs AC, Kuksenok O, Dayal P, Bhattacharya A (2012) Chemical oscillators in structured media. Acc Chem Res 45(12): 2160–2168
47. Weitz M, Kim J, Kapsner K, Winfree E, Franco E, Simmel FC (2014) Diversity in the dynamical behaviour of a compartmentalized programmable biochemical oscillator. Nat Chem 6(4):295–302
48. Guan Y, Li Z, Wang S, Barnes PM, Liu X, Xu H, Jin M, Liu AP, Yang Q (2018a) A robust and tunable mitotic oscillator in artificial cells. eLife 7:e33549
49. Guan, Wang S, Jin M, Xu H, Yang Q (2018b) Reconstitution of cell-cycle oscillations in microemulsions of cell-free Xenopus egg extracts. J Vis Exp JoVE 139:58240
50. Sun M, Li Z, Wang S, Maryu G, Yang Q (2019a) Building dynamic cellular machineries in droplet-based artificial cells with single-droplet tracking and analysis. Anal Chem 91(15):9813–9818
51. Günther A, Jensen KF (2006) Multiphase microfluidics: from flow characteristics to chemical and materials synthesis. Lab Chip 6(12):1487–1503
52. Whitesides GM, Ostuni E, Takayama S, Jiang X, Ingber DE (2001) Soft lithography in biology and biochemistry. Annu Rev Biomed Eng 3(1):335–373
53. Sun M, Li Z, Yang Q (2019b) μdropi: a hand-held microfluidic droplet imager and analyzer built on raspberry pi. J Chem Educ 96(6):1152–1156
54. Sun M, Maryu G, Wang S, Yang Q, Bailey RC (2021) Plug-in tubes allow tunable oil removal, droplet packing, and reaction incubation for time-controlled droplet-based assays. Biomicrofluidics 15(2)
55. Hazel J, Krutkramelis K, Mooney P, Tomschik M, Gerow K, Oakey J, Gatlin JC (2013) Changes in cytoplasmic volume are sufficient to drive spindle scaling. Science 342(6160):853–856
56. Good MC, Vahey MD, Skandarajah A, Fletcher DA, Heald R (2013) Cytoplasmic volume modulates spindle size during embryogenesis. Science 342(6160): 856–860
57. Hara Y, Merten CA (2015) Dynein-based accumulation of membranes regulates nuclear expansion in Xenopus laevis egg extracts. Dev Cell 33(5):562–575
58. Brownlee C, Heald R (2019) Importin α partitioning to the plasma membrane regulates intracellular scaling. Cell 176(4):805–815
59. Newport J, Kirschner M (1982) A major developmental transition in early Xenopus embryos: I. characterization and timing of cellular changes at the midblastula stage. Cell 30(3):675–686
60. Piñeros L, Frolov N, Ruiz-Reynés D, Van Eynde A, Cavin-Meza G, Heald R, Gelens L (2025) The nuclear-cytoplasmic ratio controls the cell-cycle period in compartmentalized frog egg extract. Curr Biol 35(18):4426–4441
61. Jeon NL, Dertinger SKW, Chiu DT, Choi IS, Stroock AD, Whitesides GM (2000) Generation of solution and surface gradients using microfluidic systems. Langmuir 16(22):8311–8316
62. Rombouts J, Tavella F, Vandervelde A, Phong C, Ferrell J, Yang Q, Gelens L (2025) Mechanistic origins of temperature scaling in the early embryonic cell cycle. Nat Commun 16(1):8045
63. Genot AJ, Baccouche A, Sieskind R, Aubert-Kato N, Bredèche N, Bartolo JF, Taly V, Fujii T, Rondelez Y (2016) High-resolution mapping of bifurcations in nonlinear biochemical circuits. Nat Chem 8(8): 760–767
64. Muñoz-Espín D, Cañamero M, Maraver A, Gómez-López G, Contreras J, Murillo-Cuesta S, Rodríguez-Baeza A, Varela-Nieto I, Ruberte J, Collado M, et al (2013) Programmed cell senescence during mammalian embryonic development. Cell 155(5):1104–1118
65. Strogatz SH (2018) Nonlinear dynamics and chaos: with applications to physics, biology, chemistry, and engineering. CRC Press, New York
66. Martinez-Corral R, Liu J, Süel GM, Garcia-Ojalvo J (2018) Bistable emergence of oscillations in growing bacillus subtilis biofilms. Proc Natl Acad Sci 115(36):E8333–E8340
67. Rata S, MFSP Rodriguez, Joseph S, Peter N, Iturra FE, Yang F, Madzvamuse A, Ruppert JG, Samejima K, Platani M, et al (2018) Two interlinked bistable switches govern mitotic control in mammalian cells. Curr Biol 28(23):3824–3832
68. Zlotek-Zlotkiewicz E, Monnier S, Cappello G, Le Berre M, Piel M (2015) Optical volume and mass measurements show that mammalian cells swell during mitosis. J Cell Biol 211(4):765–774
69. Jin M, Tavella F, Wang S, Yang Q (2022) In vitro cell cycle oscillations exhibit a robust and hysteretic response to changes in cytoplasmic density. Proc Natl Acad Sci 119(6):e2109547119
70. Li Z, Wang S, Sun M, Jin M, Khain D, and Yang Q (2024) Comprehensive parameter space mapping of cell cycle dynamics under network perturbations. ACS Synth Biol 13(3):804–815
71. Kane DA, Kimmel CB (1993) The zebrafish midblastula transition. Development 119(2):447–456
72. Samoilov MS, Price G, Arkin AP (2006) From fluctuations to phenotypes: the physiology of noise. Sci STKE 2006(366):re17–re17
73. Anderson GA, Gelens L, Baker JC, Ferrell JE (2017) Desynchronizing embryonic cell division waves reveals the robustness of Xenopus laevis development. Cell Rep 21(1):37–46
74. Lan G, Sartori P, Neumann S, Sourjik V, Tu Y (2012) The energy–speed–accuracy trade-off in sensory adaptation. Nat Phys 8(5):422–428
75. Govern CC and Pieter Rein ten Wolde (2014) Energy dissipation and noise correlations in biochemical sensing. Phys Rev Lett 113(25):258102
76. Cao Y, Wang H, Ouyang Q, Tu Y (2015) The free-energy cost of accurate biochemical oscillations. Nat Phys 11(9):772–778

77. Rust MJ, Golden SS, O'Shea EK (2011) Light-driven changes in energy metabolism directly entrain the cyanobacterial circadian oscillator. Science 331(6014):220–223
78. Phong C, Markson JS, Wilhoite CM, Rust MJ (2013) Robust and tunable circadian rhythms from differentially sensitive catalytic domains. Proc Natl Acad Sci 110(3):1124–1129
79. Wang S, Yourston L, Maryu G, et al. (2026) Energy-Precision Trade-off in Mitotic Oscillators Revealed by ATP Modulation in Artificial Cells. bioRxiv. https://doi.org/10.64898/2026.03.02.709190
80. Chang JB, Ferrell JE Jr (2013) Mitotic trigger waves and the spatial coordination of the Xenopus cell cycle. Nature 500(7464):603–607
81. Bischof J, Brand CA, Somogyi K, Májer I, Thome S, Mori M, Schwarz US, Lénárt P (2017) A Cdk1 gradient guides surface contraction waves in oocytes. Nat Commun 8(1):849
82. Vergassola M, Deneke VE, Di Talia S (2018) Mitotic waves in the early embryogenesis of drosophila: bistability traded for speed. Proc Natl Acad Sci 115(10):E2165–E2174
83. Nolet FE, Vandervelde A, Vanderbeke A, Piñeros L, Chang JB, Gelens L (2020) Nuclei determine the spatial origin of mitotic waves. Elife 9:e52868
84. Afanzar O, Buss GK, Stearns T, Ferrell JE Jr (2020) The nucleus serves as the pacemaker for the cell cycle. Elife 9:e59989
85. Cebrián-Lacasa D, Piñeros L, Vanderbeke A, Ruiz-Reynés D, Wouters T, Goryachev AB, Frolov N, Gelens L (2024) Spiral waves speed up cell cycle oscillations in the frog cytoplasm. arXiv preprint arXiv:2412.16094
86. Puls O, Ruiz-Reynés D, Tavella F, Jin M, Kim Y, Gelens L, Yang Q (2024) Spatial heterogeneity accelerates phase-to-trigger wave transitions in frog egg extracts. Nat Commun 15(1):10455
87. Maryu G, Yang Q (2022) Nuclear-cytoplasmic compartmentalization of cyclin B1-Cdk1 promotes robust timing of mitotic events. Cell Rep 41(13)
88. Huang J-H, Chen Y, Huang WYC, Tabatabaee S, Ferrell JE Jr (2024) Robust trigger wave speed in Xenopus cytoplasmic extracts. Nat Commun 15(1): 5782
89. Cheng X, Ferrell JE Jr (2018) Apoptosis propagates through the cytoplasm as trigger waves. Science 361(6402):607–612
90. Co HKC, Wu C-C, Lee Y-C, Chen S-H (2024) Emergence of large-scale cell death through ferroptotic trigger waves. Nature 631(8021):654–662
91. Orlando DA, Lin CY, Bernard A, Wang JY, Socolar JES, Iversen ES, Hartemink AJ, Haase SB (2008) Global control of cell-cycle transcription by coupled CDK and network oscillators. Nature 453(7197): 944–947
92. Papagiannakis A, Niebel B, Wit EC, Heinemann M (2017) Autonomous metabolic oscillations robustly gate the early and late cell cycle. Mol Cell 65(2): 285–295
93. Liu X, Oh S, Peshkin L, Kirschner MW (2020) Computationally enhanced quantitative phase microscopy reveals autonomous oscillations in mammalian cell growth. Proc Natl Acad Sci 117(44): 27388–27399
94. Aydogan MG, Steinacker TL, Mofatteh M, Wilmott ZM, Zhou FY, Gartenmann L, Wainman A, Saurya S, Novak ZA, Siu-Wong S, et al (2020) An autonomous oscillation times and executes centriole biogenesis. Cell 181(7):1566–1581
95. Bakshi A, Iturra FE, Alamban A, Rosas-Salvans M, Dumont S, Aydogan MG (2023) Cytoplasmic division cycles without the nucleus and mitotic CDK/cyclin complexes. Cell 186(21):4694–4709
96. Cross FR, Buchler NE, Skotheim JM (2011) Evolution of networks and sequences in eukaryotic cell cycle control. Philos Trans R Soc B Biol Sci 366(1584):3532–3544
97. Lande-Diner L, Boyault C, Kim JY, Weitz CJ (2013) A positive feedback loop links circadian clock factor clock-bmal1 to the basic transcriptional machinery. Proc Natl Acad Sci 110(40):16021–16026
98. Stewart-Ornstein J, Ho Cheng WJ, Lahav G (2017) Conservation and divergence of p53 oscillation dynamics across species. Cell Syst 5(4):410–417
99. Wagner A (2005) Circuit topology and the evolution of robustness in two-gene circadian oscillators. Proc Natl Acad Sci 102(33):11775–11780
100. Elowitz MB, Leibler S (2000) A synthetic oscillatory network of transcriptional regulators. Nature 403(6767):335–338
101. Stricker J, Cookson S, Bennett MR, Mather WH, Tsimring LS, Hasty J (2008) A fast, robust and tunable synthetic gene oscillator. Nature 456(7221): 516–519

Leveraging Autonomous Biological Oscillators for Directed Evolution of Switchable Proteins

8

Vojislav Gligorovski, Marco Labagnara, Lorenzo Scutteri, and Sahand Jamal Rahi

Abstract

Engineering proteins that can switch between states is of strong interest to biotechnology. These include multi-state transcription factors, membrane receptors, kinases, and synthetic molecular logic gates. While traditional directed evolution approaches excel at optimizing steady-state protein functionalities, they require substantial adaptation to treat transitions in states of proteins. This chapter describes how the emerging regulatory logic of autonomous biological oscillators can be used to impose time-varying selection pressure for evolving switchable proteins. We first review the challenges of evolving switchable proteins and examine past strategies, including fluorescence-activated cell sorting (FACS) screens, alternating chemical selection and counterselection methods, and phage display techniques. We then compare these with "optovolution," an approach recently developed by our lab, which couples a protein's activity to host cell cycle progression and leverages programmable light inputs to drive continuous evolution of proteins. To this end, we explain the design principles underlying the use of an autonomous oscillator for evolution and how it promotes selection for both "on" and "off" states, as well as switching between them. In a direct comparison with a conventional directed evolution method, we elucidate how optovolution proves more reliable by avoiding evolutionary routes that hinder counterselection. In conclusion we discuss how optovolution opens new avenues for creating proteins with complex logic and dynamic control, complementing latest advances in computational protein design.

Keywords

Directed evolution · Synthetic biology · Cell cycle control · Protein engineering · Optogenetics · Switchable proteins · Multi-state proteins · Light-inducible transcription factors · LOV domain · EL222 · PhyB/PIF3 · Tet-On system · AND gate · Budding yeast

8.1 Introduction

Living systems rely on proteins not only as catalysts or structural components but also as switches that toggle between functional states in response to changes in the environment. Such switchable, multi-state proteins are used for reversible, on-demand control in synthetic biology. However, when designed rationally, or placed in a new biological context, protein switches may not be immediately useful and

V. Gligorovski · M. Labagnara · L. Scutteri · S. J. Rahi (✉)
Laboratory of the Physics of Biological Systems, Institute of Physics, EPFL, Lausanne, Switzerland
e-mail: sahand.rahi@epfl.ch

M. G. Aydogan (ed.), *Biological Time Control*, Advances in Experimental Medicine and Biology 1521,
https://doi.org/10.1007/978-3-032-36189-9_8

require optimization. Directed evolution, based on iterative mutation and selection, has been one of the key techniques for generating or improving protein properties. In traditional directed evolution, one typically selects for a single property under fixed conditions; for instance, an enzyme's catalytic efficiency [1–5], binding affinity of an antibody [6, 7], or different properties of fluorescent proteins [8, 9]. In contrast, multi-state, allosterically regulated protein switches have conformation-dependent activity.

Examples of the latter include allosteric enzymes, transcription factors regulated by metabolites or light, kinases, and membrane receptors. The evolution of such switchable proteins, which have at least two functional states ("on" and "off") poses additional challenges for the design of evolutionary experiments [10, 11]. For a given switchable protein of interest (SPOI), one must select for high activity in one state and low activity in another, while preserving the protein's capacity to switch between states reliably and to respond to exogenous input. The central challenge is that applying selection to only one state at a time can corrupt the other state. For example, evolving switchable proteins by only selecting for the "on" state poses the risk that mutations which lock the protein in the "on" conformation will be fixed, thus compromising switching.

Even more challenging is applying selection pressure toward specific switching kinetics. Existing directed evolution methods, which often rely on endpoint assays after hours of induction, are ill-suited for directly selecting for fast switching kinetics. A mutant might pass endpoint assays for "on" and "off" states, but its slow transition between the two states will make its use impractical. Thus, an ideal evolutionary approach would also apply pressure on the speed of transitions between states.

Various approaches have been developed to evolve multi-state proteins with steady-state selection mechanisms: A common strategy is to alternate selection for protein states; that is, to repeatedly toggle environmental conditions to favor one state first, then the other (Fig. 8.1a). For example, to evolve a ligand-responsive receptor, one alternates between the presence and absence of ligand: in the presence of ligand, high protein activity is selected; and in its absence, the low protein activity [12–46]. This strategy has been implemented using different selection methods, e.g., fluorescence-activated cell sorting (FACS), chemical selection, and in vitro selection (Fig. 8.1b–d).

In this chapter, building on our recent work [10], we explore a novel solution to these challenges: using the emerging logic of autonomous biological oscillators to drive the directed evolution of switchable proteins with a focus on leveraging the cell cycle oscillator that drives DNA replication and segregation [10] as a selection mechanism. In doing so, the central goal is to couple the proliferation of an organism to the switching behavior of a SPOI. We first review in greater detail how previous methods have evolved switchable proteins, and what limitations these techniques face. We then describe how to overcome these limitations by enforcing a rhythmic selection that continuously tests both output states of the SPOI in each cycle of an autonomous oscillator.

8.2 Traditional Methods for Evolving Protein Switches

Several experimental strategies have been devised to direct the evolution of switchable proteins. These strategies can be classified by how they impose selection for the "on" and "off" states of the SPOI. Below and in Fig. 8.1, we highlight three categories of techniques:

8.3 FACS-Based Alternating Screening

Fluorescence-activated cell sorting (FACS) can enable rapid isolation of protein variants by coupling the activity of the switchable protein to a fluorescent readout. Populations of cells are subjected to alternating sorting: one for high fluorescence upon growth under conditions that induce the SPOI's "on" state, and the other for

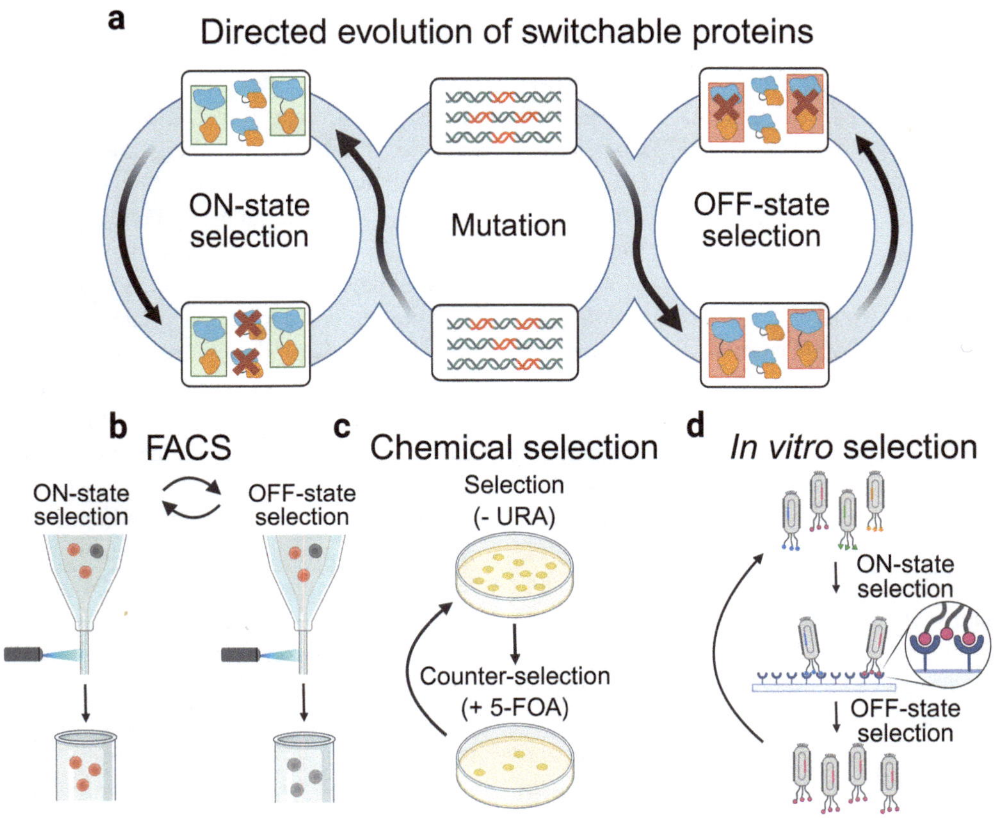

Fig. 8.1 (**a**) Existing approaches for evolving switchable proteins are based on alternating selection conditions for high on-state and low off-state activity. In practice, this can be achieved by (**b**) alternating FACS screening, (**c**) chemical selection and counterselection, or (**d**) in vitro selection and elution

low fluorescence corresponding to the "off" state upon removing the inducer. Unlike chemical selection, FACS permits quantitative tuning of selection stringency by adjusting the gating threshold for the sorted population.

In a sophisticated version of this method, Wang et al. [20] evolved a light- and calcium-gated transcription factor ("FLARE") by six rounds of FACS sorting. A yeast-display library of $>10^7$ AsLOV2 variants fused to a TEV cleavage site and flanked by HA/FLAG epitopes was alternately enriched for light-triggered cleavage and dark-state resistance. In the positive sorts, yeast cells were incubated with TEV protease under illumination, upon which cells with low FLAG/HA fluorescence ratios (corresponding to efficient cleavage by TEV) were selected. In the negative sorts, the library was treated with the protease in the dark, upon which cells with high FLAG/HA ratios (indicative of intact TEVcs) were retained. Six screening rounds yielded "eLOV," a five-mutation LOV variant that, when transplanted into FLARE, increased the light/dark signal in HEK cells as well as in neurons.

FACS-based methods have the advantage of speed of selection: millions of cells can be sorted in minutes, and multiple selection rounds can be completed in a day [47]. However, FACS may not be easily accessible. Another limitation of the system is that one must couple the SPOI's activity to fluorescent output. This is relatively straightforward, e.g., for transcription factors, but evolving other types of proteins might require additional engineering steps to create a suitable reporter. For instance, evolving protein–protein interactions requires building a two-hybrid system [48]. Despite these caveats, the tunable, high-dimensional selection that FACS offers has been successfully used to evolve a variety of biosensors and synthetic gene circuits with improved dynamic characteristics [12, 16, 17, 20, 30, 33–35, 40, 43].

8.4 Chemical Selection/Counterselection

Another directed evolution approach uses markers that can be selected for and against. In *E. coli*, such dual selectors are the herpes simplex virus thymidine kinase (hsvTK) and the tetracycline efflux pump TetA. hsvTK can rescue cells in the presence of 5FdU but renders cells susceptible to nucleoside analog pro-drugs [32]. TetA is essential for tetracycline resistance but also increases uptake of toxic metals, making cells more sensitive to $NiCl_2$ [37].

In budding yeast, a widely used selection/counterselection mechanism is based on the *URA3* gene [49–51]. *URA3* encodes orotidine-5′-phosphate decarboxylase, an enzyme required for uracil biosynthesis. Cells lacking *URA3* cannot grow in the absence of uracil (−Ura medium). Conversely, cells expressing *URA3* die on media containing 5-fluoroorotic acid (5-FOA), which Ura3 converts into 5-fluorouracil, which is toxic [50].

By placing *URA3* under the control of a SPOI, one can alternate between −Ura and +5-FOA conditions to evolve a SPOI. For instance, Taslimi et al. [22] used a yeast two-hybrid *URA3* reporter to evolve CRY2-CIB optogenetic switch variants with different photocycle lifetimes. Their alternating −Ura (positive) and +5-FOA (negative) selection conditions, paired with illumination regimes allowed screening for both fast-cycling and slow-cycling variants. This approach yielded two key CRY2 mutants: (a) L348F, a slow-cycling variant that prolonged the light-activated state time (4.5× slower than WT, $t\frac{1}{2} \approx 24$ min), and (b) W349R, a fast-cycling variant that switches off faster than WT (~50% faster than WT, $t\frac{1}{2} \approx 2.5$ min). Transplanting L348F into a split Cre showed a ~5-fold higher light/dark dynamic range and robust recombination after a single brief pulse with low dark background, illustrating how *URA3*/5-FOA schemes can select for desired switching kinetics and functional performance in vivo.

While effective, this approach has potential pitfalls. In particular, the system is sensitive to leaky expression of *URA3*, which can prevent kick-starting an evolutionary campaign, as we show below. Additionally, the selection pressure that can be encoded using this system cannot be easily tuned. Lastly, as shown by others [52] and confirmed by us below, the probability of obtaining beneficial mutations in the SPOI versus mutations elsewhere in the genome is not optimized in this system, meaning "cheater" variants (i.e., mutations outside SPOI) can dominate the evolved population. To illustrate these challenges in practice, we performed tests of the *URA3*/5-FOA method using a light-inducible transcription factor (El222) as a SPOI (see the Sect. 8.8).

In summary, *URA3*/5-FOA selection is useful for enforcing the selection of two opposite states, but it requires tight control of basal activity and tends to be slow, with each round taking approximately 2–3 days for colonies to form. Nonetheless, it has been one of the few practical ways to evolve switchable systems and has yielded improved optogenetic dimers [22] and other ligand biosensors [15].

8.5 Phage Display and In Vitro Selection

A third approach uses phage-based systems to select for binding differences between SPOI states. Phage display can be adapted for protein switches by performing selection for binding under one condition and then counterselection using elution under another condition.

A notable example is the evolution of the improved light-inducible dimer (iLID) by Guntas et al. [24]. This system is based on the binding of the SsrA peptide to the SspB protein, which was made light-inducible by fusing SsrA to the terminal Jα helix of a LOV domain [53]. Using Rosetta-guided design, the authors built a 743-point-mutation library inside the LOV domain, displayed the variants on M13 phage, and alternated positive selection (binding under blue light) and negative selection (dark-state elution), followed by ELISA ranking and recombination of top hits. This campaign yielded improved iLID variants with a >50-fold dynamic range. The switch enabled second-scale recruitment and reversible control of small-GTPase signaling in mammalian cells.

Phage display and ribosome display are extremely powerful in terms of the number of variants that can be screened in a single run [54, 55]. Their in vitro nature also allows precise control over biochemical conditions and rapid switching between buffer with and without ligand. One limitation of display methods is that they are best suited for evolving binding affinity [23–26]. Also, translating in vitro results to in vivo functionalities may face challenges, as the protein might behave differently in an intracellular context.

An emerging directed evolution method at the interface of in vivo and in vitro approaches is compartmentalized partnered replication (CPR) [56, 57]. In CPR, one links the activity of interest to the replication of genes within micro-compartments (water-in-oil droplets). This technique couples selection to gene amplification, enabling enormous variant populations to be processed in liquid format, though it requires a bespoke setup and careful design. To evolve protein switches, CPR has been used as a positive selection step coupled with another method for negative selection [31]. Alternatively, Ellefson et al. [46] created an inverter circuit that also allows negative selection based on CPR. In their implementation, TrpR activity was wired to Taq DNA polymerase so that, in positive CPR, ligand-activated variants derepressed a λPR–Taq cassette and were exponentially amplified within droplets. Whereas in negative CPR, placing Taq under the PTrp promoter prevented amplification of constitutively active or l-tryptophan–responsive variants.

In summary, existing selection methods have tradeoffs that need to be considered when designing directed evolution campaigns for switchable proteins. Selection by FACS requires a suitable reporter and specialized equipment; *URA3*/5-FOA counterselection imposes strong on/off pressure but limited tunability; and phage/in vitro systems can explore larger libraries than *in cellulo* systems but are so far limited to evolving binders and evolved variants might not function inside eukaryotic cells without further modification. Although each approach has yielded successes, all are constrained in their capacity to continuously enforce selection for dynamic switching properties at short timescales.

Other key variables in the experimental evolution of switchable proteins include the selection scheme (Fig. 8.2a), mutagenesis strategy (Fig. 8.2b), the choice of host organism (bacterial, yeast, or phage/in vitro systems, Fig. 8.2c), and the format of evolution (liquid culture vs. solid medium screening, Fig. 8.2d). We summarize these design choices from past directed evolution campaigns for protein switches [12–46] below. In terms of selection modality, studies have used a broad range of schemes (Fig. 8.2a): chemical marker-based survival selection is most common (used in ~40% of campaigns), followed by fluorescence-based selection by FACS (~30%), with a smaller number of experiments using manual selection, phage display, or CPR. Among mutagenesis approaches (Fig. 8.2b), error-prone PCR has been the dominant strategy (roughly half of published campaigns that focused on evolving switches), with the remainder relying on site-directed mutagenesis or rationally designed libraries. The most frequent hosts for in vivo evolution (Fig. 8.2c) were *E. coli*, used in ~65% of studies, and budding yeast in ~30% of articles, alongside a small number of published phage or cell-free systems.

Each of these factors influences the trajectory and outcome of evolution and, akin to the selection schemes, reflects the tradeoffs that one faces when designing evolutionary experiments. For example, using yeast as a host can be preferable for the evolution of eukaryotic proteins that might require post-translational modifications

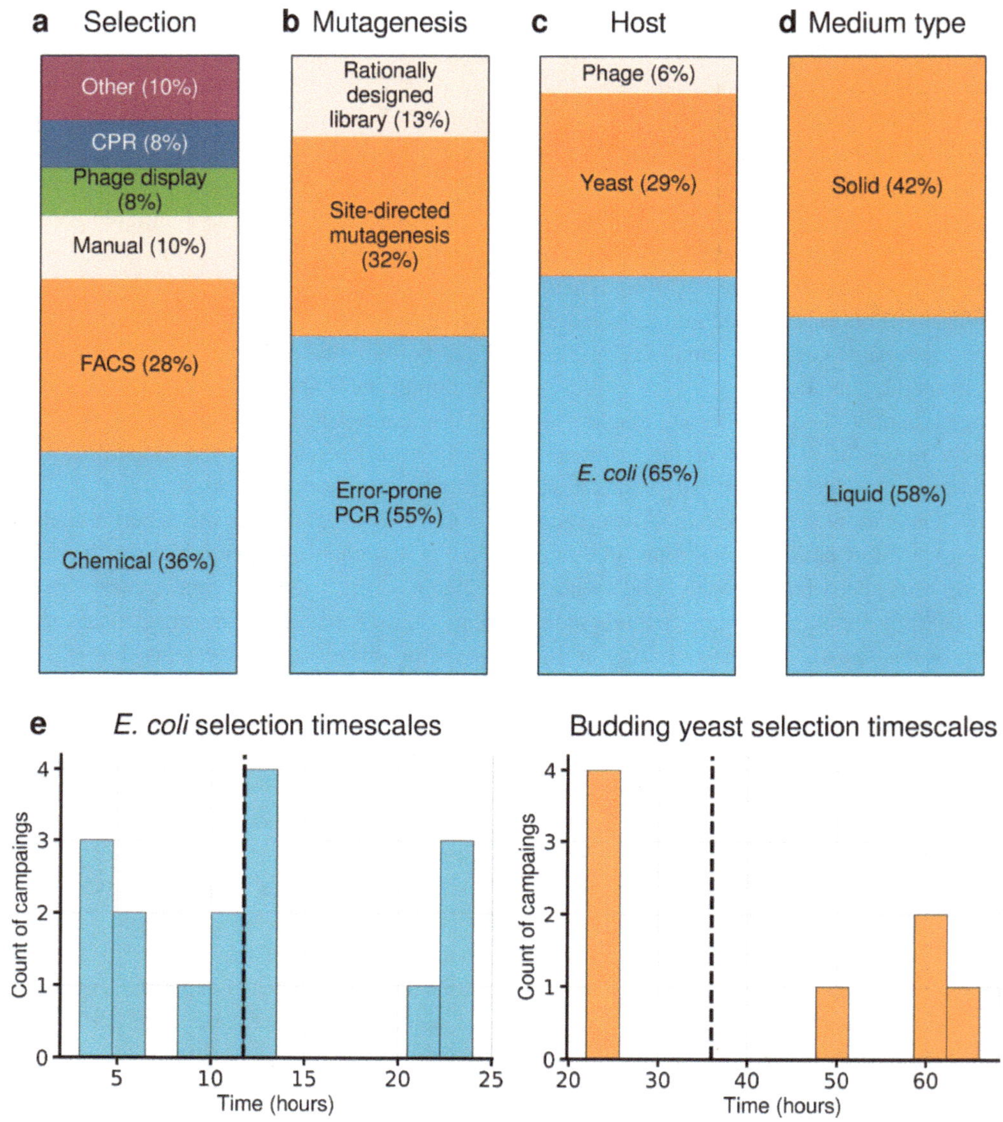

Fig. 8.2 Key parameters in previous directed evolution campaigns of switchable proteins [12–46]. (**a**) Selection schemes used across studies, including marker-based survival selection (~40%), fluorescence-based FACS (~30%), and less frequent approaches: manual selection, phage display, and compartmentalized partnered replication (CPR). (**b**) Mutagenesis strategies employed, with error-prone PCR used in roughly half of campaigns, followed by site-directed mutagenesis and rationally designed libraries. (**c**) Host organisms used for evolving switches: *E. coli* (~65%), budding yeast (~30%), and a small number of phage or cell-free systems. (**d**) Experimental format: most campaigns (~60%) were performed in liquid culture, while others used solid media (colony screening on plates). (**e**) Timescales of selection rounds in *E. coli* versus *S. cerevisiae* (averaged when studies used multiple cycle times). Yeast rounds (orange, n = 8; 22–66 h; median ~42 h) are typically longer than *E. coli* rounds (blue, n = 18; 3–30 h; median ~12 h), consistent with slower growth of budding yeast. Dashed lines indicate medians

[58]. However, the slower growth of yeast compared to bacteria can lengthen each on–off cycle, thereby constraining the maximal switching speed that can be imposed on the switchable protein of interest (SPOI). This difference is apparent when comparing typical cycle times (Fig. 8.2e): selection rounds in *S. cerevisiae* often require one or more days per round (22–66 h range; median ~42 h), whereas in *E. coli* many rounds can be completed within a day (3–30 h range; median ~12 h). Similarly, the experimental format imposes its own compromises (Fig. 8.2d): Liquid culture is readily amenable to continuous or automated evolution (such as in PACE [59]), in which mutants compete and the best variant can sweep the population; by contrast, using solid media allows comprehensive fitness-agnostic isolation of individual mutants but is more labor-intensive. Error-prone PCR is straightforward but introduces frequent deleterious mutations, which increases the required throughput of the screen.

8.6 Leveraging Autonomous Biological Oscillators for Evolution

We hypothesized that a biological clock [60–62] can serve as a dynamic selection mechanism for the evolution of switchable proteins. Classic examples of such clocks include the circadian oscillator [63], which runs with ~24-h cycles, and the cell division cycle, which can oscillate with a period on the order of hours [64]. In the context of evolving a SPOI, an oscillator can impose intrinsically alternating selection conditions without human intervention. If the SPOI state is linked to oscillator phases, then the cycles will give a fitness advantage to those cells in which the SPOI switches on time. Thus, alternating selection pressure would be built into the operation of the oscillator. The timing of the oscillator defines the range of evolvable switching dynamics.

To utilize an autonomous biological oscillator for evolving switchable proteins, several design requirements must be met:

- *Tight Coupling of the SPOI to the Oscillator*: The SPOI output must be directly entrained to the oscillator, such that cellular fitness depends on the SPOI's correct state at the correct time. In practice, this means that if the oscillator is held in one state (i.e., constant "on" or constant "off" input), cells should proliferate markedly worse.
- *Mutational Robustness*: The system must be engineered so that redundant or compensatory pathways that could bypass the selection are removed. Anticipating such pathways might not always be trivial and might require iterative optimization of the genomic background used for evolution [10, 52, 65].
- *Reliable Oscillator Timing and Input Synchronization*: The oscillator and the external input driving the SPOI must be coordinated to enable continuous selection. The SPOI input should be controlled at a timescale shorter than the oscillator period.
- *Tolerance for Initially Weak Performers*: At the start of evolution, the wild-type SPOI might perform poorly. If selection is too strict initially, even the starting genotype cannot survive long enough to diversify. Therefore, a practical design includes a temporary "buffer" in selection to allow the initial population to grow slowly despite suboptimal switching. This could be a helper gene that is lost over time, or permissive selection conditions that are gradually tightened.
- *Modularity and Generalizability*: The system should be designed so that it can adapt to SPOIs with different biological functions. Ideally, the core oscillator coupling mechanism must be agnostic to the specific SPOI function.
- *Genetic Stability During Continuous Culture*: Long-term evolution experiments can lead to mutations in the host chassis, unrelated to the SPOI, such as polyploidy, or chromosomal rearrangements, which could change the trajectory of SPOI evolution (e.g., by changing the gene copy number). Genome engineering and the choice of selection mechanisms can help ensure integrity and prevent the emergence of variants that undermine the evolutionary campaign.

By meeting these criteria, an oscillator-driven evolution system can achieve continuous selection of dynamic, multi-state proteins, with states being iteratively selected as the oscillator transitions through different phases. Every oscillation becomes a test of the SPOI's ability to switch correctly on the required timescale.

In principle, any biological process whose oscillations influence cellular fitness could be used for this purpose (cell cycle, circadian rhythms, developmental cycles, or synthetic gene oscillators). We chose the cell division cycle for our implementation [10] because it has relatively fast cycles and is strongly linked to fitness. By linking the SPOI's activity to the cell cycle, we imposed a requirement that the SPOI be active at one step but inactive at another.

8.7 The Cell Cycle as a Dynamic Selection Mechanism

The eukaryotic cell cycle is a robust and well-characterized biological cycle, with a typical period of 90–120 min in budding yeast under optimal conditions [66, 67]. Its core events are driven by interactions between cyclins, the cyclin-dependent kinase Cdk1, and the anaphase-promoting complex APC/C, which regulate sequential cell cycle events such as budding, DNA replication, and mitosis [66]. Cell cycle control coordinates and temporally segregates mutually exclusive functional states, ensuring that each of the central cell cycle events happens exactly once per cycle [66]. We leveraged this property as a selection mechanism for a continuous directed evolution paradigm termed optovolution [10].

We first surveyed core regulators of *S. cerevisiae* to find one that is required for one cell-cycle transition but harmful if expressed continuously (Fig. 8.3a). Specifically, cells should require pulsatile activity for proliferation (Fig. 8.3b) but die if the SPOI is continuously off (Fig. 8.3c) or on (Fig. 8.3d). To this end, G1/S and mitotic cyclins proved unsuitable as continuous induction failed to arrest or led to irreversible blocks. On the other hand, Clb5 (an S-phase cyclin) fulfilled two of the requirements: pulsed expression enabled periodic cycles (Fig. 8.3b), sustained expression impeded growth and became lethal over multiple cycles (Fig. 8.3d).

To make growth strictly contingent on SPOI output, we further engineered the genomic background. We removed endogenous cyclins that are partially redundant with Clb5, rendering Clb5 both essential (Fig. 8.3c) and dosage-sensitive. Specifically, we deleted ***CLN1–3*** and ***CLB3–6***, thereby preventing trivial bypass mutations in the cell cycle control system. Next, because pulsed Clb5 alone initiates Start unreliably, we added a construct in which the G1/S cyclin *CLN2* is controlled by the same SPOI. This preserves Clb5 essentiality for S-phase and mitotic entry while increasing the likelihood that a brief pulse of SPOI activity triggers a cycle.

To initiate evolution even when the starting SPOI can only be forced "on" and "off" insufficiently, we use a "training-wheel" plasmid that carries copies of *CLB5* and *CLN2*; the campaign begins by counterselecting against this plasmid in 5-FOA medium while imposing permissive pulses, then tightening the pulse regime to create strong selection. The mutator allele *pol3-L523D* boosts genome-wide mutation rates, increasing mutational diversity above background levels.

During evolution using this strain, we selected against both the mutations that lock the SPOI in the "on" or "off" states and the mutations that impair the timing or level of activity. This tight coupling satisfies the key requirement we set forth: each cell division event imposes selection for proper SPOI switching and mutations elsewhere in the genome should not bypass the SPOI control.

In our recent work implementing this paradigm, we demonstrated that optovolution can improve several classes of protein switches [10]:

1. Starting with the wild-type El222 optogenetic transcription factor [68–72], the oscillator-driven evolution produced mutants with increased sensitivity, reduced leakiness, or altered spectral properties. Specifically, we

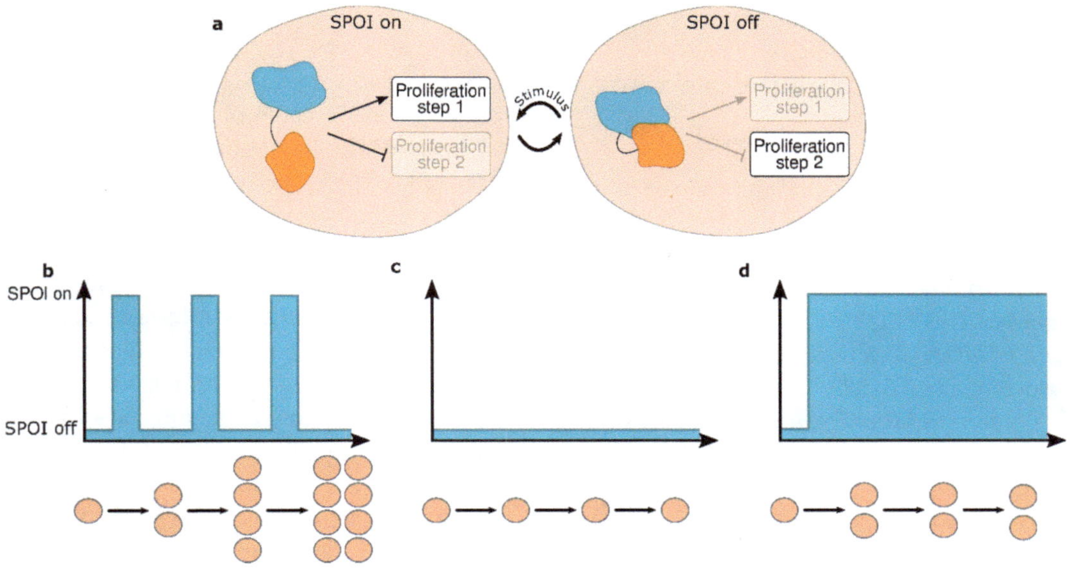

Fig. 8.3 (**a**) In optovolution, the SPOI controls a cell cycle regulator that is required at one stage of proliferation but toxic at another. The key idea underlying the continuous self-selection of proteins is that engineered cells proliferate only in response to alternating SPOI activity (**b**), not under sustained off (**c**) or on (**d**) states

discovered multiple El222 variants that displayed up to 16-fold higher transcriptional output under low light compared to wild type, mutants with dramatically lower leakiness (basal activity in the dark nearly indistinguishable from background), and green-light-responsive variants which can be multiplexed with blue-responsive LOV domains.

2. The PhyB-Pif3 optogenetic system requires the exogenous supplementation of the chromophore PCB [73]. Using optovolution, we discovered that deleting the ABC transporter *YOR1* obviates PCB supplementation in yeast. Further work showed that Yor1 exports biliverdin or a biliverdin derivative out of the cell so that in its absence, endogenous levels suffice for PhyB-Pif3 function.
3. To generalize beyond light sensors, we evolved a logic gate in which a degron-tagged rtTA—a part of the widely used Tet-On system—requires both transcriptional induction (via El222/*LIP*) and doxycycline. Starting from weak output, optovolution yielded rtTA variants with substantially higher ON activity while maintaining low leakiness, demonstrating that the same cell cycle-coupled scheme evolves computational functions, not just single-input switches.

In summary, using the cell cycle as a dynamic selection engine proved to be a powerful way to maintain selection pressure on both states of the switch. The cell's own oscillator enforced the timing, and the experimenter simply had to program a periodic light signal and keep the culture fed. This approach effectively intertwines evolutionary selection with cell physiology: the selective agent is not a static drug or a manual step, but the ticking of the cell cycle clock. We call this approach optovolution to highlight the combination of optical control and continuous evolution. It opens the door to evolving proteins that respond on the timescale of biological processes, and that potentially can handle more complex input–output logic than binary on/off.

8.8 Comparison of Optovolution to URA3/5-FOA Selection/Counterselection Method

In order to compare optovolution with a traditional alternating selection method, we carried out parallel evolution experiments with the same SPOI. Specifically, we tried to evolve the blue light-inducible El222 transcription factor using *URA3*/5-FOA selection, and compare it with optovolution of *EL222*.

For the *URA3*/5-FOA method, we constructed a yeast strain where El222 drove the expression of *URA3* (mimicking the design of Taslimi et al. [22]). In principle, under blue light, this strain should express *URA3* and grow on −Ura medium. In darkness, El222 is off, so *URA3* should not be expressed and cells should survive in medium containing 5-FOA.

We found this system to be highly sensitive to leakiness of El222 in darkness. Even a small amount of background *URA3* expression in the "off" state was presumably enough to kill cells (Fig. 8.4). This occurred despite having single copies of *EL222* and *URA3* integrated in the genome to minimize expression. As a result, the counterselection step was not possible, and we could not complete a full alternating selection cycle. This illustrates a general issue that, if the selection thresholds are not tunable, one state's selection can dominate and wipe out the population before the other state can even be tested.

We next attempted selection on 5-FOA, reasoning that we could obtain initial survivors and thus reduce the leakiness of El222. We performed a "dark survival" selection, looking for low-leakiness variants that could survive on 5-FOA medium plates. However, upon sequencing 200 evolved colonies, we discovered that none had mutations in *EL222*. Instead, some had mutations in *URA3* that presumably inactivated the enzyme (data not shown), allowing cells to evade selection without improving the switch.

We next compared the approach to optovolution. An optovolution strain running under the control of El222 was used to evolve light-sensitized variants by growing cells on a low-light pulsing regime [10]. In contrast to the *URA3*/5-FOA method, the optovolution experiment readily yielded improved *EL222* variants.

One could argue that our implementation of the *URA3*/5-FOA method was not fully optimized and that it could be improved by using a weaker *URA3* variant, tuning the enzyme expression level, employing an alternative light-inducible promoter backbone with reduced leakiness, or adjusting the concentration of 5-FOA. However, these optimizations involve trial and error. Optovolution, on the other hand, needed no such tweaking. The requirements to run the cell cycle balanced the evolutionary pressure: if a mutant

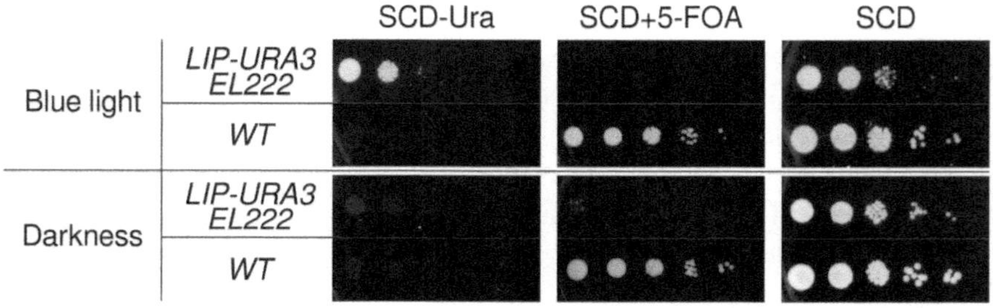

Fig. 8.4 Challenges of using the *URA3*/5-FOA method for the directed evolution of El222. The test *ura3Δ0* strain carried single copies of *LIP-URA3* and *PGK1pr-EL222* constructs each. The control strain was wild-type W303, which is auxotrophic for uracil. Because the test strain did not proliferate in darkness on +5-FOA, we could not perform a light-sensitization campaign. Further, none of 200 mutant colonies that emerged under these conditions had mutations in *EL222*. Therefore, the opposite campaign for low-leakiness mutants also failed. SCD, synthetic complete medium with 2% glucose

was too leaky, it would suffer in the dark phase; if it was too weak under light, it could not trigger S-phase.

Another comparison point is the speed of evolutionary improvement. In an alternating plate method, after each round, researchers typically isolate clones, test them, mutagenize again, etc. After several rounds, which might take weeks, they may obtain a desired variant. Optovolution, being continuous, gave us significantly improved variants in a few days without intervention.

Lastly, optovolution is well-suited to complex selection patterns or multi-input logic. The *URA3*/5-FOA method, FACS, or phage display can, in principle, be extended to more complicated switching (e.g., a protein that responds to two signals) by adding more alternating conditions at the cost of increased labor. In contrast, leveraging an oscillator allows adding complexity by simply programming the input signal differently, all while cells keep cycling. In principle, one could evolve a protein that integrates several different wavelengths of light by cycling through a sequence of light combinations in each cell cycle (e.g., a cycle of no light, red light, blue light, and both lights to select for proper response in each of four states). Only cells whose SPOI gives the correct output in all of those phases would successfully divide. While we have not demonstrated multi-color input evolution, the principle is a natural extension of our previous work.

8.9 Conclusions and Outlook

Directed evolution enables fitness landscape explorations that are infeasible with rational design by sampling vast mutant libraries and allowing beneficial mutations to accumulate iteratively. Recent advances in the field have enabled non-continuous directed evolution of SPOIs. In contrast, optovolution represents a novel paradigm for continuous directed evolution of SPOIs. By coupling budding yeast's internal cell cycle clock with an external control signal to select for protein switching, the method evolves dynamic, multi-state protein properties [10].

Several challenges can in principle now be addressed. One is the creation of more sophisticated protein logic gates [74–78]. Many cellular decisions involve proteins and circuits that integrate multiple inputs, e.g., the amount of DNA damage and arrest times jointly determining the override decision of the DNA damage checkpoint [79]. Using multiple orthogonal light inputs (i.e., different wavelengths), researchers can now select for logic gating that responds correctly to a predefined desired combination of inputs. Recent expansions of the optogenetic toolbox provide independent control of various light-responsive domains (UV, blue, green, and red), making their multiplexing increasingly feasible [80–84]. An optovolution scheme could, for example, evolve a single protein to perform a Boolean logic function, which has hitherto only been achieved with semi-automated alternating selection schemes or clever circuit design [12–46]. Continuous evolution ought to streamline this process by allowing all conditions to be imposed in a single culture.

We also foresee integration with laboratory automation and protein design driven by artificial intelligence (AI) [85–88]. Much of the evolution process can be automated by using continuous culture devices and computer-controlled light input. Algorithms can be used to adjust the light schedule or mutagenesis rate on the fly based on real-time feedback of population growth. Additionally, optovolution can complement the AI design of protein switches. Computational design can be used to propose an initial pool of protein switches that are then optimized by optovolution. The results of selection can then be fed back into the AI model, establishing an iterative platform for switchable protein engineering.

In summary, autonomous biological oscillators offer a powerful new way to impose complex, time-dependent selection pressures for protein evolution. Our case study comparing it to more established methods highlights its robustness and speed benefits. Looking ahead, the ability to evolve proteins with complex control logic and dynamic behavior will be invaluable for building the next generation of synthetic biology devices, ranging from biosensors and controllers for cell biology applications to reversible gene switches for cell-based therapies.

Acknowledgments We acknowledge support for VG, ML, LS, and SJR provided by SNSF grants CRSK-3_190526, 310030_204938, and CRSK-3_221036 awarded to SJR. LS is also supported by the EPFLglobaLeaders doctoral fellowship.

References

1. Mills DR, Peterson RL, Spiegelman S (1967) An extracellular Darwinian experiment with a self-duplicating nucleic acid molecule. Proc Natl Acad Sci USA 58:217–224
2. Kuchner O, Arnold FH (1997) Directed evolution of enzyme catalysts. Trends Biotechnol 15:523–530
3. Schmidt-Dannert C, Arnold FH (1999) Directed evolution of industrial enzymes. Trends Biotechnol 17: 135–136
4. Packer MS, Liu DR (2015) Methods for the directed evolution of proteins. Nat Rev Genet 16:379–394
5. Wang Y et al (2021) Directed evolution: methodologies and applications. Chem Rev 121: 12384–12444
6. Li Y et al (2005) Directed evolution of human T-cell receptors with picomolar affinities by phage display. Nat Biotechnol 23:349–354
7. Xu L et al (2002) Directed evolution of high-affinity antibody mimics using mRNA display. Chem Biol 9: 933–942
8. Crameri A, Whitehorn EA, Tate E, Stemmer WPC (1996) Improved green fluorescent protein by molecular evolution using DNA shuffling. Nat Biotechnol 14: 315–319
9. Campbell RE et al (2002) A monomeric red fluorescent protein. Proc Natl Acad Sci USA 99:7877–7882
10. Gligorovski V et al (2024) Light-directed evolution of dynamic, multi-state, and computational protein functionalities. bioRxiv. https://doi.org/10.1101/2024.02.28.582517
11. Jang J, Woolley GA (2021) Directed evolution approaches for optogenetic tool development. Biochem Soc Trans 49:2737–2748
12. Groseclose TM, Rondon RE, Herde ZD, Aldrete CA, Wilson CJ (2020) Engineered systems of inducible anti-repressors for the next generation of biological programming. Nat Commun 11:4440
13. Sanchez MI, Ting AY (2020) Directed evolution improves the catalytic efficiency of TEV protease. Nat Methods 17:167–174
14. Tominaga M, Nozaki K, Umeno D, Ishii J, Kondo A (2021) Robust and flexible platform for directed evolution of yeast genetic switches. Nat Commun 12:1846
15. Klauser B, Atanasov J, Siewert LK, Hartig JS (2015) Ribozyme-based aminoglycoside switches of gene expression engineered by genetic selection in *S. cerevisiae*. ACS Synth Biol 4:516–525
16. Skjoedt ML et al (2016) Engineering prokaryotic transcriptional activators as metabolite biosensors in yeast. Nat Chem Biol 12:951–958
17. Snoek T et al (2020) Evolution-guided engineering of small-molecule biosensors. Nucleic Acids Res 48:e3
18. Usherenko S et al (2014) Photo-sensitive degron variants for tuning protein stability by light. BMC Syst Biol 8:128
19. Lu H et al (2019) A yeast system for discovering optogenetic inhibitors of eukaryotic translation initiation. ACS Synth Biol 8:744–757
20. Wang W et al (2017) A light- and calcium-gated transcription factor for imaging and manipulating activated neurons. Nat Biotechnol 35:864–871
21. Woloschuk RM, Reed PMM, McDonald S, Uppalapati M, Woolley GA (2020) Yeast two-hybrid screening of photoswitchable protein–protein interaction libraries. J Mol Biol 432:3113–3126
22. Taslimi A et al (2016) Optimized second-generation CRY2–CIB dimerizers and photoactivatable Cre recombinase. Nat Chem Biol 12:425–430
23. Huang Z et al (2020) Creating red light-switchable protein dimerization systems as genetically encoded actuators with high specificity. ACS Synth Biol 9: 3322–3333
24. Guntas G et al (2015) Engineering an improved light-induced dimer (iLID) for controlling the localization and activity of signaling proteins. Proc Natl Acad Sci USA 112:112–117
25. Reis JM et al (2018) Discovering selective binders for photoswitchable proteins using phage display. ACS Synth Biol 7:2355–2364
26. Wang H et al (2016) LOVTRAP: an optogenetic system for photoinduced protein dissociation. Nat Methods 13:755–758
27. Kawano F, Aono Y, Suzuki H, Sato M (2013) Fluorescence imaging-based high-throughput screening of fast- and slow-cycling LOV proteins. PLoS One 8: e82693
28. Yokobayashi Y, Weiss R, Arnold FH (2002) Directed evolution of a genetic circuit. Proc Natl Acad Sci USA 99:16587–16591
29. Maranhao AC, Ellington AD (2017) Evolving orthogonal suppressor tRNAs to incorporate modified amino acids. ACS Synth Biol 6:108–119
30. Baumschlager A, Weber Y, Cánovas D, Dionisi S, Khammash M (2022) Enhancing the performance of Magnets photosensors through directed evolution. bioRxiv. https://biorxiv.org/lookup/doi/10.1101/2022.11.14.516313, https://doi.org/10.1101/2022.11.14.516313
31. Meyer AJ, Segall-Shapiro TH, Glassey E, Zhang J, Voigt CA (2019) Escherichia coli "Marionette" strains with 12 highly optimized small-molecule sensors. Nat Chem Biol 15:196–204
32. Tashiro Y, Fukutomi H, Terakubo K, Saito K, Umeno D (2011) A nucleoside kinase as a dual selector for

genetic switches and circuits. Nucleic Acids Res 39: e12
33. Taylor ND et al (2016) Engineering an allosteric transcription factor to respond to new ligands. Nat Methods 13:177–183
34. Tang S-Y, Fazelinia H, Cirino PC (2008) AraC regulatory protein mutants with altered effector specificity. J Am Chem Soc 130:5267–5271
35. Lee SK et al (2007) Directed evolution of AraC for improved compatibility of arabinose- and lactose-inducible promoters. Appl Environ Microbiol 73: 5711–5715
36. Shong J, Huang Y-M, Bystroff C, Collins CH (2013) Directed evolution of the quorum-sensing regulator EsaR for increased signal sensitivity. ACS Chem Biol 8:789–795
37. Nomura Y, Yokobayashi Y (2007) Dual selection of a genetic switch by a single selection marker. Biosystems 90:115–120
38. Muranaka N, Sharma V, Nomura Y, Yokobayashi Y (2009) An efficient platform for genetic selection and screening of gene switches in Escherichia coli. Nucleic Acids Res 37:e39–e39
39. Sharma V, Nomura Y, Yokobayashi Y (2008) Engineering complex riboswitch regulation by dual genetic selection. J Am Chem Soc 130:16310–16315
40. Richter F et al (2013) Engineering of temperature- and light-switchable Cas9 variants. Nucleic Acids Res 44: gkw930
41. Tominaga M, Ike K, Kawai-Noma S, Saito K, Umeno D (2015) Rapid and liquid-based selection of genetic switches using nucleoside kinase fused with aminoglycoside phosphotransferase. PLoS One 10: e0120243
42. Saeki K, Tominaga M, Kawai-Noma S, Saito K, Umeno D (2016) Rapid diversification of BetI-based transcriptional switches for the control of biosynthetic pathways and genetic circuits. ACS Synth Biol 5: 1201–1210
43. Ike K et al (2015) Evolutionary design of choline-inducible and -repressible T7-based induction systems. ACS Synth Biol 4:1352–1360
44. Guntas G, Mansell TJ, Kim JR, Ostermeier M (2005) Directed evolution of protein switches and their application to the creation of ligand-binding proteins. Proc Natl Acad Sci USA 102:11224–11229
45. Tashiro Y et al (2016) Directed evolution of the autoinducer selectivity of *Vibrio fischeri* LuxR. J Gen Appl Microbiol 62:240–247
46. Ellefson JW, Ledbetter MP, Ellington AD (2018) Directed evolution of a synthetic phylogeny of programmable Trp repressors. Nat Chem Biol 14:361–367
47. Picot J, Guerin CL, Le Van Kim C, Boulanger CM (2012) Flow cytometry: retrospective, fundamentals and recent instrumentation. Cytotechnology 64:109–130
48. Fields S, Song O (1989) A novel genetic system to detect protein–protein interactions. Nature 340:245–246
49. Boeke JD, La Croute F, Fink GR (1984) A positive selection for mutants lacking orotidine-5′-phosphate decarboxylase activity in yeast: 5-fluoro-orotic acid resistance. Mol Gen Genet 197:345–346
50. Boeke JD, Trueheart J, Natsoulis G, Fink GR (1987) 5-Fluoroorotic acid as a selective agent in yeast molecular genetics. Methods Enzymol 154:164–175
51. Stojković L, Gligorovski V, Geramimanesh M, Labagnara M, Rahi SJ (2025) Automated plasmid design for marker-free genome editing in budding yeast. G3 15:jkae297
52. Armstrong JO et al (2024) *URA6* mutations provide an alternative mechanism for 5-FOA resistance in *Saccharomyces cerevisiae*. bioRxiv. https://doi.org/10.1101/2024.06.03.597250
53. Lungu OI et al (2012) Designing photoswitchable peptides using the AsLOV2 domain. Chem Biol 19: 507–517
54. Mattheakis LC, Bhatt RR, Dower WJ (1994) An in vitro polysome display system for identifying ligands from very large peptide libraries. Proc Natl Acad Sci USA 91:9022–9026
55. Jaroszewicz W, Morcinek-Orłowska J, Pierzynowska K, Gaffke L, Węgrzyn G (2022) Phage display and other peptide display technologies. FEMS Microbiol Rev 46:fuab052
56. Abil Z, Ellefson JW, Gollihar JD, Watkins E, Ellington AD (2017) Compartmentalized partnered replication for the directed evolution of genetic parts and circuits. Nat Protoc 12:2493–2512
57. Ellefson JW et al (2014) Directed evolution of genetic parts and circuits by compartmentalized partnered replication. Nat Biotechnol 32:97–101
58. Gonzalez-Perez D, Garcia-Ruiz E, Alcalde M (2012) Saccharomyces cerevisiae in directed evolution: an efficient tool to improve enzymes. Bioengineered 3: 174–179
59. Esvelt KM, Carlson JC, Liu DR (2011) A system for the continuous directed evolution of biomolecules. Nature 472:499–503
60. Mofatteh M et al (2021) Autonomous clocks that regulate organelle biogenesis, cytoskeletal organization, and intracellular dynamics. elife 10:e72104
61. Papagiannakis A, Niebel B, Wit EC, Heinemann M (2017) Autonomous metabolic oscillations robustly gate the early and late cell cycle. Mol Cell 65:285–295
62. Lu Y, Cross FR (2010) Periodic cyclin-Cdk activity entrains an autonomous Cdc14 release oscillator. Cell 141:268–279
63. March S et al (2024) Autonomous circadian rhythms in the human hepatocyte regulate hepatic drug metabolism and inflammatory responses. Sci Adv 10: eadm9281

64. Rahi SJ, Pecani K, Ondracka A, Oikonomou C, Cross FR (2016) The CDK-APC/C oscillator predominantly entrains periodic cell-cycle transcription. Cell 165: 475–487
65. Azizoglu A, Bijman EY, Stelling J, Brent R (2024) Evolverator: an engineered *in cellulo* yeast system to drive rapid continuous evolution of proteins. bioRxiv. https://doi.org/10.1101/2024.09.01.610536
66. Morgan DO (2007) The cell cycle: principles of control. New Science Press in association with Oxford University Press; Distributed inside North America by Sinauer Associates, London/Sunderland
67. Rahi SJ et al (2017) Oscillatory stimuli differentiate adapting circuit topologies. Nat Methods 14:1010–1016
68. Motta-Mena LB et al (2014) An optogenetic gene expression system with rapid activation and deactivation kinetics. Nat Chem Biol 10:196–202
69. Zhao EM et al (2018) Optogenetic regulation of engineered cellular metabolism for microbial chemical production. Nature 555:683–687
70. Benzinger D, Khammash M (2018) Pulsatile inputs achieve tunable attenuation of gene expression variability and graded multi-gene regulation. Nat Commun 9:3521
71. Gligorovski V, Sadeghi A, Rahi SJ (2023) Multidimensional characterization of inducible promoters and a highly light-sensitive LOV-transcription factor. Nat Commun 14:3810
72. Gligorovski V, Rahi SJ (2024) Construction and characterization of light-responsive transcriptional systems. In: Marchisio MA (ed) Synthetic promoters, vol 2844. Springer US, New York, pp 261–275
73. Shimizu-Sato S, Huq E, Tepperman JM, Quail PH (2002) A light-switchable gene promoter system. Nat Biotechnol 20:1041–1044
74. Chen Z et al (2020) De novo design of protein logic gates. Science 368:78–84
75. Gao XJ, Chong LS, Kim MS, Elowitz MB (2018) Programmable protein circuits in living cells. Science 361:1252–1258
76. Fink T et al (2019) Design of fast proteolysis-based signaling and logic circuits in mammalian cells. Nat Chem Biol 15:115–122
77. Omersa N, Aden S, Kisovec M, Podobnik M, Anderluh G (2020) Design of protein logic gate system operating on lipid membranes. ACS Synth Biol 9:316–328
78. Vishweshwaraiah YL, Chen J, Chirasani VR, Tabdanov ED, Dokholyan NV (2021) Two-input protein logic gate for computation in living cells. Nat Commun 12:6615
79. Sadeghi A, Dervey R, Gligorovski V, Labagnara M, Rahi SJ (2022) The optimal strategy balancing risk and speed predicts DNA damage checkpoint override times. Nat Phys 18:832–839. https://doi.org/10.1038/s41567-022-01601-3
80. Tous C, Kinstlinger IS, Rice MEL, Deng J, Wong WW (2025) Multiplexing light-inducible recombinases to control cell fate, Boolean logic, and cell patterningin mammalian cells. Sci Adv 11:eadt1971
81. Müller K et al (2013) Multi-chromatic control of mammalian gene expression and signaling. Nucleic Acids Res 41:e124
82. Kramer MM, Mühlhäuser WWD, Weber W, Radziwill G (2021) Multichromatic control of signaling pathways in mammalian cells. Adv Biol 5:2000196
83. Jang J, McDonald S, Uppalapati M, Woolley GA (2019) Green, orange, red, and far-red optogenetic tools derived from cyanobacteriochromes. bioRxiv. https://biorxiv.org/lookup/doi/10.1101/769422, https://doi.org/10.1101/769422
84. Kainrath S, Stadler M, Reichhart E, Distel M, Janovjak H (2017) Green-light-induced inactivation of receptor signaling using cobalamin-binding domains. Angew Chem Int Ed 56:4608–4611
85. DeBenedictis EA et al (2022) Systematic molecular evolution enables robust biomolecule discovery. Nat Methods 19:55–64
86. Yu T, Boob AG, Singh N, Su Y, Zhao H (2023) In vitro continuous protein evolution empowered by machine learning and automation. Cell Syst 14:633–644
87. Rapp JT, Bremer BJ, Romero PA (2024) Self-driving laboratories to autonomously navigate the protein fitness landscape. Nat Chem Eng 1:97–107
88. Scutteri L et al (2025) De novo design of light-regulated dynamic proteins using deep learning. bioRxiv. https://doi.org/10.1101/2025.08.12.669910

9 Physical Frameworks and Theories on How Autonomous Biological Oscillators Synchronize

Fabio Echegaray-Iturra and Evelyn Tang

Abstract

Rhythms and oscillations are widely observed in biological systems. However, it is unclear how such noisy systems can produce robust and consistent timing, or how individual molecules or time-keeping systems couple together to produce macroscopically observable rhythms. Here, we focus on the synchronization of oscillatory processes at different scales: from the sub-cellular to the cellular and intercellular levels. Using examples from the circadian rhythms in cyanobacteria, cell cycle in eukaryotes, the firing of neurons, and from other emerging findings on the frontier of autonomous clocks, we discuss different paradigms of coupling or entrainment. We then highlight different methods of analysis from deterministic to stochastic approaches. Specifically, we discuss methods to quantify synchronization and its possible physical mechanisms, as well as prevailing mathematical models. Lastly, we close with a discussion of suggestions for new experiments or theoretical investigation.

Keywords

Autonomous oscillators · Biological clocks · Synchronization · Entrainment · Coupled oscillators · Stochastic oscillations · Deterministic oscillations · Cell cycle oscillator · Circadian rhythms · Phase locking · Chemical coupling · Mechanical coupling · Kuramoto model · Stochastic modeling

9.1 Introduction

Rhythms and oscillations are everywhere in biological systems, so it is evident that time must be kept. But, given that underlying reactions are noisy or with small copy numbers, how could time be kept consistently and robustly? How are individual molecules in a cell coupled together to produce a macroscopically observable rhythm?

Researchers have wondered about how organisms account for the progression of time for centuries now, yet a quantitative understanding of this problem began to emerge only over the last century. Initial observations naturally focused on the correlations between physiological responses to environmental cues such as light in day/night cycles or temperature [92, 152]. By the

F. Echegaray-Iturra
Department of Biochemistry and Biophysics, University of California, San Francisco, San Francisco, CA, USA

Department of Genetics and Development, Columbia University, New York, NY, USA

E. Tang (✉)
Center for Theoretical Biological Physics and the Department of Physics and Astronomy, Rice University, Houston, TX, USA
e-mail: e.tang@rice.edu

M. G. Aydogan (ed.), *Biological Time Control*, Advances in Experimental Medicine and Biology 1521,
https://doi.org/10.1007/978-3-032-36189-9_9

mid-twentieth century, empirical evidence suggested that the dynamics of several processes that correlated with these environmental cues were driven by something different: endogenous molecular feedback mechanisms that can display dynamics independently of the environment [152]. When a process in biology embeds a molecular mechanism to keep track of time, we refer to it as a *biological clock*. These molecular "time keepers" can operate as oscillators (although, in general, their dynamics can be different from oscillatory), and interact with the environment. Our discussion will be centered on exploring the different ways clocks of the oscillatory kind could synchronize and how they would reach synchronization through collective behavior. In particular, we will focus on synchronization of autonomous oscillators.

There are currently several distinctions regarding biological clocks in the field. For instance, Gliech and Holland [70] calls a clock *directed* when the response in time is faithfully respected and the outcome is predetermined; much like the fuse on a firework. Conversely, the clock is *reactive* when the mechanism executes a decision based on the timing of its external inputs. A clock is linear if it does not oscillate (hourglass clocks, Rensing et al. [158]). The monotonic temporal progression of this group of clocks is usually involved in processes such as organismal development or aging [158]. A classic example of these clocks is the decreasing capacity of somatic cells to proliferate, a process mediated by the shortening of telomeres with each replication cycle [19, 35]. It stands to reason that clocks with a linear regime do not have the capacity to synchronize (at most they could wait for a different clock but never go back in phase). Since we are interested in the synchronization of autonomous oscillators, we will not discuss the linear group but focus exclusively on the mechanisms of synchronization in oscillatory phenomena.

Previous studies on synchronization have carefully considered specific systems, e.g. coupled neurons or excitable media. Still, autonomous biological oscillators are observed at different scales in biology: sub-cellular, cellular, and inter-cellular levels. To address mechanisms that underlie the synchronization of oscillators at these different scales of biology, we selected the circadian rhythms in cyanobacteria, cell cycle in eukaryotes, the firing of neurons, and other examples of emerging autonomous oscillators as case examples, respectively.

After presenting examples of these clocks, we will go through different paradigms of coupling or entrainment. We highlight different methods of analysis to quantify synchronization and its resulting behavior. Later, we delve into possible physical mechanisms as well as computational models and approaches. We close with a discussion of open questions and possibilities for experimental detection or theoretical investigation.

9.2 Autonomous Biological Oscillations in Cellular Systems

9.2.1 Inter-Cellular Scale: The Case of Neurons

Synchronization of biological oscillators has been explored in various systems, and is especially studied in the coupling of neuron excitations (Fig. 9.1a). In this regard, key contributions have been made by Ermentrout and Kopell [55], where they used the phase response curve to study weakly coupled oscillators. Later on, Wilson and Moehlis [182] and Wilson and Ermentrout [181] developed isostable response curves to help extend coupled oscillator analysis beyond the weak coupling regime. As there exist extensive books and reviews on these well-studied systems, our chapter will focus instead on the two cellular systems below.

9.2.2 Cellular Scale: The Case of the Cell Cycle

A classic example of autonomous oscillations in biology has been studied within the context of the

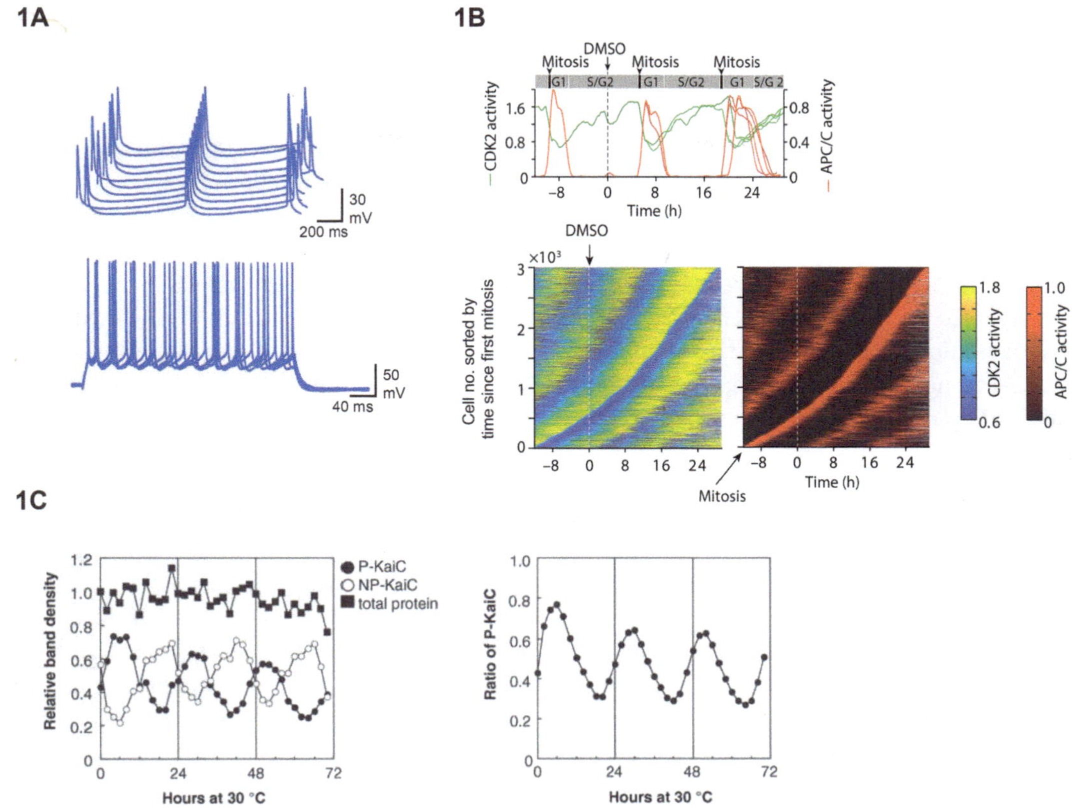

Fig. 9.1 Different cellular oscillations and clocks. (**a**) Trains of action potentials in nerve cells. (Adapted from Caillard [23]). (**b**) Kinase activity and proteasome activity during the cell cycle. The time series shows Cdk2 activity and APC/C activity signals from independent cells. The heatmap kymographs below show this activity represented in individual rows per each cell, collectively sorted by the time of their first mitosis. Thus, the heatmaps give a sense of the isochrons in the collection of cells sampled. (Adapted from Cornwell et al. [34]). (**c**) In vitro oscillations of KaiC phosphorylation that governs the circadian rhythms in cyanobacteria. These oscillations provided a model biological system for self-sustainable oscillations consisting of just proteins and ATP, without external cues. (Adapted from Nakajima et al. [132] where they were first observed)

cell cycle. See Fig. 9.1b for an example of how enzymatic oscillations in the cell cycle look like in cultured mammalian cells. Historically, the determination of the factors that govern DNA replication and cellular division has been at the kernel of cell biology research [138]. In this regard, a key experiment showed that a centrifuged fraction from the lysate of *Xenopus* eggs could trigger mitosis, hence these extracts were hypothesized to contain the maturation-promoting factor (MPF) needed for egg maturation [114]. The determination of MPF from these fractions had suggested that the process was heavily mediated by the phosphorylation of biochemical substrates through the activity of cyclin-dependent kinases (Cdks) [56, 64, 65, 123, 126, 137]. The idea that some steps in the cell cycle could be rate limiting for the progression of the cell cycle, and that a subset of enzymes can prevent cell cycle progression (otherwise known as checkpoints), supported the notion that the cell cycle could be temporally *controlled*, therefore qualifying for the status of a biological oscillator [77, 136]. Subsequent in vitro biochemical and phospho-proteomic experiments have provided evidence that the number of phosphorylations these kinases exert at mitotic entry are in the order of a thousand of substrates [36, 42, 47, 126, 137]. While other mechanisms of cell cycle progression have been pointed out more recently—e.g. metabolic pathways [96, 106, 112], changes in the state of the cytosol such as pH [99], and the role of forces in the process [22, 39, 40, 171, 172]—it is unclear if and how these mechanisms feedback on each other to ultimately converge on cell cycle progression. As we understand it today, the progression of cell cycle is largely mediated by the post-translational effect of kinases. We discuss in this chapter how such a post-translational modification (PTM) could potentially lead to coupling in Sect. 9.4.1.

In an effort to understand mathematically the role of kinase phosphorylation and its effects on cellular division, several models based on the post-translational action of kinases over substrates have followed. Early models were primarily based on biochemical kinetics, making heavy use of ordinary differential equations (ODEs) to search for regions in the phase-space that would lead to oscillatory regimes [71, 173]. Initially, the analysis faced criticism mainly from two fronts: First, there was an insufficient theoretical framework to support the use of ODEs to model the complexity of molecular systems in the context of cell cycle progression [67, 122], and; second, the solution of a system of ODEs requires to indicate the initial conditions—otherwise known as the initial value problem (IVP)- and the parameters of the system [37, 175]. To this day, the latter issue remains challenging. The former has been addressed in simplified situations, in which this type of formalism can adequate experimental data (see Sect. 9.3.1 for a justification of the methodology). Interestingly, experimental results concur with rate-based ODE models of the cell cycle in some cases. In particular, the models capture the inherent oscillatory regime of the cell cycle oscillator (CCO) and the bistability between S-phase and mitosis, observable in a great range of cells despite the stringent assumptions of the formulation [27, 46, 80, 124, 135, 157]. The models are also consistent with experimental data when reaction products are diffused in space, and can serve as a tool to understand other kinds of phenomena such as the emergence of waves in development [27, 45, 46].

Models based on ODEs also have predictive power. For example, in Rata et al. [157] the authors experimentally confirm a steady state in the prophase stage of the cell cycle that was predicted from their model when the levels of Cdk1 activity were low in prophase (orange solid line in Fig. 9.2a). In this state, the cells show chromosomes partially condensed, with Cyclin B translocated to the nucleus (and not cytosolic), and with a rounded-up cell morphology [157]. A similar state on interphase-cells, where Cyclin B is in the nucleus but no mitotic entry, can be induced if cells are subjected to external pressure while having low levels of PLK1 activity [41]. In summary, the experiments regarding cell division and its inherent oscillatory cycles have led to the advent of mathematical models which have, on the one hand, increased our understanding of how cells replicate and, on the other hand, extended their purpose to inspire the modeling of other distinct biological phenomena.

9.2.3 Sub-Cellular Scale: The Case of Circadian Rhythms

While the circadian rhythm has been long studied and is evidently of vital interest for health and regulation, the discovery that the daily rhythms in cyanobacteria could be studied in vitro allowed great progress. The KaiC phosphorylation cycle—the daily rhythms that drive the cyanobacterial circadian clock—persists without transcription or even translation. Nakajima et al. [132] reconstituted the self-sustainable oscillation of KaiC phosphorylations in vitro by incubating KaiC with KaiA, KaiB, and ATP. The period of the in vitro oscillation was found to be stable (Fig. 9.1c), even despite changes in temperature, indicating that these oscillations are temperature compensated. Corroborating these results, the circadian periods observed in KaiC mutant strains in vivo were consistent with those measured in vitro [132]. This discovery opened the path for a careful molecular examination and the

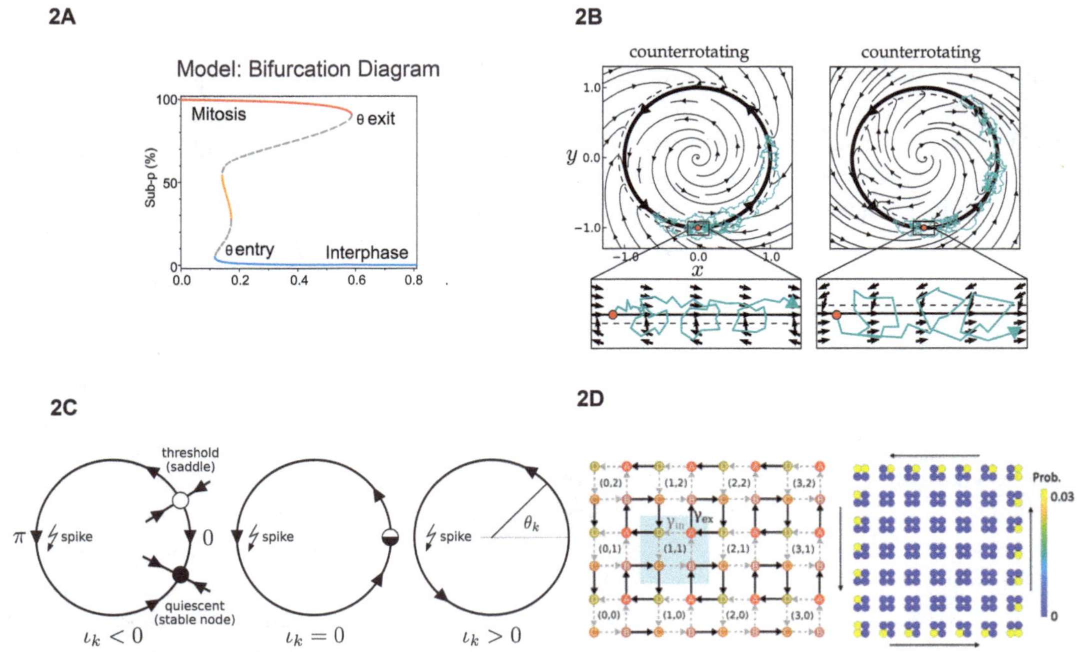

Fig. 9.2 Different paradigms for emergent oscillations. (**a**) Deterministic oscillators or limit cycles. In this example, the authors present a deterministic model of cell cycle progression in the form of a phase diagram. The system can oscillate through three different subsets of fixed points: the first two are the canonical states of mitosis and interphase in red and blue, respectively. A proposed intermediate state of fixed points is drawn with an orange line. (Adapted from Rata et al. [157]). (**b**) Deterministic limit cycles with noise added to them. (Adapted from Newby and Schwemmer [134]). (**c**) Limit cycles that only oscillate with noise, e.g. this saddle node bifurcation on an invariant circle (SNIC). (Adapted from Bick et al. [17]). (**d**) Stochastic oscillators that can oscillate at the single molecule level, but which require coupling to create a coherent oscillation that is macroscopically detectable. (Adapted from Agudo-Canalejo and Tang [2])

theoretical study of the cyanobacterial circadian clock in the following two decades [33, 141, 163, 192]. In subsequent sections, we will examine more closely mechanisms for synchronization and computational approaches for this system.

9.2.4 Beyond cells

Oscillations in macroscopic systems are also common, e.g., quasicycles in predator-prey systems [151]. However, cycles on the large organismal scale are outside the scope of this chapter.

9.3 How Do Coherent Oscillations Emerge?

9.3.1 Deterministic Oscillations

Synchronization was first observed by Christiaan Huygens when analyzing the motion of the pendulum clock he had invented in 1657 [86, vol. 5, p. 246]. When two pendulums were hung by one common wooden beam, he noticed after a while that the frequencies ultimately reached equilibrium. Since this seminal observation, collective synchronization of oscillators has been found in almost every corner of science [150]. From a physical perspective, an oscillator is represented by a dynamical system

$$\dot{x} = f(x), x \in \mathrm{R}^n \tag{9.1}$$

when the hypotheses of the Poincaré Recurrence Theorem are met. It is often assumed that the system evolves in a vector space that has finite dimensions, that it is smooth and has an inner product with Euclidean metrics like R^n, or Riemannian such as C^n. It should be noted, however, that the analysis of dynamical systems can be extended to other spaces with a sense of metric such as Banach Spaces. In the study of biological clocks, the vector space usually models the number and amount of chemical species in the system, their activity (e.g. by considering the phosphorylated state of the molecule as active or inactive), or some coarse-grained state variable such as bending force or pressure. Although this representation could be compelling at the meso and macro scales, the molecular complexity of biological systems tends to surpass this type of formulation [118, 185]. Therefore, these systems present several challenges to take into account, such as a broad distribution of different timescales, the determination of precise concentrations in biochemical reactions, and the possibility that the observed phenomena could be the manifestation of an emergent property [5, 159]. In particular, the determination of parameters that make an ODE model of a biological system representative, alongside their initial conditions, has been a matter of debate since the inception of these techniques [5, 48, 107, 113, 149]. We will present specific examples of common ODE models in Sect. 9.5.

Synchronization is often achieved through a *coupling element*. For example, to synchronize many pendulums, the free move of a common base provides the necessary coupling for these oscillators to synchronize [142]. In biology, the synchronization of cilia that extrude from the plasma membrane is attributed to hydrodynamic forces [76]. A current challenge in biological systems is to *identify the sources of coupling* when synchronization is observed. These sources can vary from energetic requirements, chemical constraints, time constraints, thermodynamic configurations and self-organization, mechanical forces and, more generally, the topology of the system when the coupling cannot be attributed solely to one source.

Even more challenging is to determine if an observed oscillation is the outcome of a single source or the interference of different sources. In the simple example of the *repressilator*, where the oscillator is synthetically designed to be orthogonal to any biological clock in the system, it is still impossible to attribute basic characteristics such as its frequency solely to the synthetic circuit. Instead, the high complexity that is found in *Escherichia coli* bacteria translates to high noise readouts of gene expression and gene product fluctuations influenced by other cellular processes [51, 52]. Thus, the elements that can potentially interfere with the transcription factors governing the system can be potentially large and are currently not completely known. Techniques for the causal determination of synchronization or independence in oscillatory systems will greatly help us elucidate the dynamical nature of this and other examples in biology.

9.3.2 Stochastic Oscillations

The molecular basis of biological clocks emerges from stochastic cycles, but they become more regular and reliably observable often through collective behavior. As such, theoretical frameworks are needed to describe the stochasticity that is inherent in these systems. To this end, new methods are called for, because classical deterministic definitions do not apply to stochastic systems, as the latter does not display precise periodic orbits. In other words, the return times (from "isochron" to "isochron") in these systems are stochastic. In this regard, we review below some emerging models and methods for the study of stochastic oscillators.

The most basic paradigm of a stochastic oscillation would be to simply add noise to the deterministic limit cycle described in the previous section (see Fig. 9.2b adapted from Newby and Schwemmer [134]). However, such limit cycles would still oscillate robustly in the limit when noise is reduced to zero. Thus, this paradigm is not qualitatively different from the canonical case of deterministic oscillators.

To go beyond just adding noise to existing deterministic limit cycles, attention should be directed to cycles without an underlying deterministic limit and that only oscillate in the presence of noise. Examples include quasicycles [20] and the different forms of excitable systems, e.g. a saddle-node bifurcation on an invariant circle (SNIC) in the presence of noise (Fig. 9.2c). In the latter, the system is near the bifurcation point and yet does not oscillate in the deterministic case. In the presence of noise, the system is pushed past the bifurcation point, when oscillations ensue. Other examples that require noise include a spiral sink with noise, such as in Pérez-Cervera et al. [147], or the damped harmonic oscillator as described in Pérez-Cervera et al. [148]. In biology, there also exist limit cycles that emerge from adding the contributions of many noisy molecules, such as from ion channels as studied in Anderson et al. [6].

Lastly, there can also be stochastic oscillators that demonstrate oscillations on the single molecule level (Fig. 9.2d). For instance, a topological model supports emergent oscillations in single molecules of the KaiC circadian clock that was shown in Fig. 9.1c [192]. The stochastic nature is evident as coupling between molecules is needed to prevent their phases from drifting apart. Still, it is unusual for a single molecule to show a stochastic limit cycle; rather, stochastic switches such as the ion channels mentioned above that open and close randomly, are more common.

As we see in these examples, there are new phenomena produced by stochastic oscillators that would be relevant for biological clocks at large. The ongoing challenge is to relate new mathematical descriptions to experimentally studied biological clocks. Notably, the differences in these models and paradigms inspire the development of experimental tests that disambiguate between them. In the last section, we provide suggestions on possible experimental probes.

9.3.3 Synchronization: Entrainment Versus Coupling Between Oscillators

When considering synchronization between or among biological oscillators, we observe an intricacy of rhythms that pass from one cellular system to another, e.g., from a pacemaker to observed oscillations elsewhere in the body. Such interacting rhythms are often poorly captured by mathematical models that typically describe groups of identical oscillators, or at most a few different oscillators. These simple isolated models are rarely observed in biology, where many different rhythms can affect each other, e.g. how the circadian rhythm can couple to the cell cycle (Fig. 9.3a). However, recent experiments in vitro have successfully isolated oscillations that can persist independently of any external cues (Fig. 9.3b), which provide systems that can be easier to tackle using quantitative and modeling approaches.

To distinguish between these different cases, we use *entrainment* to describe the first case, in which a rhythm from one biological system drives or affects another. In the second case, we use *coupling* to describe interactions between similar biological systems, such as a set of molecules in vitro, that can collectively produce a coherent oscillation. As our chapter is on the various mechanisms of entrainment and coupling, we explore concrete pathways for such mechanisms in the next section.

9.4 Physical Mechanisms of Synchronization

The formal models of synchronization are, in general, an abstraction of lower-level phenomena clumped in the representation of differences in phase. The collective aggregation and eventual convergence of phases is accomplished through a (set of) coupling parameter(s) between different oscillators. Experimentally, the coupling elements can be physical objects. For example, in synchronizing pendulums, it is the freely

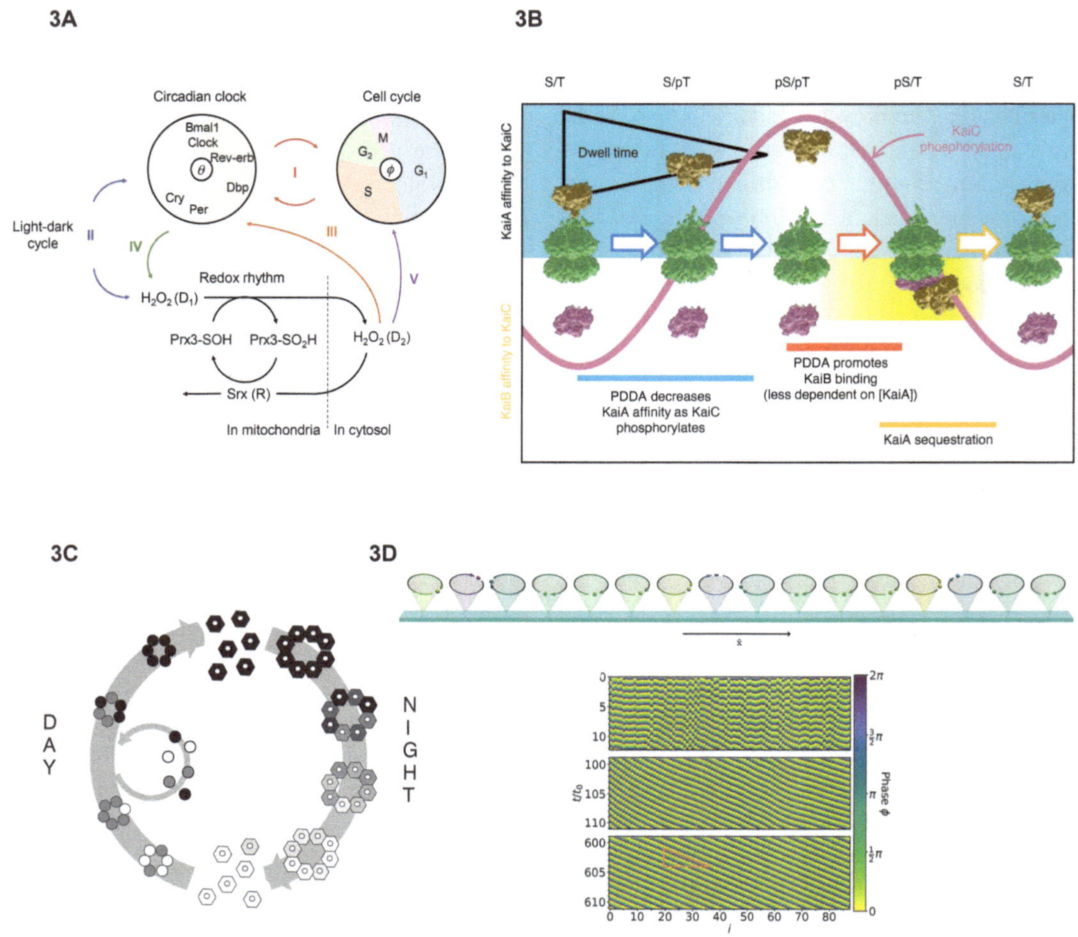

Fig. 9.3 Mechanisms for synchronization. (**a**) Example of a model that accounts for the coupling between the circadian clock, the cell cycle and redox rhythms through a system of ODEs. These equations in turn describe a three coupled oscillator model. Each arrow represents a coupling between the different oscillators. (Figure adapted from Masuda et al. [117]). (**b**) Chemical coupling of KaiC molecules, that regulate the circadian rhythm of cyanobacteria. Here, individual molecules are expected to be coupled using sequestration of KaiA molecules, where competition over limited KaiA makes front runners slow down and laggards speed up. (Figure adapted from Mori et al. [128]). (**c**) An alternative mechanism for the synchronization of the previous system has been hypothesized monomer exchange between KaiC hexamers. This could provide a route towards averaging of the phase of different hexamers. (Figure adapted from Emberly and Wingreen [53]). (**d**) An underlying underlying mechanical substrate can also couple oscillators, e.g. in this example of cilia carpets that perform coherent waves. Here, the coupling mechanism could be hydrodynamic interactions or elastic stretching. (Adapted from Hickey et al. [81])

moving base onto which the oscillators are placed that provides the mechanism of coupling [85]. Moreover, it has been shown that this degree of freedom (DOF) is sufficient to derive a formulation that follows the Kuramoto model for the synchronizing phases of the pendulums [142]. As such, one key aspect in the analysis of synchronization is to study the different sources of coupling (i.e. the couplers). The coupling can manifest in some physical feature, or some mechanism of communication and, in general, it is something that enables the oscillators to interact. What follows in this section is a deductive approach to enumerating such possible mechanisms of coupling with no attempt to make them exhaustive or complete; instead, our efforts aim to inspire avenues of thinking to find the potential ways in which autonomous oscillations can couple.

9.4.1 Chemical Coupling

Chemical coupling often arises due to a scarcity, or competition, for key molecules necessary to complete a biological cycle. This arises in various contexts, from the circadian rhythm to glycolytic oscillations in yeast cells and to the synchronization of cAMP oscillations. In the circadian rhythm, this can happen due to competition over limited KaiA, also known as KaiA sequestration or differential affinity [128, 141, 163, 164, 174] (see Fig. 9.3b). For glycolytic oscillations, the signaling molecule acetaldehyde diffuses between cells and can slow down or speed up oscillations [184]. Lastly, in cAMP oscillations, the diffusion of extracellular cAMP out of a cell could similarly perform this function [95].

Chemical competition on substrates can synchronize oscillations. One classic example is in the activity of kinases and phosphatases that promote progression through the cell cycle via phosphorylation/dephosphorylation of a common set of substrates (see Sect. 9.2.2 for an introduction to the cell cycle). For instance, it has been shown that the kinase cyclin-dependent kinase 1 (Cdk1) partnered to cyclin A or B promotes entry into mitosis [127]. Although the regulation of Cdk1 is complex and involves the interaction with activators, inhibitors and PTMs [36, 50, 127], activation of Cdk1 happens through the phosphorylation at a conserved T-loop residue (Thr160 in human Cdk1). However, the inhibitory phosphorylations at Thr14 and Tyr15 by Wee1/Myt1 kinases prevent cell cycle progression, and it is only after the phosphatase Cdc25 removes these phosphorylations that the cell can enter mitosis (see Crncec and Hochegger [36] for a review of mitotic entry). A similar argument can be put forward for the entry into S-phase when Cdk2 makes a complex with cyclins A or E [15]. Thus, the competition of kinases and phosphatases on the cellular substrate that is amicable to Cdk phosphorylation determines progression of the cell cycle. The cell cycle oscillator (CCO) can, in turn, entrain other key processes in cell biology. For example, in centriole biogenesis, oscillatory regimes are usually observed in the formation of the cartwheel, which occur concomitantly with the CCO [11]. A key determinant in the formation of the cartwheel is the concentration and activity of the kinase PLK4 on centrioles [14, 61]. However, it has been shown in the syncytial cycles of *Drosophila* that PLK4 oscillations can still happen when the CCO is halted by RNAi against the main cyclins (i.e. cyclin A-B-B3; [10, 120]). In a similar manner, nuclear divisions can also run independently from metaphase furrow divisions if the CCO is halted by RNAi against cyclins A-B-B3 [12]. Thus, the CCO dominates regimes that can be otherwise independently executed through mechanisms orthogonal to cyclin expression and, therefore, Cdk activity [10, 12]. A related entrainment of the CCO with nuclear conformation has been demonstrated in the case of the translocation of the phosphatase Cdc14 from the cytosol to the nucleus (and vice versa, Lu and Cross [115]).

Beyond the understanding of DNA replication and mitosis, cell cycle Cdks and cyclin-dependent kinase inhibitors (CKIs) act as regulatory factors in other classes of mechanisms such as cell differentiation, inflammation, wound repair, and metabolism [130, 140]. These effects can be independent of checkpoint activation, e.g. as evidenced by the p53-independent increase of CKI $p21^{CIP1}$ expression in terminally differentiated cells [144]. Connection to metabolism was inferred in experiments performed in *S. cerevisiae* in which, by virtue of modulating key nutrient metabolites (e.g. pyruvate) with cell cycle progression (by the addition of the mating pheromone α-factor), a strong coupling of metabolism with early and late stages of the cell cycle was observed [140]. Thus, cell cycle activity can manifest itself in the collective growth or inhibition of cell proliferation in processes such as the healing of tissue, in cell differentiation, or by coupling metabolic cues through the combined action of Cdks, CKIs and nutrients [130, 140]. This seems to be in stark contrast as to when the cells are manipulated in culture, where the phases of the cell cycle have been observed to be independent of each other and couple only after the modification of molecules in the cytosol such as the inhibition of Cdk2 or the overexpression of cyclin D [28]. An interesting avenue forward then is the determination of the precise factors that lead to the coupling and synchronicity of these large populations of cells acting in tandem to account tissue remodeling.

Allosteric chemical competition can also synchronize oscillations. For example, during mitotic spindle formation, oscillation of astral microtubule (MT) polymerization and depolymerization can be synchronized with the progression of the cell cycle to accomplish a reliable mitosis. This is clearly evidenced in the case of kinetochore attachments errors, in which stringent mechanisms of control take place at the sites of error to delay MT polymerization and promote their depolymerization in an Aurora B dependent manner [24, 98, 109, 121]. MT polymerization fluctuates from low lengths to elongated bundles, and from radially organized MT-asters, to subsequent spindles as mitotic progression takes place. The mechanisms by which this and other MT-derived properties (e.g. persistent length and flexural rigidity) materialize are chiefly mediated by the local configuration of microtubule-associated proteins (MAPs) loaded onto the MT lattice [89]. One of these MAP proteins is the (+)-end directed kinesin mitotic centromere-associated kinesin (MCAK), which promotes the depolymerization of MTs upon reaching the end of the MT (+) ends [63]. Since there is an ample spectrum of other proteins prepared to bind to the tips and lattices of MTs, the localization of MCAK to the (+)-end is in direct competition with other enzymes that are ready promote opposite effects such as rescues, polymerization and branching of MTs [3]. In consequence, MT oscillations can be synchronized with the cell cycle through the allosteric competition of MAP proteins on MT-lattices to accomplish objectives such as the correction of errors in spindle formation and to conduct mitosis reliably.

9.4.2 Mechanical Coupling

As previously stated, a degree of coupling is often necessary to induce synchronization of oscillations. Mechanical forces have been implicated in several key biological processes such as cell cycle progression [16, 84, 97, 105] or cell differentiation [38, 83, 166, 189]. Thus, it is interesting from a synchronization perspective how various forms of mechanical force could potentially serve as couplers. The following section will attempt to allude to biochemical processes that are prompted by internal forces in the cell, which can then translate to external forces at the level of tissues, to be then fed back into their own internal cell state.

Internal forces can synchronize oscillations. Although there are no direct efforts exploring the biochemical interdependence between forces at the cellular and tissue scales, there are results that can be integrated toward this goal and further provide inspiration for future research. For example, the kinase Aurora B has been largely documented in its role of kinetochore attachment correction during chromosome segregation [4, 21, 29, 32, 43, 75, 129]. This activity, in turn, is actively synchronized by the CCO at the final step of mitotic exit to aid the formation of the cytokinetic ring and to further complete cytokinesis [60, 74]. The proliferation of cells obtained in the latter step, in turn, has mechanical consequences in terms of intercellular forces such as the pressure that is induced in tissues [26, 62, 94, 116]. How can a kinase have a role at the molecular and meso scale in such a short time window? In anaphase, chromosomes are actively aligned prior to their segregation by forces exerted from and on the spindle. Oftentimes, oscillations of the kinetochores between the sister chromatids are observed, and several lines of evidence have suggested that this oscillatory phenomenon could be driven by the tension that the chromatids exert to each other [90, 169, 177]. In turn, these oscillations are correlated with error correction in kinetochore attachments in a process mediated through Aurora B kinase to destabilize improper kinetochore attachments and assure that the pairs of sister chromatids are attached to opposite poles of the spindle [4, 21, 29, 32, 43, 75, 129]. Further evidence has suggested that kinetochores undergo a conformational change when under tension, a state that prevents detachment from MTs by phosphorylation from Aurora B kinase [43, 104] This process has not only been proposed to maintain chromosome stability by preventing lagging chromosomes in anaphase, but has also been implicated in the resolution of merotelic attachments to correctly form the metaphase plate of the spindle [31, 32, 98, 167]. Thus, tension-dependent Aurora B activity can synchronize progression of cell cycle events through the phosphorylation of kinetochores [43], MT depolymerization by the regulation of depolymerizing kinesins such as MCAK [54, 98, 121] and by the activation of the spindle checkpoint (SC) [111]. In turn, the expression levels and activity of Aurora B directly mediate actomyosin response through the Rac pathway and directly correlate with cellular responses at the mesoscale; e.g. in tissue generation of the salivary glands and in the uncontrolled proliferation of cells [26, 156, 168]. At longer timescales, the effect of Aurora B on tissues accounts for the build-up of local pressure through the increase of cell density and inflammation [26, 62, 94, 116]. These responses can feed back to other cells, thus closing a loop through mechanical forces at the intercellular level.

In the case of the circadian rhythms, it has been suggested that monomer exchange between KaiC hexamers could provide a means for averaging the phase differences that could otherwise arise among the hexamers [53, 191] (see Fig. 9.3c). In contrast, there can also be mechanical feedback provided by an underlying substrate, e.g., through hydrodynamic interactions or elastic stretching, which can cause the synchronization of nearby oscillators [81, 93]. An example of such a possibility can be found in the synchronization of cilia carpets (Fig. 9.3d).

Oscillations driven by external forces feedback to internal processes in the cell. Cells respond to external physical cues such as pressure or plasma membrane tension. This, in turn, can translate into the synchronization of events through a biochemical pathway; e.g. metabolic modulation and nutrient consumption through the Yes-associated protein (YAP)-Hippo signaling pathway [87]. A potential place for this to happen is in the intercellular tension between cell boundaries promoted by actomyosin contractility. This intercellular tension is often manifested as surface contractions of cells in tissue or individual eggs [18, 78, 82, 100, 110, 165].

Plasma membrane contractility is mediated by changes in membrane curvature. Evidence suggests that the underlying force-generating mechanism is largely provided by actin polymerization and further actomyosin contractility [13, 101, 110, 186]. Since myosin II has also been reported to reduce tensile stress upon mechanical perturbation [49], the evidence amounts to a model in which actin bundles polymerize and stabilize when under tension or high concavity [30, 59, 79, 110, 186]. The effect that forces and curvature have at the membrane subsequently induce molecules to react upon these changes by either activation of ion channels [176] or by changing their conformational state, often revealing hidden binding sites that trigger downstream effects such as in the case of the binding of Talin and Vinculin [44, 143], the creation of focal adhesions [25, 145], or in metabolism through the activation of YAP proteins by Rho GTPases [87]. In the latter example, YAP activity acts as a hub alongside the Hippo signaling network to coordinate the activity of different internal processes such as glucose metabolism, lipid metabolism, amino acid & nucleotide metabolism, and regulation of transcription through transcription factors [87]. In the context of mitosis, it has been demonstrated that Cyclin B readily translocates to the nucleus when induced by external forces transmitted through the LINC-complex, subsequently licensing cell cycle progression into mitotic entry [41]. In consequence, external forces and membrane curvature can serve as a coupler to many internal cellular processes.

9.4.3 Entrainment by Emergent Properties: The Case of Temperature

From a thermodynamic perspective, oscillations are understood as *temporal self-organization*, and as such, they are enunciated under the concept of *dissipative structures* [153, 154]. This generalization accounts for open systems which accomplish their self-organization by the dissipation of energy and thus operate far from thermodynamic equilibrium. A dissipative structure forms when the system changes its constitution, and a stable fixed point becomes unstable at a critical bifurcation point. When such a transition occurs, the system can manifest not only in oscillatory regimes, but also in stationary spatial patterns known as *Turing patterns* and propagating waves [72]. Moreover, the transition from a population of incoherent oscillators to collective synchrony was quickly recognized as one of the chief examples of Pirigogine's theory [103]. Thus, the evolution of the state of a thermodynamic system can account for the progression of time [154, 155].

Such system-state-driven time progression can be observed in the process of *phase separation*. Here, the biological system transitions from one macro state to another (e.g. liquid phases of different densities) as a function of an aggregated parameter such as substrate phosphorylation [178]. In mammalian models, phase separation through phosphorylation has been implicated in cell cycle regulation through the hyperphosphorylation of the nucleolar phosphoprotein Ki-67 [188]. Furthermore, the dynamics of phase separation can provide a theoretical framework for rounds of cyclic division in protocells from their constituent droplets [194]. Finally, as elaborated next in this section, a thermal source can also show the capacity to synchronize cycles of cell division in protocells by inducing phase separation and micelle formation through protonation [162].

Energy is the main macro parameter that determines the overall dynamic of system. In classical mechanics, this parameter can bring a system to the onset of bifurcations and chaos as exemplified in the case of a double pendulum [91]. One of the manifestations of this concept is in the internal kinetic energy of molecules in the form of *temperature*.

At the cellular scale, thermal cues have shown the capacity to synchronize events such as cycles of cell division in protocells. An in vitro study showed how thermal oscillations can drive the growth and division of protocells, providing evidence of a primitive cell cycle being driven chiefly by thermal fluctuations [162]. In their study, the main parameter that drives the formation of micelles and coalescence to droplets is the protonation free energy pK_a of prebiotic fatty acids. At high pH, deprotonated fatty acids form micelles, whereas at low pH the now protonated fatty acids degenerate into an oil phase. To make the connection to temperature, a key piece of the argument is that several physiologically-relevant buffers changed their pH when heat was applied and temperature raised. For example, in the case of Tris-HCl, pH decreased two units when temperature was increased from 25 to 95 °C [162]. Thus, the cycle of cellular formation and dissolution can be driven in an elementary way by the change of pH levels as a function of temperature. Interestingly, an orthogonal study in the yeast *Saccharomyces cerevisiae* also concludes a link between acidification of the cytoplasm and cell cycle progression [139]. Here, the authors conclude that the alkalinization of vacuoles (the equivalent of lysosomes in cerevisiae) precedes mitosis and reacidifies as cells divide [139]. Thus, thermal fluctuations can drive responses in organisms which can, in turn, account for time progression and provide a mechanism of synchronization.

In biological systems, energy consumption seems to be tightly regulated. It has been shown in *Drosophila embryos* that key steps of the cell cycle can deviate from their canonical regimes under changes of temperature [58]. This is of high interest, because calorimetry studies have shown that the energy consumption of several organisms do not increase massively over the course of embryogenesis [131, 160, 170]. This suggests that heat is meticulously distributed to different processes inside the developing embryo through potentially different mechanisms. One such beneficiary process could be the assembly of the nucleolus, which has been shown to be partly dependent on temperature [57]. In broader terms, given the constricted energy observations that follow embryogenesis in these organisms and the small increase of heat from start to finish of the process, it stands to reason that energy is strictly controlled in the developmental program. This energy constraint could then induce synchronization by regulating the availability of energy to the independent processes in the developing embryo. Theoretical results have shown thus far the sufficiency of simple thermodynamic rules to support the existence of systems that can organize around consuming or avoiding energy sources [180]. However, a lack of experimental evidence makes it difficult to build a strong case for biological organisms. Hence, it would be interesting to gather experimental proof to either falsify or support the hypothesis of energetic constraints in the synchronization of autonomous clocks in development.

Embryogenesis in oviparous organisms presents an ideal system to study energy fluxes since no work is exerted on the egg from external sources and, crucially, because the egg does not move. Consequently, one can attribute the heat Q measured during calorimetry experiments entirely to the change in the internal energy U of the organism, which enables analyzing the change of heat according to $\Delta U = Q$. Rodenfels et al. [160] observed that heat oscillations measured by calorimetry fully followed cell cycle oscillations in the development of zebrafish embryos. In a similar study, a stoichiometric calculation of the energy consumption during the embryogenesis of *Drosophila melanogaster* yielded 13.4 mJ attributed to the oxidation of glucose and triacylglycerol (TAG) to create adenosine triphosphate (ATP) [170]. The same study measured this energy consumption using calorimetry, obtaining a value of 10 mJ, which in turn equates to 110 nmol of ATP. Interestingly, when a similar calorimetry experiment was performed in *Xenopus* specimens, the authors found that the heat liberated in the embryogenesis of the frog was independent of temperature [131]. Thus, it is likely that the main source of heat in these embryos comes from cell cycle activity, such as in the case of zebrafish embryos [131, 160]. This in turn could be a fantastic starting point in the exploration of synchronicity of autonomous clocks that may correlate to the cell cycle, since the energy constraint necessarily implicates restrictions on the phase space that these systems can accommodate, further narrowing down the search space of key elements such as coupling molecules.

9.5 Mathematical Models

9.5.1 Deterministic Models

Mass-action kinetics often provides a set of deterministic equations with chemical concentration as variables. This is sufficient to produce oscillations or limit cycle in biology, including the circadian clocks in both cyanobacteria [53, 163, 164, 174] and mammalian cells [73], as well as glycolytic oscillations in yeast cells [184].

How can we assure that ODEs of chemical kinetics is a valid formulation to model biological phenomena? The thermodynamic argument that supports this premise hitherto resides in the definition of the *propensity function* [67, 68, 122]. Consider that a set of $N \geq 1$ chemical species $\{S_1, \ldots, S_N\}$ interact inside a "well stirred" and fixed volume Ω, at a constant temperature and capable to react into $M \geq 1$ different reactants known as reaction channels—$\{R_1, \ldots, R_M\}$. Then given the state of the system $X(t) \equiv (X_1(t), \ldots, X_N(t))$, the *propensity function* a_j is defined as [122]

$a_j(x)dt \equiv$ probability, given $X(t) = x$, that a random combination of R_j reactions will occur somewhere inside Ω at time t in the next infinitesimal time interval $[t, t + dt)$,

$$j = 1, \ldots, M. \tag{9.2}$$

When some constraints are applied to this function, the chemical master equation (CME) looks exactly like the rate equations from classic chemical kinetics [68]. These conditions are:

1. The system is well mixed, concentrations remain uniform in Ω.
2. The number of reactions that are allowed to happen in the time interval is exactly one.
3. The population of all molecules in the state vector $X(t)$ are very large compared to 1.

Together, these axioms model the behavior of several well-mixed chemical species reacting in a dilute fluid. The form of the CME is then expressed in terms of the *mean* value of the random variable $X(t)$ follows [68]

$$\frac{d\langle X_i(t)\rangle}{dt} = \sum_{j=1}^{M} v_{ij}\langle a_j(X_i(t))\rangle \quad (i=1,\dots,N), \tag{9.3}$$

which in turn is in clear resemblance to the heuristic of classic chemical kinetics, dictating that the species of the population of molecules in $X(t)$ evolve *deterministically*. It is even pointed out in Gillespie [68] that, under the hypothesis of existence of a *macroscopically infinitesimal timescale*, the stochastic master equation can be further approximated by the sum of a deterministic component and a fluctuating component, a result that is formally akin to the Langevin equation. The main caveat in the presented formulation is that the result is not derived from Liouville's equation; however, it does present a clear connection between the CME and thermodynamics [67]. Situations in which this type of formulation deviates from empirical experiments include when the molecular counts are low, when the discreteness of molecules in the volume plays a role, where noise is a determinant factor of the outcome, or where the phenomena could be part of an emergent property [5, 8, 52, 119, 159, 179]. In these cases, stochastic approaches are more suitable to provide in silico predictions [66, 69].

Oscillatory regimes can emerge as a result of their mathematical modeling. However, the analysis of whether the phase of two oscillatory variables is synchronized can be cryptic. Can independent albeit coupled systems synchronize themselves? And if so, can we mathematically analyze them? In this regard, Arthur Winfree discovered that an ensemble of weakly coupled oscillators can go through the analog of a phase transition and collectively synchronize [183].

In general, the analysis of f in Eq. (9.1) needs to take into account both amplitude and phase of the state variable in phase space. However, in the simpler case when oscillations are exponentially stable, the state of the system can be transformed in a way in which the phase of the system ϑ is preserved but the amplitude disregarded [88]. In this situation, we will introduce definitions in this section that come to aid in the understanding of different regimes in an ensemble of oscillators, and where the concept of synchronization naturally arises from them. Formally, the phase representation of a network of coupled oscillators will have the form of Eq. (9.4c) in the following system of equations

$$\dot{X} = f\left(x, \vec{\Omega}, \vartheta\right) \tag{9.4a}$$

where f can be further decomposed in

$$\dot{x}_i = g_i(x_i) + \in h_i(x_1,\dots,x_n,\in) \tag{9.4b}$$

$$\dot{\vartheta}_i = \Omega_i + \in O_i(\vartheta_1,\dots,\vartheta_n,\in), \vartheta_i \in S^1 \tag{9.4c}$$

The network of Eq. (9.4a) is weak when $\epsilon \ll 1$. The distinctions between phase and frequency in systems of this form have stricter constraints when the system is phase locked as opposed to frequency locked. Following Izhikevich and Kuramoto [88], the oscillators are frequency locked when Eq. (9.4c) has a periodic stable orbit $\theta(t) = (\vartheta_1(t), \dots, \vartheta_n(t))$ on the n-torus T^n. The oscillators are entrained when they are $1:1:\dots:1$ frequency locked. The oscillators are phase locked when there is an $(n-1)\times n$ integer matrix M that has linearly independent rows such that $M\vartheta(t) = \text{const}$. The oscillators are synchronized when they are entrained and phase locked. The oscillators are phase trapped when they are locked in frequency but not in phase. Ordered from most strict to less strict: synchronized, entrained, {frequency locked, phase locked}, phase trapped. Curly braces indicate the concepts that share some degree of equivalence.

A vast majority of mathematical studies performed on synchronous systems stems from the study of the simplified Eq. (9.4a). A further simplification was put forward by Kuramoto when he assumed that the oscillators not only are weakly coupled, but almost identical. In this case, the phases of the system can be represented by [102]:

$$\dot{\theta}_i = \omega_i + Kr\sin(\phi - \theta_i), \tag{9.5}$$

in this model, each phase θ_i has a corresponding natural frequency ω_i, and collectively synchronize over the mean-field phase ϕ with a coupling strength Kr. Increasingly heterogeneous models can be accounted if this latter term is changed for an interaction coefficient between every pair of oscillators $(\theta_i, \theta_j)i/ = j$ [161, 187, 190].

There are several ways of synchronizing the phases represented in Eq. (9.4c). Key results are derived from Graph Theory. Here, the interaction among a collection of oscillators is considered as a network, with every oscillator represented as a node and their connections representing their coupling. The topology of the network has the potential to drive a collection of oscillators from being inherently incoherent to phase locked by changes in the connection of a few edges [9]. In a similar line, when the connectivity of the network is high, synchronicity tends to emerge spontaneously [1]. Other ways to synchronize a network is to change the topology between connectivity states as a function of time [146] or by means of an optimization process [7].

9.5.2 Stochastic Models

In contrast to deterministic equations, another approach is to use stochastic simulations, often based on the Chemical master equation. From simulated trajectories, one can see the effects of microscopic mechanisms and stochastic trajectories, for example in the circadian rhythm where stochastic asynchronous oscillators align their phase upon coupling [133, 141]. Such modeling has also been used to study the synchronization of cAMP oscillations [95].

Another way to introduce stochasticity is by adding noise to differential equations. The effects of stochasticity here can be contrasted to the mechanism above using stochastic simulations directly, where noise comes from underlying Poisson processes of microscopic reactions, which can lead to phase diffusion. Here, noise instead comes from fluctuations in molecular concentrations in processes such as the production or degradation of molecules. Examples of these systems are in the study of repressilators or relaxation oscillators [108, 193].

9.6 Future Directions

The recent developments we have highlighted prompt new directions for exploration such as experimental tests or new modeling for understanding the coupling of autonomous oscillators. For instance, it will be critical to disambiguate experimentally between different paradigms of coupled oscillators, such as models where oscillations observed in ensembles only emerge through the coupling of many molecules [141, 174], or an alternative paradigm describing noisy oscillations inherent to individual molecules [192].In particular, we expect each model to predict differing levels of fluctuations along the oscillation period or demonstrate a synchronization phase transition, which could also be tested with experiments. Lastly, the possibility of changing the number of molecules or other concentration levels to see their effects on the oscillation coherence, affords good possibility for comparison between theory and experiment.

Additional to the requirement of independence in autonomous clocks, a major requirement for a circuit to qualify as a *biological clock* is its capability to account for the progression of time in an *irreversible manner* [70, 125, 152].

In the In the case of the cell cycle, reversibility has been justified through different mechanisms such as the existence of rate limiting steps and of checkpoints to slow down or prevent cell cycle progression, respectively. This set of results indeed are consistent with the notion of monotonicity in cell cycle progression [77, 136, 138]. However, pharmacological, mechanical and genetic manipulation to cells have shown that some of the steps in the cell cycle can be reversed [41, 57, 58, 157]. It would be an interesting avenue of research to elucidate in which capacity these induced states could be accomplished by physiological processes in the cell; in this way, we would further determine if and how much variation does the cell cycle have in its internal progression. One thing is clear though: if we accept the hypothesis that multiple biological clocks can couple, in a way that could be represented to some capacity by Eq. (9.4c), then one must assume these oscillators will move back and forth in phase while trying to converge to a common frequency. There could be multiple ways to account for this family of regimes in a way that is consistent with current results. For example, the regimes of clocks that go bidirectionally in time could happen at a higher frequency than the observed progression and thus look like they are irreversible at the mesoscale while having high fluctuations below the microscale. This in turn would have profound implications to how we understand these processes in terms of the definitions we currently hold and how their activities enact the regulation of the state of biological systems at the tissue and organismal level.

Acknowledgments E.T. would like to thank Chongbin Zheng, Peter Thomas and Max Kreider for helpful suggestions and references. E.T. also acknowledges support from the NSF Center for Theoretical Biological Physics and CAREER Award (PHY-2019745 and DMR-2238667), as well as the Chan-Zuckerberg Foundation. F.E. would like to thank Mustafa Aydogan, Andrew Alamban and Helfrid Hochegger for the stimulating discussions and kind suggestions to the manuscript. F.E. received support from National Institutes of Health (DP2GM154328 through Mustafa Aydogan) while working on this manuscript.

References

1. Abdalla P et al (2022) Expander graphs are globally synchronising. https://doi.org/10.48550/arXiv.2210.12788 (visited on 11/14/2023)
2. Agudo-Canalejo J, Tang E (2024) Topological phases in discrete stochastic systems. Rep Prog Phys 88:102601
3. Akhmanova A, Steinmetz MO (2010) Microtubule +TIPs at a glance. J Cell Sci 123(20):3415–3419
4. Akiyoshi B et al (2010) Tension directly stabilizes reconstituted kinetochore-microtubule attachments. Nature 468(7323):576–579
5. Alfieri R et al (2011) Modeling the cell cycle: from deterministic models to hybrid systems. Biosystems 105(1):34–40
6. Anderson DF, Ermentrout B, Thomas PJ (2015) Stochastic representations of ion channel kinetics and exact stochastic simulation of neuronal dynamics. J Comput Neurosci 38(1):67–82
7. Ansarinasab S et al (2024) Optimized multi-variable coupling can improve synchronization in complex networks. Nonlinear Dyn 112(20):18491–18500
8. Arkin A, Ross J, McAdams HH (1998) Stochastic kinetic analysis of developmental pathway bifurcation in phage λ-infected Escherichia coli cells. Genetics 149:1633
9. Arola-Fernández L et al (2022) Emergence of explosive synchronization bombs in networks of oscillators. Commun Phys 5(1):264
10. Aydogan MG et al (2020) An autonomous oscillation times and executes centriole biogenesis. Cell 181:44
11. Aydogan MG et al (2018) A homeostatic clock sets daughter centriole size in flies. J Cell Biol 217(4): 1233–1248
12. Bakshi A et al (2023) Cytoplasmic division cycles without the nucleus and mitotic CDK/cyclin complexes. Cell 186(21):4694–4709.e16
13. Banerjee S, Gardel ML, Schwarz US (2020) The actin cytoskeleton as an active adaptive material. Annu Rev Condens Matter Phys 11:421–439
14. Banterle N, Gönczy P (2017) Centriole biogenesis: from identifying the characters to understanding the plot. Annu Rev Cell Dev Biol 33:23–49
15. Barr AR et al (2016) A dynamical framework for the all-or-none G1/S transition. Cell Syst 2(1):27–37
16. Benham-Pyle BW, Pruitt BL, Nelson WJ (2015) Mechanical strain induces E-cadherin–dependent Yap1 and β-catenin activation to drive cell cycle entry. Science 348(6238):1024–1027
17. Bick C et al (2020) Understanding the dynamics of biological and neural oscillator networks through exact mean-field reductions: a review. J Math Neurosci 10(1):1–43
18. Bischof J et al (2017) A Cdk1 gradient guides surface contraction waves in oocytes. Nat Commun 8(1):849
19. Blackburn EH (1991) Structure and function of telomeres. Nature 350(6319):569–573

20. Boland RP, Galla T, McKane AJ (2008) How limit cycles and quasi-cycles are related in systems with intrinsic noise. J Stat Mech 2008(9):P09001
21. Broad AJ, DeLuca KF, DeLuca JG (2020) Aurora B kinase is recruited to multiple discrete kinetochore and centromere regions in human cells. J Cell Biol 219(3):e201905144
22. Brooker HR et al (2018) A novel live-cell imaging system reveals a reversible hydrostatic pressure impact on cell-cycle progression. J Cell Sci 131(15): jcs212167
23. Caillard O (2011) Pre & postsynaptic tuning of action potential timing by spontaneous GABAergic activity. PLoS One 6(7):e22322
24. Cao LG, Wang YL (1996) Signals from the spindle midzone are required for the stimulation of cytokinesis in cultured epithelial cells. Mol Biol Cell 7(2): 225–232
25. Carisey A et al (2013) Vinculin regulates the recruitment and release of core focal adhesion proteins in a force-dependent manner. Curr Biol 23(4):271–281
26. Carmena M, Earnshaw WC (2003) The cellular geography of Aurora kinases. Nat Rev Mol Cell Biol 4(11):842–854
27. Chang JB, Ferrell JE Jr (2013) Mitotic trigger waves and the spatial coordination of the Xenopus cell cycle. Nature 500(7464):603–607
28. Chao HX et al (2019) Evidence that the human cell cycle is a series of uncoupled, memoryless phases. Mol Syst Biol 15(3):e8604
29. Cheeseman IM et al (2006) The conserved KMN network constitutes the core microtubule-binding site of the kinetochore. Cell 127(5):983–997
30. Chen T et al (2019) Large-scale curvature sensing by directional actin flow drives cellular migration mode switching. Nat Phys 15(4):393–402
31. Cimini D et al (2001) Merotelic kinetochore orientation is a major mechanism of aneuploidy in mitotic mammalian tissue cells. J Cell Biol 153(3):517–528
32. Cimini D et al (2006) Aurora kinase promotes turnover of kinetochore microtubules to reduce chromosome segregation errors. Curr Biol 16(17): 1711–1718
33. Cohen SE, Golden SS (2015) Circadian rhythms in cyanobacteria. Microbiol Mol Biol Rev 79(4): 373–385
34. Cornwell JA et al (2023) Loss of CDK4/6 activity in S/G2 phase leads to cell cycle reversal. Nature 619(7969):363–370
35. Counter CM (1996) The roles of telomeres and telomerase in cell life span. Mutat Res 366(1):45–63
36. Crncec A, Hochegger H (2019) Triggering mitosis. FEBS Lett 593(20):2868–2888
37. Csikasz-Nagy A (2009) Computational systems biology of the cell cycle. Brief Bioinform 10(4):424–434
38. Dado D et al (2012) Mechanical control of stem cell differentiation. Regen Med 7(1):101–116
39. Dantas M, Lima JT, Ferreira JG (2021) Nucleus-cytoskeleton crosstalk during mitotic entry. Front Cell Dev Biol 9:649899
40. Dantas M et al (2021) Mechanosensitive control of mitotic entry. bioRxiv. https://doi.org/10.1101/2021.09.23.461473
41. Dantas M et al (2022) Nuclear tension controls mitotic entry by regulating cyclin B1 nuclear translocation. J Cell Biol 221(12):e202205051
42. Daub H et al (2008) Kinase-selective enrichment enables quantitative phosphoproteomics of the kinome across the cell cycle. Mol Cell 31(3):438–448
43. De Regt AK et al (2022) Tension can directly suppress Aurora B kinase-triggered release of kinetochore-microtubule attachments. Nat Commun 13(1):2152
44. Del Rio A et al (2009) Stretching single talin rod molecules activates vinculin binding. Science 323(5914):638–641
45. Deneke VE, Di Talia S (2018) Chemical waves in cell and developmental biology. J Cell Biol 217(4): 1193–1204
46. Deneke VE et al (2016) Waves of Cdk1 activity in S phase synchronize the cell cycle in drosophila embryos. Dev Cell 38(4):399–412
47. Dephoure N et al (2008) A quantitative atlas of mitotic phosphorylation. Proc Natl Acad Sci 105(31):10762–10767
48. DiStefano J III (2015) Dynamic systems biology modeling and simulation. Academic Press. 886 pp
49. Duda M et al (2019) Polarization of Myosin II refines tissue material properties to buffer mechanical stress. Dev Cell 48(2):245–260.e7
50. Elledge SJ, Winston J, Harper JW (1996) A question of balance: the role of cyclin-kinase inhibitors in development and tumorigenesis. Trends Cell Biol 6(10):388–392
51. Elowitz MB, Leibler S (2000) A synthetic oscillatory network of transcriptional regulators. Nature 403(6767):335–338
52. Elowitz MB et al (2002) Stochastic gene expression in a single cell. Science 297(5584):1183–1186
53. Emberly E, Wingreen NS (2006) Hourglass model for a protein-based circadian oscillator. Phys Rev Lett 96(3):038303
54. Ems-McClung SC et al (2013) Aurora B inhibits MCAK activity through a phosphoconformational switch that reduces microtubule association. Curr Biol 23(24):2491–2499
55. Ermentrout GB, Kopell N (2006) Frequency plateaus in a chain of weakly coupled oscillators, I. SIAM J Math Anal 15(2):215–237. https://doi.org/10.1137/0515019
56. Evans T et al (1983) Cyclin: a protein specified by maternal mRNA in sea urchin eggs that is destroyed at each cleavage division. Cell 33(2):389–396
57. Falahati H, Wieschaus E (2017) Independent active and thermodynamic processes govern the nucleolus

assembly in vivo. Proc Natl Acad Sci 114(6): 1335–1340
58. Falahati H et al (2021) Temperature-induced uncoupling of cell cycle regulators. Dev Biol 470: 147–153
59. Fernandez-Gonzalez R et al (2009) Myosin II dynamics are regulated by tension in intercalating cells. Dev Cell 17(5):736–743
60. Ferreira JG et al (2013) Aurora B spatially regulates EB3 phosphorylation to coordinate daughter cell adhesion with cytokinesis. J Cell Biol 201(5): 709–724
61. Fırat-Karalar EN, Stearns T (2014) The centriole duplication cycle. Philos Trans R Soc Lond B Biol Sci 369(1650):20130460
62. Francisco-Recuero I et al (2013) HCV mediated Aurora kinase B activity inhibition regulates anti-inflammatory response and HCV initial extracellular infectivity: 2019. Hepatology 58:1185A
63. Gardner MK et al (2011) Depolymerizing kinesins Kip3 and MCAK shape cellular microtubule architecture by differential control of catastrophe. Cell 147(5):1092–1103
64. Gautier J et al (1990) Cyclin is a component of maturation-promoting factor from Xenopus. Cell 60(3):487–494
65. Gavet O, Pines J (2010) Activation of cyclin B1–Cdk1 synchronizes events in the nucleus and the cytoplasm at mitosis. J Cell Biol 189(2):247–259
66. Gillespie DT (1977) Exact stochastic simulation of coupled chemical reactions. J Phys Chem 81(25): 2340–2361
67. Gillespie DT (1992) A rigorous derivation of the chemical master equation. Physica A 188(1–3): 404–425
68. Gillespie DT (2000) The chemical Langevin equation. J Chem Phys 113(1):297–306
69. Gillespie DT (2007) Stochastic simulation of chemical kinetics. Annu Rev Phys Chem 58(1):35–55
70. Gliech CR, Holland AJ (2020) Keeping track of time: the fundamentals of cellular clocks. J Cell Biol 219(11):e202005136
71. Goldbeter A (1991) A minimal cascade model for the mitotic oscillator involving cyclin and Cdc2 kinase. Proc Natl Acad Sci 88(20):9107–9111
72. Goldbeter A (2017) Dissipative structures and biological rhythms. Chaos 27(10):104612
73. Gonze D et al (2005) Spontaneous synchronization of coupled circadian oscillators. Biophys J 89(1): 120–129
74. Goto H et al (2003) Aurora-B regulates the cleavage furrow-specific vimentin phosphorylation in the cytokinetic process*. J Biol Chem 278(10):8526–8530
75. Gregan J et al (2011) Merotelic kinetochore attachment: causes and effects. Trends Cell Biol 21(6): 374–381
76. Hamilton E et al (2019) Motile cilia hydrodynamics: entrainment versus synchronization when coupling through flow. Philos Trans R Soc Lond B Biol Sci 375(1792):20190152
77. Hartwell LH, Weinert TA (1989) Checkpoints: controls that ensure the order of cell cycle events. Science 246(4930):629–634
78. Harvey EN, Fankhauser G (1933) The tension at the surface of the eggs of the salamander, triturus (diemyctylus) viridescens. J Cell Comp Physiol 3(4):463–475
79. Hayakawa K, Tatsumi H, Sokabe M (2011) Actin filaments function as a tension sensor by tension-dependent binding of cofilin to the filament. J Cell Biol 195(5):721–727
80. Hégarat N et al (2020) Cyclin a triggers mitosis either via the Greatwall kinase pathway or cyclin B. EMBO J 39(11):e104419
81. Hickey DJ, Golestanian R, Vilfan A (2023) Nonreciprocal interactions give rise to fast cilium synchronization in finite systems. Proc Natl Acad Sci USA 120(40):e2307279120
82. Houliston E et al (1993) Axis establishment and microtubule-mediated waves prior to first cleavage in Beroe Ovata. Development 117(1):75–87
83. Hu D et al (2023) Mechanical force directs proliferation and differentiation of stem cells. Tissue Eng Part B Rev 29(2):141–150
84. Huang S, Chen CS, Ingber DE (1998) Control of cyclin D1, p27Kip1, and cell cycle progression in human capillary endothelial cells by cell shape and cytoskeletal tension. Mol Biol Cell 9(11):3179–3193
85. Hugenii C (1673) Horoloquim oscilatorium. Muguet, Paris. Reprinted in English as: The pendulum clock. Iowa State UP, Ames
86. Huygens C (1893) Œuvres complètes de Christiaan Huygens: correspondance 1664–1665, vol 5. Nijhoff
87. Ibar C, Irvine KD (2020) Integration of Hippo-YAP signaling with metabolism. Dev Cell 54(2):256–267
88. Izhikevich EM, Kuramoto Y (2006) Weakly coupled oscillators. Encycl Math Phys 5:448
89. Janke C (2014) The tubulin code: molecular components, readout mechanisms, and functions. J Cell Biol 206(4):461–472
90. Jaqaman K et al (2010) Kinetochore alignment within the metaphase plate is regulated by centromere stiffness and microtubule depolymerases. J Cell Biol 188(5):665–679
91. Jiménez-López J, García-Garrido VJ (2024) Chaos and regularity in the double pendulum with Lagrangian descriptors. Int J Bifurcat Chaos 34(16):2450201
92. Kalmus H (1940) New research in the diurnal periodicity of animals. Acta Med Scand 103(S108): 227–233
93. Kanale AV et al (2022) Spontaneous phase coordination and fluid pumping in model ciliary carpets. Proc Natl Acad Sci USA 119(45):e2214413119
94. Katsha A et al (2013) Aurora kinase a promotes inflammation and tumorigenesis in mice and human gastric neoplasia. Gastroenterology 145(6): 1312–1322.e8

95. Kim J et al (2007) Stochastic noise and synchronisation during dictyostelium aggregation make cAMP oscillations robust. PLoS Comput Biol 3(11):e218
96. Kirova DG et al (2022) A ROS-dependent mechanism promotes CDK2 phosphorylation to drive progression through S phase. Dev Cell 57(14): 1712–1727.e9
97. Klein EA et al (2009) Cell-cycle control by physiological matrix elasticity and in vivo tissue stiffening. Curr Biol 19(18):1511–1518
98. Knowlton AL, Lan W, Stukenberg PT (2006) Aurora B is enriched at merotelic attachment sites, where it regulates MCAK. Curr Biol 16(17):1705–1710
99. Koch LM et al (2020) Cytosolic pH regulates proliferation and tumour growth by promoting expression of cyclin D1. Nat Metab 2(11):1212–1222
100. Kojima MK (1962) Cyclic changes of the cortex and the cytoplasm of the fertilized and the activated sea urchin egg. Embryologia 7(1):81–94
101. Kuo Y-C et al (2015) Oscillatory shear stress mediates directional reorganization of actin cytoskeleton and alters differentiation propensity of mesenchymal stem cells. Stem Cells 33(2):429–442
102. Kuramoto Y (1984) Chemical oscillations, waves, and turbulence. Springer, Berlin, Heidelberg, 178 pp
103. Kuramoto Y, Nishikawa I (1987) Statistical macrodynamics of large dynamical systems. Case of a phase transition in oscillator communities. J Stat Phys 49(3):569–605
104. Lampson MA et al (2004) Correcting improper chromosome–spindle attachments during cell division. Nat Cell Biol 6(3):232–237
105. Lancaster OM et al (2013) Mitotic rounding alters cell geometry to ensure efficient bipolar spindle formation. Dev Cell 25(3):270–283
106. Lee IH, Finkel T (2013) Metabolic regulation of the cell cycle. Curr Opin Cell Biol 25(6):724–729
107. Levine WS (ed) (2018) The control systems handbook: control system advanced methods, 2nd edn. CRC Press, Boca Raton, 1798 pp
108. Li C, Chen L, Aihara K (2007) Stochastic synchronization of genetic oscillator networks. BMC Syst Biol 1(1):1–11
109. Li S, Kasciukovic T, Tanaka TU (2024) Kinetochore–microtubule error correction for biorientation: lessons from yeast. Biochem Soc Trans 52(1):29–39
110. Liu AP et al (2008) Membrane-induced bundling of actin filaments. Nat Phys 4(10):789–793
111. Liu D et al (2009) Sensing chromosome bi-orientation by spatial separation of Aurora B kinase from kinetochore substrates. Science 323(5919):1350–1353
112. Liu W et al (2023) Lactate regulates cell cycle by remodeling the anaphase promoting complex. Nature 616:790–797
113. Ljung L (2010) Perspectives on system identification. Annu Rev Control 34(1):1–12
114. Lohka MJ, Hayes MK, Maller JL (1988) Purification of maturation promoting factor, an intracellular regulator of early mitotic events. Proc Natl Acad Sci 85(9):3009–3013
115. Lu Y, Cross FR (2010) Periodic cyclin-Cdk activity entrains an autonomous Cdc14 release oscillator. Cell 141(2):268–279
116. Madejón A et al (2015) Hepatitis C virus-mediated Aurora B kinase inhibition modulates inflammatory pathway and viral infectivity. J Hepatol 63(2): 312–319
117. Masuda K, Sakurai T, Hirano A (2024) A coupled model between circadian, cell-cycle, and redox rhythms reveals their regulation of oxidative stress. Sci Rep 14(1):15479
118. Mazzocchi F (2008) Complexity in biology: exceeding the limits of reductionism and determinism using complexity theory. EMBO Rep 9(1):10–14
119. McAdams HH, Arkin A (1997) Stochastic mechanisms in gene expression. Proc Natl Acad Sci 94(3):814–819
120. McCleland ML, O'Farrell PH (2008) RNAi of mitotic cyclins in drosophila uncouples the nuclear and centrosome cycle. Curr Biol 18(4):245–254
121. McHugh T et al (2019) The depolymerase activity of MCAK shows a graded response to Aurora B kinase phosphorylation through allosteric regulation. J Cell Sci 132(4):jcs228353
122. McQuarrie DA (1967) Stochastic approach to chemical kinetics. J Appl Probab 4(3):413–478
123. Minshull J, Blow JJ, Hunt T (1989) Translation of cyclin mRNA is necessary for extracts of activated Xenopus eggs to enter mitosis. Cell 56(6):947–956
124. Mochida S et al (2016) Two bistable switches govern M phase entry. Curr Biol 26(24):3361–3367
125. Mofatteh M et al (2021) Autonomous clocks that regulate organelle biogenesis, cytoskeletal organization, and intracellular dynamics. eLife 10:e72104
126. Morgan DO (1995) Principles of CDK regulation. Nature 374(6518):131–134
127. Morgan DO (1997) Cyclin-dependent kinases: engines, clocks, and microprocessors. Annu Rev Cell Dev Biol 13:261–291
128. Mori T et al (2018) Revealing circadian mechanisms of integration and resilience by visualizing clock proteins working in real time. Nat Commun 9(1): 3245
129. Murata-Hori M, Wang Y-l (2002) The kinase activity of Aurora B is required for kinetochore-microtubule interactions during mitosis. Curr Biol 12(11): 894–899
130. Nabel EG (2002) CDKS and CKIS: molecular targets for tissue remodelling. Nat Rev Drug Discov 1(8): 587–598
131. Nagano Y, Ode KL (2014) Temperature-independent energy expenditure in early development of the African clawed frog *Xenopus laevis*. Phys Biol 11(4): 046008

132. Nakajima M et al (2005) Reconstitution of circadian oscillation of cyanobacterial KaiC phosphorylation in vitro. Science 308(5720):414–415
133. Nandi A, Ramaswamy R (2008) Synchronization of coupled stochastic oscillators: the effect of topology. Pramana 70(6):1165–1174
134. Newby JM, Schwemmer MA (2014) Effects of moderate noise on a limit cycle oscillator: counterrotation and bistability. Phys Rev Lett 112(11):114101
135. Novak B, Tyson JJ (2022) Mitotic kinase oscillation governs the latching of cell cycle switches. Curr Biol 32:2780
136. Nurse P (1975) Genetic control of cell size at cell division in yeast. Nature 256(5518):547–551
137. Nurse P (1990) Universal control mechanism regulating onset of M-phase. Nature 344(6266): 503–508
138. Nurse P (2000) A long twentieth century of the cell cycle and beyond. Cell 100(1):71–78
139. Okreglak V et al (2023) Cell cycle-linked vacuolar pH dynamics regulate amino acid homeostasis and cell growth. Nat Metab 5(10):1803–1819
140. Özsezen S et al (2019) Inference of the high-level interaction topology between the metabolic and cell-cycle oscillators from single-cell dynamics. Cell Syst 9(4):354–365.e6
141. Paijmans J, Lubensky DK, ten Wolde PR (2017) A thermodynamically consistent model of the post-translational Kai circadian clock. PLoS Comput Biol 13(3):e1005415
142. Pantaleone J (2002) Synchronization of metronomes. Am J Phys 70(10):992–1000
143. Papagrigoriou E et al (2004) Activation of a vinculin-binding site in the talin rod involves rearrangement of a five-helix bundle. EMBO J 23(15):2942–2951
144. Parker SB et al (1995) P53-independent expression of $p21^{Cip1}$ in muscle and other terminally differentiating cells. Science 267(5200):1024–1027
145. Pasapera AM et al (2010) Myosin II activity regulates vinculin recruitment to focal adhesions through FAK-mediated paxillin phosphorylation. J Cell Biol 188(6):877–890
146. Pereti C, Fanelli D (2020) Stabilizing Stuart-Landau oscillators via time-varying networks. Chaos, Solitons Fractals 133:109587
147. Pérez-Cervera A, Lindner B, Thomas PJ (2021) Isostables for stochastic oscillators. Phys Rev Lett 127(25):254101
148. Pérez-Cervera A et al (2023) A universal description of stochastic oscillators. Proc Natl Acad Sci USA 120(29):e2303222120
149. Pia Saccomani M et al (2019) Calculating all multiple parameter solutions of ODE models to avoid biological misinterpretations. Math Biosci Eng 16(6):6438–6453
150. Pikovsky A, Popovych O, Maistrenko Y (2001) Resolving clusters in chaotic ensembles of globally coupled identical oscillators. Phys Rev Lett 87(4): 044102
151. Pineda-Krch M et al (2007) A tale of two cycles—distinguishing quasicycles and limit cycles in finite predator–prey populations. Oikos 116(1):53–64
152. Pittendrigh CS (1958) Perspectives in the study of biological clocks. University of California Press, pp 239–268
153. Prigogine I (1969) Structure, dissipation and life. In: Theoritical physics and biology. Wiley Interscience
154. Prigogine I (2017) Non-equilibrium statistical mechanics, Dover books on physics, Dover edn. Dover Publications, Inc, New York, 319 pp
155. Prigogine I, Stengers I, Prigogine I (1984) Order out of chaos: man's new dialogue with nature. Bantam Books, Toronto/New York, 349 pp
156. Qi G et al (2007) Aurora-B expression and its correlation with cell proliferation and metastasis in oral cancer. Virchows Arch 450(3):297–302
157. Rata S et al (2018) Two interlinked bistable switches govern mitotic control in mammalian cells. Curr Biol 28(23):3824–3832.e6
158. Rensing L, Meyer-Grahle U, Ruoff P (2001) Biological timing and the clock metaphor: oscillatory and hourglass mechanisms. Chronobiol Int 18(3): 329–369
159. Ricard J (2003) Qu'entendons-nous par complexite biologique? C R Biol 2(326):133–140
160. Rodenfels J, Neugebauer KM, Howard J (2019) Heat oscillations driven by the embryonic cell cycle reveal the energetic costs of signaling. Dev Cell 48(5): 646–658.e6
161. Rodrigues FA et al (2016) The Kuramoto model in complex networks. Phys Rep 610:1–98
162. Rubio-Sánchez R et al (2021) Thermally driven membrane phase transitions *enable* content *reshuffling in primitive* cells. J Am Chem Soc 143(40):16589–16598
163. Rust MJ et al (2007) Ordered phosphorylation governs oscillation of a three-protein circadian clock. Science 318(5851):809–812
164. Sasai M (2021) Mechanism of autonomous synchronization of the circadian KaiABC rhythm. Sci Rep 11(1):4713
165. Sawai T (1982) Wavelike propagation of stretching and shrinkage in the surface of the Newt's egg before the first cleavage. J Exp Zool 222(1):59–68
166. Schwartz L, da Veiga Moreira J, Jolicoeur M (2018) Physical forces modulate cell differentiation and proliferation processes. J Cell Mol Med 22(2):738–745
167. Sen O et al (2021) Kinetochore life histories reveal an Aurora-B-dependent error correction mechanism in anaphase. Dev Cell 56(22):3082–3099.e5
168. Shaalan AK et al (2021) Inhibition of Aurora kinase B activity disrupts development and differentiation of salivary glands. Cell Death Discov 7(1):1–12
169. Skibbens RV, Skeen VP, Salmon ED (1993) Directional instability of kinetochore motility during chromosome congression and segregation in mitotic newt lung cells: a push-pull mechanism. J Cell Biol 122(4): 859–875

170. Song Y et al (2019) Energy budget of drosophila embryogenesis. Curr Biol 29(12):R566–R567
171. Stiff T et al (2020) Prophase-specific perinuclear actin coordinates centrosome separation and positioning to ensure accurate chromosome segregation. Cell Rep 31(8):107681
172. Théry M et al (2005) The extracellular matrix guides the orientation of the cell division axis. Nat Cell Biol 7(10):947–953
173. Tyson JJ (1991) Modeling the cell division cycle: Cdc2 and cyclin interactions. Proc Natl Acad Sci 88(16):7328–7332
174. van Zon JS et al (2007) An allosteric model of circadian KaiC phosphorylation. Proc Natl Acad Sci 104(18):7420–7425
175. Villaverde AF et al (2019) Benchmarking optimization methods for parameter estimation in large kinetic models. Bioinformatics 35(5):830–838
176. Volkers L, Mechioukhi Y, Coste B (2015) Piezo channels: from structure to function. Pflugers Arch Eur J Physiol 467(1):95–99
177. Wan X et al (2012) The coupling between sister kinetochore directional instability and oscillations in centromere stretch in metaphase PtK1 cells. Mol Biol Cell 23(6):1035–1046
178. Weber CA et al (2019) Physics of active emulsions. Rep Prog Phys 82(6):064601
179. Weinberger LS et al (2005) Stochastic gene expression in a lentiviral positive-feedback loop: HIV-1 tat fluctuations drive phenotypic diversity. Cell 122(2): 169–182
180. Werlang T et al (2022) Emergence of energy-avoiding and energy seeking behaviors in nonequilibrium dissipative quantum systems. Commun Phys 5(1):1–8
181. Wilson D, Ermentrout B (2019) Augmented phase reduction of (Not So) weakly perturbed coupled oscillators. SIAM Rev 61(2):277–315
182. Wilson D, Moehlis J (2016) Isostable reduction of periodic orbits. Phys Rev E 94(5):052213
183. Winfree AT (1967) Biological rhythms and the behavior of populations of coupled oscillators. J Theor Biol 16(1):15–42
184. Wolf J, Heinrich R (2000) Effect of cellular interaction on glycolytic oscillations in yeast: a theoretical investigation. Biochem J 345(Pt 2):321
185. Wolf YI, Katsnelson MI, Koonin EV (2018) Physical foundations of biological complexity. Proc Natl Acad Sci 115(37):E8678–E8687
186. Wu Z et al (2018) Membrane shape-mediated wave propagation of cortical protein dynamics. Nat Commun 9(1):136
187. Xu C et al (2016) Synchronization of phase oscillators with frequency-weighted coupling. Sci Rep 6(1):21926
188. Yamazaki H et al (2022) Cell cycle-specific phase separation regulated by protein charge blockiness. Nat Cell Biol 24(5):625–632
189. Yim EKF, Sheetz MP (2012) Force-dependent cell signaling in stem cell differentiation. Stem Cell Res Ther 3(5):41
190. Zhang H, Zhang W, Gao J (2019) Synchronization of interconnected heterogeneous networks: the role of network sizes. Sci Rep 9(1):6154
191. Zhang D et al (2020) The energy cost and optimal design for synchronization of coupled molecular oscillators. Nat Phys 16(1):95–100
192. Zheng C, Tang E (2024) A topological mechanism for robust and efficient global oscillations in biological networks. Nat Commun 15(1):1–11
193. Zhou T et al (2008) Synchronization of genetic oscillators. Chaos 18(3):37126
194. Zwicker D et al (2017) Growth and division of active droplets provides a model for protocells. Nat Phys 13(4):408–413

The Emergent Architecture of Autonomous Clocks Described by Computational Algorithms of Regulatory Networks

10

Breschine Cummins, Marcio Gameiro, Konstantin Mischaikow, and Tomas Gedeon

Abstract

Autonomous oscillators in systems biology are often mathematically expressed as regulatory network models. These models are interrogated to predict oscillator behavior under various environmental conditions. The potential suite of dynamical behaviors arising from a regulatory network model is rich and varied, as one might expect of a complex system. Therefore, a mathematical model has the capability to predict previously unobserved biological oscillator behavior that arises under untested conditions. We present a computational method that describes the full suite of dynamical behaviors of a network model, enabling both mechanistic understanding of experimental observations and prediction of unexpected dynamics. These ideas are presented through example network models, including some relevant to the yeast cell cycle.

B. Cummins (✉) · T. Gedeon
Department of Mathematical Sciences, Montana State University, Bozeman, MT, USA
e-mail: breschine.cummins@montana.edu; tgedeon@montana.edu

M. Gameiro · K. Mischaikow
Department of Mathematics, Rutgers University, New Brunswick, NJ, USA
e-mail: gameiro@math.rutgers.edu; mischaik@math.rutgers.edu

Keywords

Mathematical biology · Computational biology · Network dynamics · Genetic regulatory networks · Dynamical attractor · Multilevel discrete model · State transition graph · Parameter graph · Morse graph · DSGRN · Repressilator · Hysteresis

10.1 Introduction

Mathematical modeling is an important accessory to experimentation when assessing the functional accuracy of hypothetical regulatory networks (RNs) in systems biology. The number of experiments required to discover a complex network architecture is prohibitive in cost, time, and resources, whereas in silico experimentation permits the evaluation of various hypotheses with relatively little expense. This is particularly important in complex network structures, since, as we will show, the combination of simple network motifs into a complex network gives rise to emergent dynamics that were not predictable from the motifs themselves. Interactions between motifs exhibit unexpected diversity in dynamical behavior that could be used by cells to expand the range of their phenotypic responses.

Autonomous oscillators are complex RNs with many interacting pieces that respond flexibly to environmental or cellular conditions. For

M. G. Aydogan (ed.), *Biological Time Control*, Advances in Experimental Medicine and Biology 1521,
https://doi.org/10.1007/978-3-032-36189-9_10

example, the circadian rhythm must accelerate or slow down under changing light conditions, the cell cycle must arrest if DNA is not properly replicated, and malaria parasites must coordinate synchronized bursting from red blood cells, which may involve coupling between an intrinsic clock and the human circadian rhythm [1].

Previous work [2] by two of the authors discusses the requirement of flexible response as *controllability*, as opposed to *robustness*, which is continued RN functioning without disruption or alteration under environmental perturbation. They further remark on the need for balance between controllability and robustness—some types of perturbations must be ignored, while other perturbations, even of the same magnitude, require an appropriate and timely response. Clock-like RNs can have simple structures (see, e.g., Sect. 10.3.1), but their controllability emerges by coupling such simple motifs to other RNs (Sects. 10.3.2–10.3.4). This type of controllability coexists with measurable robustness in such RN models [2].

Dynamical systems modeling of RNs often takes one of the two forms: either parameterized systems of ordinary differential equations (ODEs) that quantitatively track and predict very precise trajectories over time [3–10] or a Boolean approach that qualitatively assesses the presence or absence of fixed points and oscillations [11–17].

The advantage of ODE models, when available, is that they produce continuous trajectories of the abundance of gene products and are therefore much easier to interpret and compare with data. ODE parameters such as decay rates, maximal transcription rates, and half-saturation rates are also readily interpretable. However, the number of parameters that need to be measured and/or estimated is a significant obstacle to constructing accurate models of large regulatory networks.

The advantage of Boolean or multilevel discrete models is the ability to consider networks with dozens or even hundreds of nodes. While experiments to measure the numerical values of parameters of ODE systems are very difficult and expensive, the assessment of what types of logical functions are appropriate to combine logical inputs potentially requires less biological information. These logical functions can describe the interaction of two or more transcription factors at a regulatory region or the effect of multiple kinases on a substrate.

The analysis of large Boolean models usually consists of finding steady states and, less often, finding complex attractors. Steady-state attractors consist of a single binary state that indicates which gene expression levels are high (and regulating all downstream targets) or low (and not regulating any downstream targets). Steady states can be found by several methods that do not explicitly construct all the states of the system and transitions between them [15, 18–23]. Conversely, a complex attractor consists of multiple binary system states instead of just one, and these states are revisited persistently over time. Complex attractors are more relevant to the topic of autonomous clocks, since a cell-cycle progression or other cyclic process would be represented by such a set of states. The description of complex attractors in large Boolean systems is more challenging than steady-state analysis [24].

While many important advances have been made using Boolean and ODE approaches, they are better suited for matching a particular experimental dataset than for exploring the range of possible dynamical behavior that an RN could exhibit under nonexperimental conditions. That is because ODE simulations depend on real-valued parameters and Boolean simulations depend on the choice of logical functions, also constrained by data, and these are choices made during model construction. However, the wiring diagram of an RN represents a set of putative interactions that do not include kinetic parameters, such as decay rates, binding, and half-saturation parameters, or even choices of logical functions. To partially remedy this restriction, bifurcation analysis is often performed in ODE models in which a single parameter is varied to identify changes in long-term dynamics. The bifurcations include Hopf bifurcations leading to the appearance or disappearance of periodic orbits or steady-state bifurcations leading to multiple steady states. Some work in this direction has been proposed in the Boolean setting as well

[25, 26]. Conceptually and mathematically, a much more difficult problem is the multidimensional description of the parameter space, which can be called a multidimensional bifurcation diagram, which shows changes in dynamical behavior when multiple parameters are simultaneously varied. Such diagrams, exhaustively computed, would permit a full exploration of emergent dynamics, controllability, and robustness that would encompass nonexperimental conditions. The focus of this chapter is a hybrid modeling approach that both characterizes dynamics over identifiable regions of parameter space and systematically organizes changes in this characterization as a function of parameters.

10.2 The DSGRN Approach

Dynamic Signatures Generated by Regulatory Networks (DSGRN) [27] is a modeling framework that can be interpreted either in terms of ODEs associated with regulatory networks [28] or as a multilevel model [29–31] that may have multiple integer states rather than binary ON/OFF states like traditional Boolean models. This hybrid nature permits a better understanding of how RN dynamics change as a function of parameters compared to a Boolean modeling framework, while avoiding the need to simulate large ensembles of precise trajectories, as in traditional investigation of ODE models. The assumptions of DSGRN are that regulatory interactions are monotone (inherited from the Boolean framework) and have both minimum and maximum expression levels (inherited from the common Hill model in ODE systems).

The starting point for the capabilities of DSGRN is that given an RN it produces a (nonlinear) decomposition of parameter space into a finite, though typically very large, number of parameter regions. The collection of DSGRN parameter regions for an RN, with additional information about proximity between parameter regions, is organized into a DSGRN *parameter graph*. Associated with each real-valued parameter region, or a DSGRN parameter node, is a combinatorial characterization of the dynamics called a *state transition graph* (STG), an analog of the STG known to the Boolean community. The STG is a graph of nodes and edges, where the nodes correspond to discretized levels of proteins or mRNA that we will henceforth refer to as gene expression (states), and edges correspond to allowable transitions between states. Regardless of the parameter choice within the DSGRN parameter region, the STG is constant. However, between regions, the STG may vary. Importantly, the STG can be reduced to an invariant acyclic directed graph called a *Morse graph* that compactly summarizes dynamical information from the STG. From a combination of features from the STG and the Morse graph, we derive a qualitative *dynamic signature* from a DSGRN network model at a parameter node. The collection of dynamic signatures over the whole parameter graph provides a novel, but informative, alternative to the classical bifurcation diagrams.

DSGRN dynamic signatures include steady states (also called fixed points) with discretized expression levels (e.g., low, medium, high, and sometimes gradations of these), collections of these steady states (i.e., bistability and multistability of equilibrium states), and both stable and unstable oscillatory orbits (or oscillations) in one or more gene products, which correspond to the complex attractors mentioned earlier in the Boolean setting. Multistability may also occur with a mixture of steady states and oscillations. Oscillations represented in an STG may either have a well-defined (consistent) order of maximum and minimum gene expression levels (peak and trough ordering [32]) or have the peak-and-trough ordering may vary over time (inconsistent ordering). But in both cases there is a finite set of states that are repeatedly revisited. We choose the relevant dynamic signature for a clock-like regulatory network to be an oscillation with consistent peak ordering. This type of dynamic signature can be matched to experimental data, providing evidence that the hypothesized RN cannot be rejected as the genetic mechanism of an observed clock.

A gene expression state in the STG is represented by a sequence of integers indicating the high or low state of each node. For example, the state 0102 tracks four gene products, the first and third of which are at their lowest expression levels, the second is expressed at a high enough level to regulate one downstream target, and the fourth is at a high enough expression level to regulate two downstream targets. The number of potential states of a gene product in an RN is the number of outgoing edges from that node plus one—the number of its downstream targets plus the lowest expression level at which no downstream targets are actively regulated. In addition to integer states, STGs are annotated with the location of peaks and troughs in the expression of each gene product. That is, the edges in the STG are labeled when gene expression switches from increasing to decreasing (a local maximum) or vice versa (a local minimum). More detail through examples is provided in the next section.

Even for moderately sized RNs, STGs tend to be large. As remarked in other works [2, 28, 33, 34], the Morse graph typically provides an extremely compact representation of the dynamics and in particular discards the STG information necessary to compare DSGRN dynamics against data. Thus, the DSGRN strategy is often to use the Morse graph to identify relevant subsets of the STG for which the abovementioned state and edge labelings are available. These subsets of the STG are then searched for the dynamic signature of interest.

Network architecture does not uniquely identify a DSGRN dynamic signature; for instance, for a two-node, two-edge toggle switch, there are nine DSGRN parameter nodes that show five different kinds of long-term dynamics: four different monostable steady states and one bistable state [33]. In Sect. 10.3.3, we present a three-node, four-edge network with 360 parameter nodes, and in Sect. 10.3.2, we show a six-node, eight-edge network with 46,656 parameter nodes. There are hundreds or thousands of distinct dynamic signatures for these systems. In theory, the ensemble of dynamic signatures generated by an RN model from DSGRN is exhaustively computable, although limitations introduced by the super-exponential growth of the parameter graph make sampling unavoidable for intermediate to large graphs. This rapid growth is due more to the number of edges than the number of nodes, so that large, sparse networks are easier to compute than small, dense networks. For example, the three-node RN with all nine directed edges including self-loops has over 22 billion parameter nodes (exactly 22,284,345,888). Sweeping through each parameter node and computing each Morse graph requires 54.5 hours using 100 CPUs. In contrast, a simple loop in 15 nodes has $3^{15} = 14,348,907$ parameter nodes, a factor of 1000 less than the complete three-node RN. However, in a network with more nodes, the computation of each STG, and therefore the Morse graph, may take somewhat longer.

DSGRN is quite general in its application. It has been used to model a number of molecular systems with cyclic behavior [2, 35–38], as well as equilibrium behavior [33–35, 39, 40], and is capable of modeling mixed transcription factor and protein interaction models [41]. DSGRN is available for use under the MIT License as a Python package through `pip`, with the code base and multiple tutorial Jupyter notebooks at [27].

10.3 Examples of Complex Dynamical Behavior in Autonomous Oscillators

We assume that an RN acting as a clock has a repeatable output. The DSGRN dynamic signature corresponding to a clock is taken to be a dynamically stable oscillation with a consistent peak order determined by experimental data; we will call this a *clock signature*. Given a fixed number of network nodes, many network models support stable oscillations at one or more parameter regions, and many of these will match peak and trough ordering at the qualitative level. Therefore, we view the presence of a clock signature in a DSGRN model as evidence that the model cannot be rejected as a hypothesis for the genetic mechanism, rather than as evidence of a unique, certain model.

In the following sections, we introduce network models that support or control clock-like behavior. In Sect. 10.3.1, we discuss the three-node repressilator motif that exhibits clock-like behavior but has limited robustness as measured by DSGRN. In the following two sections, the network models will consist of network motifs of two or three nodes that when coupled produce unexpected, emergent dynamical behaviors that permit flexible responses to changing environmental conditions. In the last section, we will discuss a model of the yeast START network that supports a dynamic signature that can control the cell-cycle clock.

10.3.1 The Repressilator

In 2000, a synthetic molecular oscillator was built in *Escherichia coli* from three repressive elements [42], schematically shown in Fig. 10.1a. It was intended to show an RN that exhibits clock-like behavior by producing replicable and steady oscillations. In order to analyze the potential dynamical behaviors of this synthetic circuit, DSGRN only requires the network schematic in Fig. 10.1a; the identities and chemistry of the molecules represented by the network nodes are not needed. For small networks, DSGRN computes the entire suite of dynamical behavior for all parameter regions. Different choices of molecular species used to build the synthetic circuit could result in a different identification between the biological system and a DSGRN parameter node. Thus, different dynamic signatures, i.e., model predictions for the behavior of the biological system, would be expected. Since the specific molecular species are not required, we will use x, y, and z variables to identify the three nodes of the repressilator.

For the repressilator architecture, DSGRN divides parameter space into 27 regions. There is only one parameter region in which all three edges in the repressilator function as intended; that is, at high expression level, the edge represses, and at low expression level, it does not. We call the one parameter region with functioning edges the *essential* parameter region. In the other 26 regions, one or more of the edges are disrupted in the sense that the downstream target is always repressed or is never repressed regardless of the expression level of the upstream regulatory node.

The construction of STGs is based on a rectangular division of phase space; see Fig. 10.1c and d. Each edge in Fig. 10.1a is associated with an expression threshold beyond which regulatory activity occurs. Each threshold separates the expression levels of x, y, and z into low and high values or OFF and ON states. This network is simple enough that the DSGRN states correspond exactly with Boolean states, where 0 represents OFF states and 1 represents ON states. Whenever every node in a network model has exactly one out-edge, then the multilevel formulation of DSGRN reduces to the Boolean case. As we will see in later examples, if a node has more than one out-edge (say, k out-edges), the node can be in $k + 1$ states that indicate how many downstream genes are currently being regulated, i.e., $0, 1, \ldots, k$.

An STG is constructed by assigning each phase box to an STG node with a corresponding sequence of integer states (e.g., 100 or 101) indicating the high or the low state of each gene product; see Fig. 10.1c and d for projections of the nodes, and see Fig. 10.1b for the full three-dimensional STG at the essential parameter region. Transition arrows in the STG only exist between immediate neighbors, where being a neighbor means that 1 changes to 0 (or vice versa) in exactly one position. The directions of the arrows are determined by the choice of DSGRN parameter region; see [28] for the formal details and [2] for a more informal presentation of edge direction assignment. After arrow directions have been assigned, oscillations and fixed points can be identified. The DSGRN dynamic signature for the essential parameter region for the repressilator is shown in Fig. 10.1b and is a clock signature. The sequence of states that is sequentially and repeatedly visited has been identified in red. We use the dynamical systems nomenclature of *periodic orbit* for such a sequence of states.

Along each edge in the orbit, a maximum or minimum expression level for exactly one gene product is possible [32]. For example, the edge

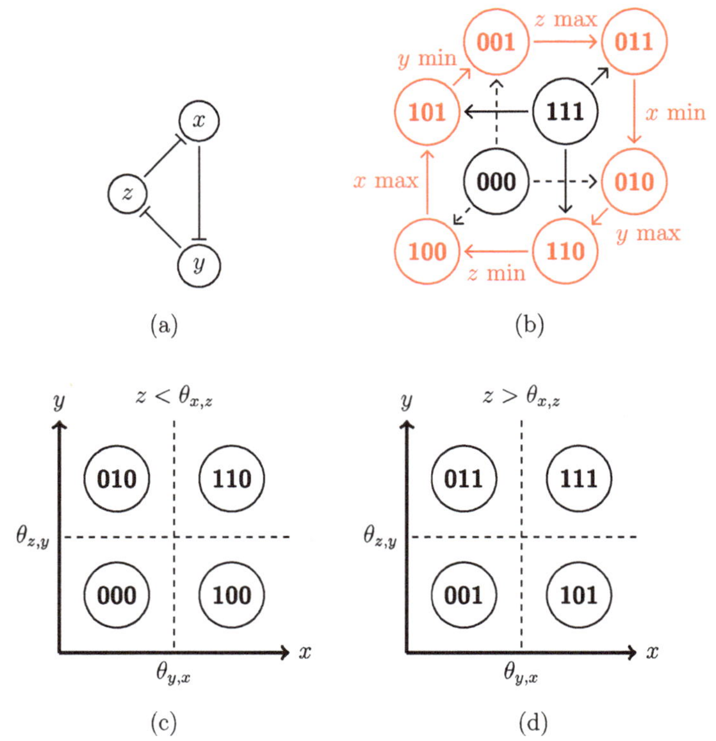

Fig. 10.1 (**a**) Network representation of the three-node repressilator. Blunt arrows indicate repression. (**b**) A clock signature of the repressilator showing oscillation in all three nodes. The x direction is forward, the y direction is horizontal, and the z direction is vertical. The states are given in the order xyz; for example, state 100 reads "x high, y low, and z low." The red nodes and edges participate in the oscillation, while the black nodes and edges do not. Dashed edges are for visual purposes only. (**c**) and (**d**) The division of $\mathbb{R}^3$ into eight boxes by three regulatory thresholds: $\theta_{y,x}$ controlling repression from x to y, $\theta_{z,y}$ controlling repression from y to z, and $\theta_{x,z}$ controlling repression from z to x. (**c**) is the projection of the eight boxes into the region where z is below its threshold, and (**d**) is the projection into the region where z is above its threshold

$100 \rightarrow 101$ in Fig. 10.1b indicates that the gene product in the third position, z, is changing from a low level to a high level. As z passes through its threshold, the node x that it regulates becomes repressed and starts decreasing from a maximum expression level. Thus the edge is labeled "x max." The order of the extrema along the periodic orbit must match experimental data in order for the repressilator in Fig. 10.1a to be a valid hypothesis for the clock mechanism. If the directions of the arrows in the repressilator were reversed, the ordering of the peak expression levels in the STG would change, and the new repressilator would not match the same datasets as the original.

The essential parameter region is the only parameter region that supports oscillations in the repressilator; that is, it is the only parameter region where the repressilator functions as a clock. In the DSGRN framework, we regard the lack of oscillations in other parameter regions as evidence of sensitivity to environmental conditions, since it is possible to disrupt the clock through many different perturbations. These adjacent parameter regions exhibit various steady states and thus cessation of oscillations. Next, we investigate how coupling two repressilators together increases the robustness of oscillations while permitting controllability.

10.3.2 Coupled Repressilators

Emergent dynamics arise when we embed a repressilator into a larger network. In this example, we explore bidirectional feedback between two repressilators, see Fig. 10.2a. The number of DSGRN parameter regions for this six-node network is quite large (46,656), and the range of potential dynamical behaviors is not clear. It seems intuitive that since two copies of the repressilator motif are present, this network should support oscillations, and indeed it does. But it also exhibits emergent, and perhaps unexpected, dynamics in which the two repressilators mutually block each other's oscillations, Fig. 10.2b.

DSGRN analysis of the network in Fig. 10.2a was performed in [36] at all essential parameter regions, of which there are 16. The essential parameter regions can be described using logical functions at the nodes y_1, y_2 and choices of threshold ordering at x_1, x_2, while the parameter region for z_1, z_2 is fixed across essential parameters. We first describe nodes y_1 and y_2, which have identical architecture, two in-edges and one out-edge each. This structure means that y_1 and y_2 integrate information from both x_1 and x_2 expression levels. The essential choices of parameters are that a y variable can be completely repressed by either x_1 or x_2 (NOR logic gate) or that the presence of both x_1 and x_2 is needed for repression to occur (NAND logic gate).

On the other hand, both x_1 and x_2 have one in-edge and two out-edges. Therefore, there is a choice of the order in which x_1 represses the downstream nodes y_1, y_2 and an independent choice for x_2. The regulator x_1 can repress y_1 at a lower expression than y_2, or vice versa, which is equivalent to choosing the order of the regulatory thresholds at the two out-edges. Since there two choices each at x_1, x_2, y_1, y_2 and these are independent of each other, there are $2^4 = 16$ essential parameter regions.

Unsurprisingly, DSGRN analysis found that 8 out of 16 essential parameter regions support oscillations in which all variables oscillate back and forth over their regulatory thresholds. Even more surprisingly, in the other eight regions, oscillations are absent, and the only attractor is an emergent steady state. The regions where the oscillations cease are those where the input integration at node y_1 is different than at node y_2; that is, if one is NAND, the other is NOR. The order of repression thresholds at nodes x_1 and x_2 is immaterial. The oscillations are disrupted by a process called *frustration*, in which the states of each repressilator are mutually blocking. To understand why this is the case, consider that NAND can be thought of as a permissive gate, in that downstream transcription is not interrupted

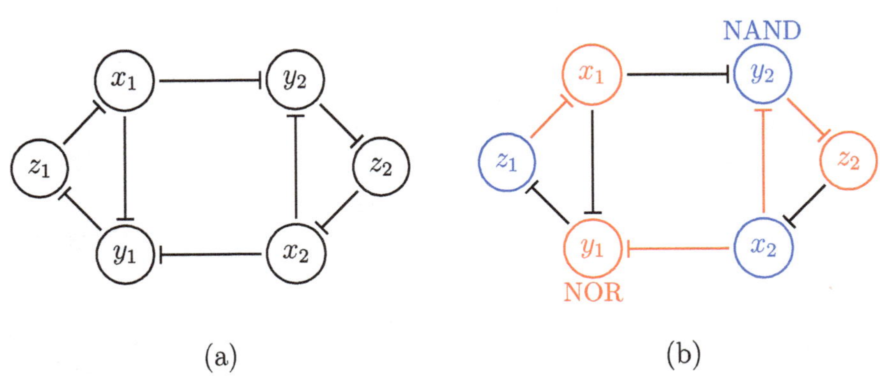

Fig. 10.2 (**a**) Two-directional coupling of repressilators. (**b**) An example of frustration leading to a steady state. Blue nodes are being transcribed, and red nodes are not being transcribed (or are being transcribed at a level too low for significant regulation). Red edges indicate active repression, while black edges indicate no repression. Note that both x_1 and x_2 must be actively repressing y_2 in order to completely suppress y_2 transcription due to the NAND logic gate, whereas y_1 transcription is blocked by the presence of either repressor as indicated by the NOR logic gate

whenever at least one of the repressors is below its regulatory expression threshold, i.e., effectively absent. On the other hand, a NOR gate only allows downstream transcription when *both* repressors are of sufficiently low expression.

As an example, consider Fig. 10.2b, where there is a NOR gate at y_1 and a NAND gate at y_2. We claim that the blue nodes are being transcribed and the red nodes are not. Indeed, transcription at y_1 occurs only if x_2 is below its regulatory threshold. Until that the desired input comes in, z_1 is actively being transcribed, and therefore x_1 is repressed. Since x_1 is repressed, y_2 is being transcribed, and therefore z_2 is repressed, ensuring that x_2 is being transcribed. Therefore, x_2 is always above its regulatory threshold, which, due to the NAND gate at y_1, ensures that y_1 can never be transcribed. In summary, the NOR gate at y_1 is waiting for y_2 expression to cease, while the NAND gate at y_2 is waiting for y_1 expression to initiate. This mutual frustration keeps the circuit at steady state and unable to oscillate.

The fact that the coupled repressilators have clock-like oscillatory behavior at 8 of 16 nodes can be interpreted as a certain amount of robustness to this clock, since some parameter perturbations will not disrupt oscillations (but they may disrupt the peak ordering). At the same time, since there are 8 of 16 regions with steady-state behavior, there is also a substantial degree of controllability built into the system. It turns out that this mechanism works for more general pairs of coupled feedback loops of different lengths and different coupling patterns, provided that the joint loop, which in this case is $x_1 \dashv y_2 \dashv z_2 \dashv x_2 \dashv y_1 \dashv z_1 \dashv x_1$, is positive (has an even number of repressive edges). If this combined loop is negative (has an odd number of repressive edges), the frustration effect never occurs.

10.3.3 Yeast Cell Cycle

The cell cycle emerges as a collection of coupled autonomous oscillators, which exhibits stereotyped patterns of gene expression that lead to a sequence of ordered cell phenotypes. In mitotic division in *Saccharomyces cerevisiae*, a cell begins with a single nucleus and genome and then begins to bud, produce spindle pole bodies, and replicate DNA. Once the DNA is replicated and separated into distinct nuclei, one nucleus travels to the bud, and mitotic division occurs. These morphological changes are accompanied by a replicable expression pattern of transcription factors and kinases [43, 44].

In Fig. 10.3a, we show an extremely simplified schematic of an autonomous transcriptional oscillator within the yeast cell division cycle derived from [2, 43]. Swi4 is a member of the complex SBF that activates Nrm1, and in turn, Nrm1 is one of the multiple regulators that repress Swi4. Swi5 is activated indirectly by Swi4 by linear chains of activators including, among others, Hcm1 and

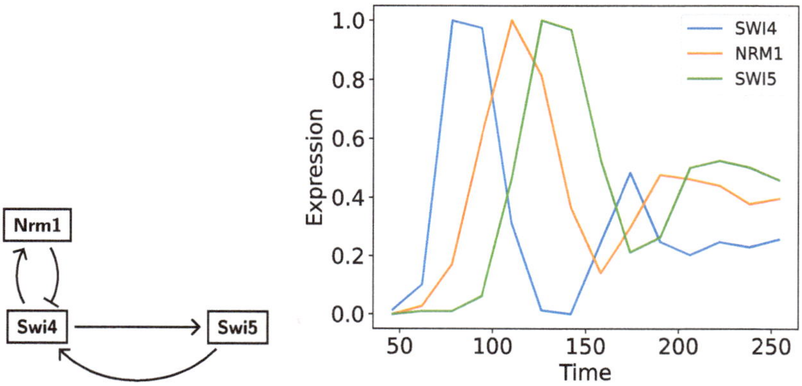

(a) (b)

Fig. 10.3 (**a**) A simplified yeast cell-cycle RN. Sharp arrows indicate activation, and blunt arrows indicate repression. Nrm1 and Swi5 have 0 and 1 integer states representing low and high expression, while Swi4 has 0, 1, and 2 integer states corresponding to its two out-edges. (**b**) Wild-type data from [46] normalized between 0 and 1

Ndd1, and Swi5 indirectly activates Swi4 via Cln3 and Whi5 [43, 45]. DSGRN divides the parameter space of this RN into 360 regions; however, similar to the repressilator examples, only 14 of these are essential where all regulatory edges are functional.

Note that the network in Fig. 10.3a consists of a negative loop between Swi4 and Nrm1 coupled to a positive loop between Swi4 and Swi5. Negative loops give rise to oscillations, and positive loops give rise to bistability [18–20, 34, 47]. If there were no emergent dynamics in the coupling between the loops, we would expect bistable steady states or stable oscillations in Swi4 and Nrm1 only. We do see the latter in four essential parameter regions; however, there are also emergent dynamics. DSGRN predicts oscillations in all three gene products in two essential parameter regions, bistability between a steady state and an oscillation in Swi4 and Nrm1 in two essential parameter regions, and monostable steady states in six essential parameter regions: three with all genes at their lowest expression level and three at their highest expression level. Other dynamics occur at nonessential parameters, where one or more disrupted edges are present.

The oscillatory time series data in Fig. 10.3b from [46] show a peak and trough ordering of Swi4 max, Nrm1 max, and Swi5 max simultaneous with Swi4 min, Nrm1 min, and Swi4 max simultaneous with Swi5 min. When peaks and/or troughs occur simultaneously, DSGRN will match any order of the simultaneous extrema.

Figure 10.4 shows the dynamic signatures for the two orbits where all three gene expression levels oscillate around at least one threshold. Since in the experimental data all genes oscillate, these are the only possible matches to wild-type data. However, they both show a different peak ordering than the experimental data in Fig. 10.3b. In both cases, by following the direction of the red arrows and examining the order of the extrema that are encountered, we see that the proposed model is too simple to capture the correct peak ordering, as either the order of Nrm1 min and Swi5 min is swapped or the order of Nrm1 max and Swi5 max is swapped. This result is not too surprising given the extreme

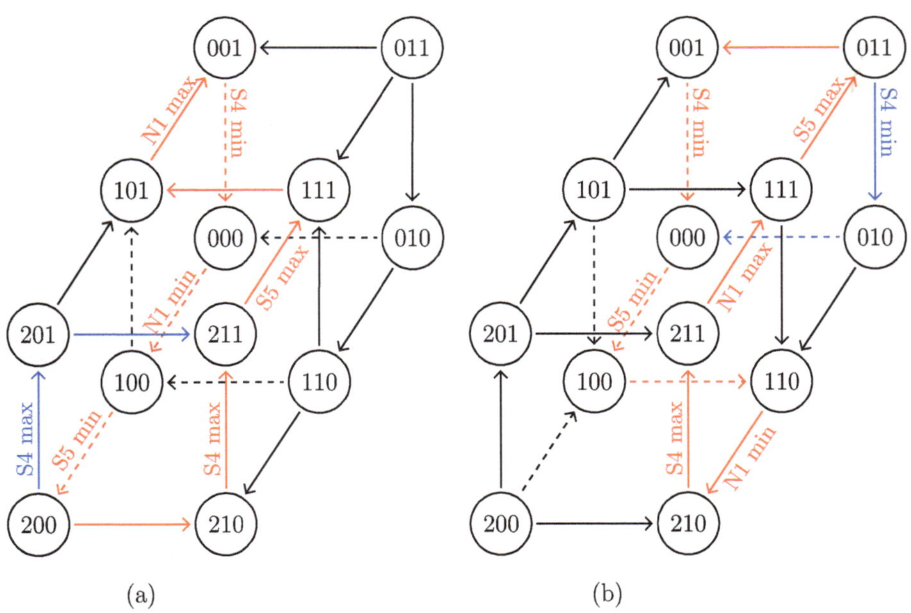

Fig. 10.4 Swi4 (S4) forward, Swi5 (S5) horizontal, and Nrm1 (N1) vertical. Dashed lines are for visual purposes only. (**a**) The red path is an orbit that is inconsistent with the peak ordering in the data. The blue edges provide an alternative path. (**b**) The red path plus alternate blue edges is an orbit at a different essential parameter, also inconsistent with the data

simplification of the RN model but demonstrates the idea for higher dimensional systems, where data and STGs are compared computationally rather than visually [2, 32, 48].

Monostable steady-state dynamics at different essential parameters, while not a match for growing wild-type cells, may describe either mutant strains or checkpoint behavior in wild-type cells. In Fig. 10.5, we see two steady states, one with all gene products low (a) and all gene products high (b).

In Fig. 10.6, we see the emergent dynamics of bistability between a stable oscillation in Swi4 and Nrm1 and a steady state where all gene products are low. Although we do not have experimental data for a cell type that exhibits this type of bistability, we remark that the peak ordering of the cycle does match the wild-type data limited to Swi4 and Nrm1 only. In this type of dynamics, sudden, large perturbations to the expression of the gene products rather than perturbations to parameter regions can cause phenotypic changes. For example, if the cell is momentarily at state 110 within the steady oscillation of Swi4 and Nrm1 but experiences an abrupt decrease in Swi5 to state 100, the model predicts that the state of the cell could decay to the steady state 000, rather than continuing the partial oscillation. The cell could use such bistability to control the phenotype. On the other hand, perturbations in parameters can change the dynamics between those described in Fig. 10.6 to those in Figs. 10.4 and 10.5 or others that are not shown.

The wealth of dynamics available to this RN architecture suggests that the cell cycle is a highly controllable system. That is, sensitivity to inputs can create a large change in the cellular phenotype. In the case of the cell cycle, this can alter the cell from a replicating state to a nonreplicating state and vice versa. In particular, a clock signature that matches peak order with wild-type data is the DSGRN equivalent of an uninterrupted wild-type cell cycle, the monostable steady states with all genes at their lowest expression level could indicate checkpoint arrest in G1, and a stable cycle in Swi4 and Nrm1 only is analogous to endocycling [49]. In the next example, we explore a control system of the yeast cell cycle

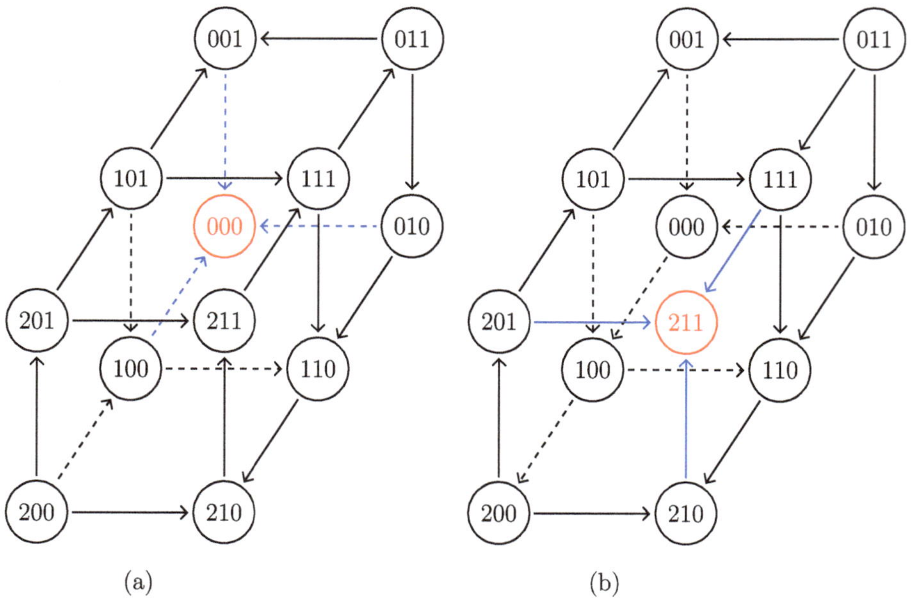

Fig. 10.5 Swi4 (S4) forward, Swi5 (S5) horizontal, and Nrm1 (N1) vertical. Dashed lines are for visual purposes only. The red nodes are steady states with (**a**) all gene products at their lowest levels, and (**b**) all gene products at their highest levels. Blue edges immediately adjacent are highlighted to show inward flow indicating stability. By inspection, there are no other steady states or complex attractors in the STGs

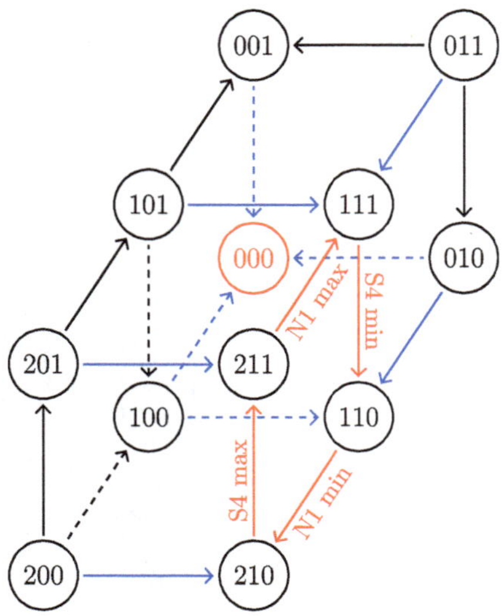

Fig. 10.6 Swi4 (S4) forward, Swi5 (S5) horizontal, and Nrm1 (N1) vertical. Dashed lines are for visual purposes only. There is bistability between a steady state with all low levels of gene products and a cycle only in Swi4 and Nrm1 (red edges on the right). The blue edges immediately adjacent are highlighted to show inward flow indicating stability

that when coupled to the cell-cycle network induces a parameter shift from a quiescent cell to a growing cell with oscillating gene expression.

10.3.4 Yeast START Network

A crucial decision for every cell is when to replicate DNA and initiate proliferation. This decision is based on multiple factors, but once the process has started DNA replication must be finished. Therefore, the influence of these factors must be uncoupled at the moment of the decision, called the *restriction point* of the cell cycle [50–52], which in *Saccharomyces cerevisiae* is controlled by the START network [53].

Several authors [54] noted that parts of the yeast START network may be analogous to the mammalian cell cycle. In [33], a simplified Rb-E2F human cell-cycle control network underwent DSGRN analysis, as well as a simplified and nearly isomorphic yeast START network, both shown in Fig. 10.7. The node S stands for an environmental signal that triggers initiation of the G1/S transition. Cln3 and Cln2 are G1 cyclins that bind to cyclin-dependent kinases that in turn disrupt binding between Whi5 and the protein complex SBF, which includes Swi4. The released SBF activates downstream targets (including Cln2) that cause the cell cycle to progress. The edges shown are reported in one of the following [45, 53, 54], except for the activating edge from Swi4 to Whi5, which is an indirect effect mediated through Hcm1 [45].

The requirement of irreversibility and decoupling that are present at the restriction point suggests that phenotypically a *bistable switch* underlies the restriction point. A bistable switch operates as a transition between two different monostable states that passes through a stage in which the system shows bistability in both states controlled by some system input S. As the input moves the system through the bistability, it will likely exhibit hysteresis. Hysteresis is asymmetry in the forward and reverse transitions between the two monostable states; that is, one direction has a higher transition threshold than the other. Experimentally, curves like those in Fig. 10.7c may be seen. In the solid line, the input signal S is increased until the expression of a gene of interest has saturated at its highest level. Once this state is firmly established, the input signal is decreased until the expression has reached low or zero levels. Hysteresis is observed when the concentration of S that triggers an increase in expression is different from that which triggers a decrease in expression.

The importance of hysteresis is that it enables noise buffering in biological processes, so that there are not rapid fluctuations between the two states when the input signal is noisy. In the context of the cell cycle, the two monostable states are proliferative (indicating commitment to transition from G1 to S phase) and quiescent (indicating rest in G1 phase). The intermediate bistable state permits the transition between these two states. This interpretation of hysteresis within the context of the mammalian Rb-E2F network is particularly important in cancer, where mutations cause repeated reentry into S phase and runaway proliferation.

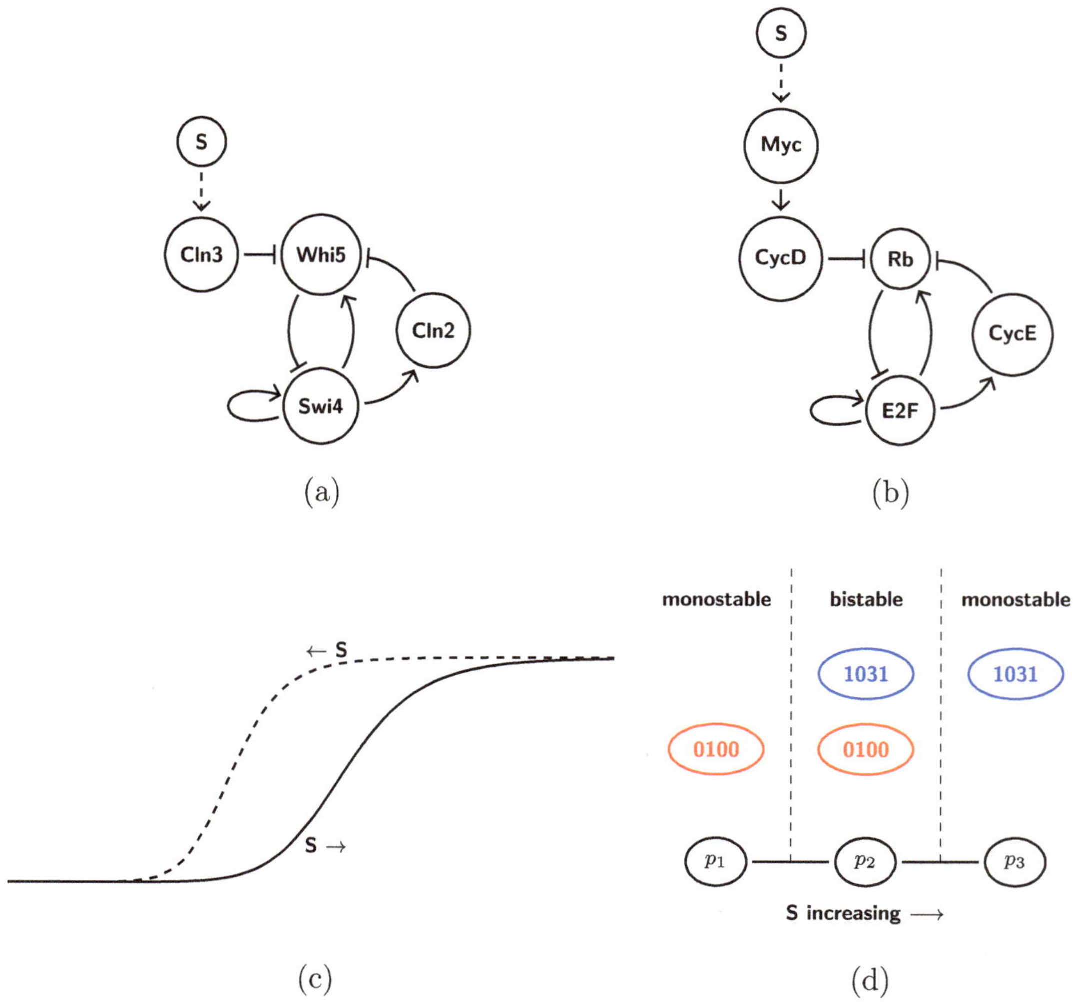

Fig. 10.7 (**a**) The four-node yeast START network analyzed in [33], where sharp arrows denote activation and blunt arrows denote repression. (**b**) The five-node Rb-E2F human cell-cycle network, isomorphic to START but for the extra node Myc, also analyzed in [33]. The prevalence of hysteresis is identical for these two networks. (**c**) Schematic of hysteresis as it would be seen in an experimental setting. (**d**) A dynamic signature that may indicate hysteresis. The quiescent state is red, and the proliferative state is blue. Hysteresis between these two states may occur if there are a minimum of three adjacent DSGRN parameter nodes with bistability between quiescence and proliferation

There is no single DSGRN dynamic signature that predicts hysteresis in an RN model. Instead, hysteresis is modeled as a sequence of dynamic signatures that occur at adjacent parameter nodes in the DSGRN parameter graph. In Fig. 10.7d, we see a DSGRN representation of hysteresis for the yeast START network, which begins with a monostable steady state in which Whi5 is active and Swi4 is inactive (quiescent state) and ends with Whi5 inactive and Swi4 active (proliferative state), with bistability in-between. The node order within the steady states is given by Cln3, Whi5, Swi4, Cln2; e.g., the state 1031 should be read as Cln3 high, Whi5 low, Swi4 high, and Cln2 high. All of Cln3, Whi5, and Cln2 have one downstream target within the simplified START network and so have binary states of 0 and 1 or OFF and ON. On the other hand, Swi4 has three downstream targets including itself and therefore has a highest activation state of 3. The parameter nodes p_1, p_2, and p_3 in Fig. 10.7d are drawn with edges between them to highlight the fact that they must be adjacent in the DSGRN parameter graph, and they must represent parameter changes in which S increases from a low to a high expression level. The full four-dimensional dynamic signature is not shown, only the collection of steady states within the dynamic signature.

The analysis in [33] to detect hysteresis was performed across all essential parameter regions for 12 hypothesized five-node networks composed of biologically realistic edges. These networks were ranked for the prevalence of hysteresis connecting the quiescent and proliferative cellular states as identified from data. The conclusion from [33] was that the network in Fig. 10.7b has the highest overall prevalence of hysteresis of the 12 networks and that the yeast START network has identical prevalence of hysteresis due to the near isomorphism between the two graphs.

This work was further extended in [34] by assessing and ranking all three-node networks in their ability to support hysteresis. The goal was to suggest a novel, robust design of a three-node switch for use in synthetic biology applications.

10.4 Importance of Emergent Dynamics

Cells need to respond to changes in their environment promptly and precisely. The responses range from relatively simple (maintenance of homeostasis in the abundance of a particular enzyme) to complex, such as those that simultaneously affect many subsystems (e.g., stress response). Small RNs may be capable of simple homeostatic control, but these networks are connected to form a cell-wide network that is capable of system-wide response. When simple networks are combined into a larger network, they need to maintain their primary functions while accommodating control mechanisms for responding appropriately to environmental signals. Therefore, a larger network must possess, in addition to the collection of primary functions of all its simple subnetworks, the ability to support new emergent phenotypes. Such emergent phenotypes are additional, unanticipated dynamics that surface as the networks connect. The ability of larger RN models to both support primary functions of motifs and exhibit unanticipated dynamics is shown in [55].

Autonomous oscillators in the cell cycle are complex phenotypes exhibiting or controlling highly coordinated sequences of events. Simple circuit motifs are capable of clock-like behavior but lack robustness in primary function and have a deficit of potential responses to environmental conditions. The combination of clock and control motifs into larger networks permits both an increase in robustness and flexibility in cellular response. For example, several cell-cycle checkpoints are capable of arresting the progression of replication, but when the key obstacle is addressed, such as DNA repair, the cell reenters the cell cycle at the precise point where it was arrested. As another example, yeast (and other organisms [56]) are capable of repeatedly doubling the genome without going through mitosis. This endocycling produces a polyploid cell that is viable and sometimes capable of reproduction.

Exploring and predicting the plasticity and robustness of complex phenotypes in RN dynamical behavior requires mathematical tools that can evaluate model behavior over parameter space. The DSGRN approach provides a method for examining the repertoire of qualitative dynamical behaviors of an RN model, as well as evaluating robustness and controllability of cellular phenotypes. This opens the door to thoroughly assessing hypothesized RN models for biological systems where multiple phenotypes have been observed, and predicting new, as-yet-unobserved phenotypes.

Acknowledgements M.G. and K.M. were partially supported by the Air Force Office of Scientific Research under awards numbered FA9550-23-1-0011 and FA9550-23-1-0400.

References

1. Smith LM, Motta FC, Chopra G, Moch JK, Nerem RR, Cummins B, Roche KE, Kelliher CM, Leman AR, Harer J, Gedeon T, Waters NC, Haase SB (2020) An intrinsic oscillator drives the blood stage cycle of the malaria parasite plasmodium falciparum. Science 368(6492):754–759
2. Fox J, Cummins B, Moseley RC, Gameiro M, Haase SB (2024) A yeast cell cycle pulse generator model shows consistency with multiple oscillatory and checkpoint mutant datasets. Math Biosci 367:109102
3. Novak B, Tyson JJ (2022) Mitotic kinase oscillation governs the latching of cell cycle switches. Curr Biol 32:1–6.
4. Jolly MK, Boareto M, Huang B, Jia D, Lu M, Ben-Jacob E, Onuchic N, Levine H (2015) Implications of the hybrid epithelial/mesenchymal phenotype in metastasis. Front Oncol 5(155)
5. Chaves M, Preto M (2013) Hierarchy of models: from qualitative to quantitative analysis of circadian rhythms in cyanobacteria. Chaos 23(2):025113
6. Jolly MK, Pillai M, et al (2023) Unraveling non-genetic heterogeneity in cancer with dynamical models and computational tools. Nat Comput Sci 3: 301–313
7. Ozbudak E, Thattai M, Lim HN, Shraiman BI, van Oudenaarden A (2004) Multistability in the lactose utilization network of *Escherichia coli*. Nature 427(737)
8. Chaves M, Gomes-Pereira L, Roux L (2021) Two-level modeling approach to identify the regulatory dynamics capturing drug response heterogeneity in single-cells. Sci Rep 11:20809
9. Balázsi A-L, Barabási G, Oltvai ZN (2005) Topological units of environmental signal processing in the transcriptional regulatory network of *Escherichia coli*. PNAS, 102(22):7841–7846
10. Aydogan MG, Steinacker TL, Mofatteh M, Wilmott ZM, Zhou FY, Gartenmann L, Wainman A, Saurya S, Novak ZA, Wong S-S, et al (2020) An autonomous oscillation times and executes centriole biogenesis. Cell, 181(7):1566–1581
11. Thomas R (1973) Boolean formalization of genetic control circuits. J Theor Biol 42:563–585
12. Thomas R, Thieffry D, Kaufman M (1995) Dynamical behaviour of biological regulatory networks-I: biological role of feedback loops and practical use of the concept of the loop-characteristic state. Bull Math Biol 57(2):247–76
13. Abou-Jaoude W, Thieffry D, and Feret J. Formal derivation of qualitative dynamical models from biochemical networks. *Biosystems*, (149):70–112, (2016)
14. Zanudo J, Albert R (2013) An effective network reduction approach to find the dynamical repertoire of discrete dynamic networks. Chaos 23:023111
15. Gan X, Albert R (2018) General method to find the attractors of discrete dynamic models of biological systems. Phys Rev E 97:042308
16. Abou-Jaoude W, Traynard P, Monteiro PT, Saez-Rodriguez J, Helikar T, Thieffry D, Chaouiya C (2016) Logical modeling and dynamical analysis of cellular networks. Front Genet 7:94
17. Chaouiya C, Naldi A, Thieffry D (2012) Logical modelling of gene regulatory networks with GINsim. Methods Mol Biol 804:463–479
18. Gouzé J-L (1998) Positive and negative circuits in dynamical systems. J Biol Syst 6(01):11–15
19. Plahte E, Mestl T, Omholt SW (1995) Feedback loops, stability and multistationarity in dynamical systems. J Biol Syst 3(02):409–413
20. Aracena J, Richard A, Salinas L (2017) Number of fixed points and disjoint cycles in monotone Boolean networks. SIAM J Discret Math 31(3):1702–1725
21. Milano M, Roli A (2000) Solving the satisfiability problem through Boolean networks. In: Lamma E, Mello P (eds) AI*IA 99: Advances in Artificial Intelligence, pp 72–83. Springer, Berlin
22. Cook SA (1971) The complexity of theorem-proving procedures. In: Proceedings of the Third Annual ACM Symposium on Theory of Computing (STOC '71), pp 151–158. Association for Computing Machinery, New York
23. Dubrova E, Liu M, Teslenko M (2012) Finding attractors in synchronous multiple-valued networks using sat-based bounded model checking. J Multiple-Valued Logic and Soft Comput 19(109)
24. Chaouya C, Monteiro P, Remy E (2024) Logical modelling, some recent methodological advances illustrated. In: AUTOMATA 2024, LNCS 14782, pp 3–22
25. Abou-Jaoudé W, Monteiro PT (2019) On logical bifurcation diagrams. J Theor Biol 466:39–63
26. Cury J, Roxo P, Manquinho V, Chaouiya C, Monteiro P (2024) Immediate neighbours of monotone Boolean functions. arXiv:2407.01337v
27. Gameiro M (2018) Dynamic signatures generated by regulatory networks. https://github.com/marciogameiro/DSGRN
28. Cummins B, Gedeon T, Harker S, Mischaikow K, Mok K (2016) Combinatorial representation of parameter space for switching networks. SIAM J Appl Dyn Syst 15(4):2176–2212
29. Crawford-Kahrl P, Cummins B, Gedeon T (2022) Joint realizability of monotone Boolean functions. J Theor Comp Sci 922:447–474
30. Gedeon T (2024) Network topology and interaction logic determine states it supports. NPJ Syst. Biol., submitted
31. Gedeon T (2024) Lattice structures that parameterize regulatory network dynamics. Math Biosci, accepted
32. Cummins B, Gedeon T, Harker S, Mischaikow K (2018) Model rejection and parameter reduction via time series. SIAM J Appl Dyn Syst 17(2):1589–1616
33. Gedeon T, Cummins B, Harker S, Mischaikow K (2018) Identifying robust hysteresis in networks. PLoS Comput Biol 14(4):e1006121

34. Gameiro M, Gedeon T, Kepley S, Mischaikow K (2021) Rational design of complex phenotype via network models. PLoS Comput Biol 17(7):e1009189
35. Hari K, Duncan W, Ibrahim MA, Jolly MK, Cummins B, Gedeon T (2023) Assessing biological network dynamics: comparing numerical simulations with analytical decomposition of parameter space. NPJ Syst Biol Appl 9:29
36. Gedeon T, Cummins B (2023) Oscillator death in coupled biochemical oscillators. Math Control Signals Syst 35(4):781–801
37. Fox E, Cummins B, Duncan W, Gedeon T (2022) Modeling transport regulation in gene regulatory networks. Bull Math Biol 84(8):89
38. Cummins B, Motta FC, Moseley RC, Deckard A, Campione S, Gameiro M, Gedeon T, Mischaikow K, Haase SB (2022) Experimental guidance for discovering genetic networks through hypothesis reduction on time series. PLoS Comput Biol 18(10):e1010145
39. Xin Y, Cummins B, Gedeon T (2020) Multistability in the epithelial-mesenchymal transition network. BMC Bioinform 21(1):71
40. Cummins B, Vrana J, Moseley RC, Eramian H, Deckard A, Fontanarrosa P, Bryce D, Weston M, Zheng G, Nowak J, Motta FC, Eslami M, Johnson KL, Goldman RP, Myers CJ, Johnson T, Vaughn MW, Gaffney N, Urrutia J, Gopaulakrishnan S, Biggers V, Higa TR, Mosqueda LA, Gameiro M, Gedeon T, Mischaikow K, Beal J, Bartley B, Mitchell T, Nguyen TT, Roehner N, Haase SB (2023) Robustness and reproducibility of simple and complex synthetic logic circuit designs using a DBTL loop. Synth Biol 8(1):ysad005
41. Cummins B, Gameiro M, Gedeon T, Kepley S, Mischaikow K, Zhang L (2022) Extending combinatorial regulatory network modeling to include activity control and decay modulation. SIAM J Appl Dyn Syst 21(3):2096–2125
42. Elowitz MB, Leibler S (2000) A synthetic oscillatory network of transcriptional regulators. Nature 403(6767):335–338
43. Cho C-Y, Motta FC, Kelliher CM, Deckard A, Haase SB (2017) Reconciling conflicting models for global control of cell-cycle transcription. Cell Cycle 16(20): 1965–1978
44. Cho C-Y, Kelliher CM, Haase SB (2019) The cell-cycle transcriptional network generates and transmits a pulse of transcription once each cell cycle. Cell Cycle 18(4):363–378
45. Teixeira MC, Viana R, Palma M, Oliveira J, Galocha M, Mota MN, Couceiro D, Pereira MG, Antunes M, Costa IV, et al (2023) YEASTRACT+: a portal for the exploitation of global transcription regulation and metabolic model data in yeast biotechnology and pathogenesis. Nucleic Acids Res 51(D1): D785–D791
46. Orlando DA, Lin CY, Bernard A, Wang JY, Socolar JES, Iversen ES, Hartemink AJ, Haase SB (2008) Global control of cell-cycle transcription by coupled CDK and network oscillators. Nature 453(7197): 944–947
47. Gardner T, Cantor C, Collins J (2000) Construction of a genetic toggle switch in *Escherichia coli*. Nature 403:339–342
48. Berry E, Cummins B, Nerem RR, Smith LM, Haase SB, Gedeon T (2020) Using extremal events to characterize noisy time series. J Math Biol 80(5): 1523–1557
49. Kraikivski P, Chen KC, Laomettachit T, Murali TM, Tyson JJ (2015) From start to finish: computational analysis of cell cycle control in budding yeast. NPJ Syst Biol Appl 1(1):1–9
50. Pardee AB (1974) A restriction point for control of normal animal cell proliferation. PNAS (71): 1286–1290
51. Blagosklonny MV, Pardee AB (2002) The restriction point of the cell cycle. Cell Cycle 2(1):102–109
52. Sears RC, Nevins JR (2002) Signaling networks that link cell proliferation and cell fate. J Biol Chem 277: 11617–11620
53. Costanzo M, Nishikawa JL, Tang X, Millman JS, Schub O, Breitkreuz K, Dewar D, Rupes I, Andrews B, Tyers M (2004) CDK activity antagonizes Whi5, an inhibitor of G1/S transcription in yeast. Cell 117(7):899–913
54. Wang H, Carey LB, Cai Y, Wijnen H, Futcher B (2009) Recruitment of cln3 cyclin to promoters controls cell cycle entry via histone deacetylase and other targets. PLoS Biol 7(9):e1000189
55. Duncan W, Cummins B, Gedeon T (2024) Subnetwork inclusion and switching of multilevel Boolean networks preserve parameter graph structure and dynamics. Front Appl Math Stat 10:1286957
56. Neiman M, Beaton MJ, Hessen DO, Jeyasingh PD, Weider LJ (2017) Endopolyploidy as a potential driver of animal ecology and evolution. Biol Rev 92(1): 234–247

Index

M. G. Aydogan (ed.), *Biological Time Control*, Advances in Experimental Medicine and Biology 1521,
https://doi.org/10.1007/978-3-032-36189-9